Color, Line, and Space

Color, Line, and Space

The Neuroscience of Spatio-Chromatic Vision

Edited by

Baingio Pinna

Spatial Vision Perspectives
Volume 1

///VSP///

LEIDEN • BOSTON - 2006

This book is printed on acid-free paper

Library of Congress Cataloging-in-Publication Data

The library of Congress Cataloging-in-Publication Data is available from the Publisher

ISBN10: 90 04 15306 3
ISBN13: 978 90 04 15306 6

Contents

Color, Line, and Space, pp. 1–7 (2006)

Spatio-chromatic interactions: An introduction

BAINGIO PINNA *

Department of Sciences of Languages, University of Sassari, via Roma 151, I-07100, Sassari, Italy

Phylogenetic and palaeontological evidence has shown that in the animal kingdom, the ability to perceive color differences has existed for at least half a billion years and evolved independently several times (see Parker, 1998). In primates, color vision advanced in a new way receiving a massive neural investment, considered independent of a luminance system that is evolutionarily older than the color system. These evolutionary changes suggest that color vision has some fundamental and specialized functions providing biological advantages.

Color and luminance are usually correlated and covary in the real world. It is not clear yet what the main role of color in spatial vision is and how it is related to and interacts with the luminance system. They cannot be easily separated; nevertheless it is clear that the color system is specifically designed to pick up information about variations in reflectance that constitute further important attributes for segregating the visual world into different objects and components. The distinctiveness and usefulness of the role of color in spatial vision and the emergent spatio-chromatic properties within visual scenes are easily demonstrated by comparing side-by-side the color photographs with the corresponding black-and-white ones (see Figs 1–10 below) and the color-color photographs of Fig. 8.

Figures 1–10 demonstrate that color has special and unique advantages that are absent or weak when only luminance variations are present. The most general and important role of the color system is the use of reflectance information to code the presence, the position and the figural properties of objects filling the visual world. More specifically, the color system facilitates our ability to segregate, locate and organize the world into objects and parts, and into lights and shades that model their volume and their illumination.

*E-mail: baingio@uniss.it

In Fig. 1(left), differences in color favor the presence, the salience and the kind of pumpkins that in the achromatic photograph have nearly the same luminance. The segregation on the basis of shape similarity can be enough to induce a figural segregation; however, color strongly increases the figurality of each pumpkin.

In Fig. 2(right), the circular shape of the fruits is not enough to induce clear object segregation. In Fig. 2(left), the color not only increases the figural salience of the fruits within the surrounding leaves and vine-branches, but also reveals which of them is ripe enough to be eaten. Color variations enhance the depth segregation between the segregated components.

Depth segregation of the flowers in Fig. 3(left) is strongly enhanced compared to those in the achromatic version. Color favors also the emergence of the illumination and of the light reflected by the flowers. Both of these phenomenal properties synergistically contribute to increase the volumetric 3D appearance and the figurality of the flowers that can be detected at long distance.

Color clearly defines illumination as shown in Fig. 4(left). Color also defines the chromatic properties of the light reflected by the Nuraghe (Paleolithic tower-like Sardinian monument) and, as a consequence, determines the right time of the day (late afternoon in the case of this picture). It is worth noting that the chromatic Nuraghe appears much brighter and lighter than the cloudy sky. The opposite is true for the dark achromatic Nuraghe. Color again enhances the 3D appearance.

The illumination and its properties, defined by colors, are even more clearly perceived in Fig. 5(left). The achromatic counterpart does not show any illumination, at least not in terms of the richness of properties perceived in Fig. 5(left). The color of the reflected light reveals not only the time but also the latitude of the region.

Color can make visible other complex and indirect objects like seasons. In Fig. 6(left), the autumn, clearly revealed by the color of the leaves, is completely invisible in the black and white picture.

From these basic and general advantages of color, the following ones may be derived. Color aids in detection and identification of things of interest like

Figure 1. (See color plate I) Colors favor the perception of the presence, the salience, the position and the nature of different kind of objects and fruits.

Figure 2. (See color plate I) Colors make the foraging for food easier: Achromatic fruits are almost invisible.

Figure 3. (See color plate I) Colors strongly contribute to enhance the figurality, the illumination and the 3D appearance of the flowers.

Figure 4. (See color plate I) Colors make objects to appear illuminated and brighter. Objects pop out due to the perceived illumination.

Figure 5. (See color plate II) Colors define the illumination, the light reflexes and their chromatic properties that, on their own, define the time of the day, the weather, the season, etc.

conspecifics (De Valois, 2004) or other species, possible partners (see Fig. 7), food, objects, preys or predators, etc. The opposite is also true. Some animals can hide themselves utilizing colors (Fig. 8(left) and 8(right)). Camouflage utilizes color in most cases. Plants evolved the use of colors both to exhibit and offer their flowers to some creatures and, at the same time, to hide them to some other creatures (Fig. 9). Mimicries are also commonly related to colors. In Batesian mimicry, a defenseless organism gains protection from predation by resembling mostly in colors a harmful or poisonous model organism. In Müllerian mimicry two organisms that derive protection from predation by tasting repellent, manifest a convergent evolution of similar coloration, so that predators avoid both mimics by tasting only one.

Color vision helps conspecifics discrimination and communication, for example through signaling sexual receptivity or gender or, as happens in humans, communicating emotions, ideas, personality characteristics or civilizations (see Fig. 10). Finally, secondary cognitive processes are derived from the main properties of color perception: colors elicit encoding, retrieving from memory, categorizing, describing and attributing new and emergent phenomenal properties or meanings to objects, animals and perceptual scenes. It is conceivable that color perception reaches its maximum expression, meaning and application in human cognition beyond perception itself. All the abstract creations of human mind, from science to art, use color as a basic component in conjunction and in relation to other visual properties, such as figural and spatial organization, shape formation, figure–ground and depth segregation (Wertheimer, 1923).

The previous phenomenal observations corroborate the largely accepted hypothesis that luminance and color systems are coded on separated pathways (De Valois, K., 2004; De Valois, R., 2004). More generally, the visual system depends on different and parallel pathways responsible for the ability to perceive different kinds of information. Vision is triggered by the absorption of photons by three different classes of cone (photopic vision) and one class of rods (scotopic vision). Color-opponency begins in the retina through cones sensitive to long (L), middle (M)

Figure 6. (See color plate II) Colors make visible complex and indirect objects like autumn, otherwise completely invisible.

Figure 7. (See color plate II) Colors help in conspecifics discrimination and communication.

Figure 8. (See color plate II) A grasshopper merges into the braches (left). A grasshopper and a cricket confront each other (right).

Figure 9. (See color plate III) Colors use reflectance information and color contrast to determine the presence of objects otherwise almost invisible. Only some target animals or insects can easily perceive these flowers, others cannot.

Figure 10. (See color plate III) Colors enable to define and to communicate human values, ideas, personality characteristics and/or civilizations.

and short (S) wavelengths of the visible spectrum and by superimposing excitatory (ON) and inhibitory (OFF) signals from different regions of the spectrum. An achromatic channel is created by parasol cells, by summing their cone contributions. A red/green opponent axis is created by midget ganglion cells that compare photon catches between L and M cones. A blue/yellow color axis originates with small and large bistratified cells that transmit ON signals from S cones and OFF signals from L + M cones (blue-ON receptive fields).

The three kinds of ganglion cells send their outputs to the LGN that comprises six layers. The upper four layers contain small cell bodies and are innervated by the midget ganglion cells. These are the parvocellular layers. The lower two layers contain large cell bodies where the parasol cells project. These are the magnocellular layers. The bistratified cells project to the intercalated layers underneath the major layers of the LGN.

The three parallel pathways (magno-, parvo- and koniocellular, Kaplan, 2004) project to V1 (layer 4 of the striate visual cortex). Neurons in magnocellular layers project the achromatic signal from parasol cells in layer $4C\alpha$ of V1, whereas parvo neurons carry the red/green signal of midget cells to layer $4C\beta$ of V1. In addition, neurons in koniocellular layers, which may receive blue-ON input from small bistratified cells, project into the cytochrome oxidase-rich zones of the superficial layers, the CO 'blobs'. The properties of other ganglion neurons that project to the K layers of the LGN and from here to the primary visual cortex are not yet known. From V1 the patterns of M-P-K signals are driven to the rest of the visual system again through parallel streams, each coding some portions of the neural image. Functional implications of the three pathways are the following: the M large cell bodies are responsible for perception of space, motion, spatial position, depth and convey information about achromatic luminance, while the color system depends on a different pathway containing P small cell bodies, best developed in primates, and responsible for the ability to perceive and analyze form and recognize in detail objects and faces. The K cells are finally involved in some aspects of color perception. The M, P and K streams provide inputs to two main visual pathways: the magno-dorsal stream defines the so-called 'where' system, while the parvo-ventral stream defines the so called 'what' system.

Despite the relatively independent underlying anatomical circuitry, the previous phenomenal observations show that colors not only represent basic information to determine and perceive *what* a specific portion of the visual world is, but also strongly influence *where* the perceived objects (3D organization and illumination) are located and organized in space and time. These implications cast doubts on a simplistic and seductive hypothesis that considers object recognition and color perception, on one side, and motion, spatial position, depth and figure–ground segregation, on the other side, as coded by totally different and independent populations of neuron. On the contrary, the previous figures suggest that the visual system and, more particularly, its particular phenomenal functions, like color perception, are rather difficult to ascribe to a unique and specific anatomical substrate pathway.

What is left to explain in color vision? This collection of papers focuses on how color appearance depends on spatio–chromatic interactions.

The article by Webster, Mizokami, Svec and Elliott studies the important issue of how prior adaptation to blurry images may calibrate visual response for spatial variations in color, in addition to the previously demonstrated visual response for spatial structure of luminance variations. Their results give important insight into the implications of adaptation effects for the subjective visual experience, and suggest dissociations in these visual adjustments both between luminance and color and different measures of spatial sensitivity.

Gordon and Shapley investigate the interesting phenomenon of chromatic induction, where a gray region appears colored if included within a colored surround. Their results confirm the third controversial law of Kirschmann (1891) that chromatic induction is at its maximum when the brightness contrast at the boundary between target and surroundings is minimal. They propose a quantitative theory of color appearance that includes mutually inhibitory interactions between color and brightness mechanisms that are sensing color and brightness contrast at visual boundaries.

Kingdom, Wong, Yoonessi and Malkoc, extending previous work, examine the remarkable problem of the influence of color contrast on shape perception in patterns combining shape-from-shading and shape-from-texture. The results show enhancement of perceived depth when a chromatic grating is added out-of-phase with luminance modulation, and suppression when the chromatic grating is added in-phase. In addition, the amount of change in perceived depth does not depend on the texture amplitude. This suggests that color influences shape-from-shading before this latter is combined with shape-from-texture.

Billock, Ditzinger, Kelso and Tsou in their paper study the fascinating Fechner-Benham subjective color that is believed to be governed by local interactions in early retinical mechanisms. The obtained results suggest the conclusion that Benham/Fechners subjective colors have their origin at a more central stage behind the site causing the McCollough effect.

The article by Daugirdiene, Murray, Vaitkevicius and Kulikowski analyzes the fascinating links between the cone-contrast rule and stability of perceived color through changes in color induced by a series of six test illuminants. They argue that color appearance is determined by cone contrasts of the matching samples computed with respect to the perceived background color rather than the actual background.

The article by Wollschläger and Faul proposes a novel filling-in phenomenon: a colored stripe pattern perceptually spreads onto an area confined by virtual contours in an apparent motion display (dynamic texture spreading). The results of their experiments suggest that sensory persistence plays an important role in the effect and indicate that information fragments are integrated over a time window of about 100 to 200 ms to form a complete object representation.

The review article by Todorović challenges the theory that lightness illusions can be accounted for by assuming that the lightness of surfaces is determined by the degree to which they are consistent with the prevailing illumination ('Illumination interpretation approach'). A number of interesting new cases inconsistent with this approach are shown, involving effects of gradients on the qualitative appearance of uniform regions.

The contribution by Grossberg and Hong proposes a neural model based on lightness computation including retinal adaptation effects, center-surround contrast interactions, boundary identification, filling-in, and anchoring. The model provides a consistent explanation through a quantitative simulation for a wide variety of quite interesting data sets and, more significantly, for anchoring effects: articulation, configuration, insulation, and area.

Von der Heydt and Pierson investigate the watercolor illusion and its two main properties of color spreading and figure–ground organization. They study whether the figure–ground effect is a consequence of the color illusion or due to an independent mechanism and provide psychophysical evidence that the figure–ground effect does not fully correlate with the strength of the color effect. They conjecture that part of the figure–ground effect of the watercolor pattern results from the double lines stimulating neurons that are selective for asymmetric edge profiles. Such neurons may signal border ownership and thus contribute to figure–ground segregation

The article by Pinna and Reeves presents new phenomena based on the watercolor illusion and on chromatic variations of the luminance profile across the boundaries. From these boundary profiles, several 'laws of figurality' that differ from the Gestalt laws of grouping and figure–ground segregation define the phenomenal appearance of what is perceived as a figure within the three-dimensional space and under a perceived illumination.

The complexities of spatio-chromatic interactions (see Switkes *et al.*, 1988) emerging from these studies and from the phenomenal attributes of Figs 1–10 suggest that (i) the color system is rather heterogeneous. It may contain additional substreams. Furthermore, (ii) the color system can interact with mechanisms coding other properties of the stimulus. Color perception reveals how visual information is nonlinearly integrated into a uniform percept. Studies based on cortical (the lingual and fusiform gyrus) lesions are difficult to interpret in terms of different parallel pathways and corroborate the idea that the visual system is a dynamic nonlinear strongly integrated system.

Acknowledgements

Supported by: Fondazione Banco di Sardegna, Alexander von Humboldt Foundation, PRIN ex 40% Cofin. es. 2005 (prot. 2005112805_002) and Fondo d'Ateneo (ex 60%).

REFERENCES

De Valois, K. K. (2004a). The role of color in spatial vision, in: *The Visual Neurosciences*, Chalupa, L. M. and Werner, J. S. (Eds), pp. 1003–1016. MIT Press, Cambridge, MA.

De Valois, R. L. (2004b). Neural coding of color, in: *The Visual Neurosciences*, Chalupa, L. M. and Werner, J. S. (Eds), pp. 1003–1016. MIT Press, Cambridge, MA.

Kaplan, E. (2004). The M, P, and K pathways of the primate visual system, in: *The Visual Neurosciences*, Chalupa, L. M. and Werner, J. S. (Eds), pp. 481–493. MIT Press, Cambridge, MA.

Kirschmann, A. (1891). Über die quantitativen Verhältnisse des simultanen Helligkeits- und Farben-Contrastes, *Philosophische Studien* **6**, 417–491.

Parker, A. R. (1988). Colour in Burgess Shale animals and the effect of light on evolution in the Cambrian, *Proc. R. Soc. Lond. B. Biol. Sci.* **265**, 967–972.

Switkes, E., Bradley, A. and De Valois, K. K. (1988). Contrast dependence and mechanisms of masking interactions among chromatic and luminance gratings, *JOSA A* **5**, 1149–1162.

Wertheimer, M. (1923). Untersuchungen zur Lehre von der Gestalt II, *Psychologische Forschung* **4**, 301–350.

Color, Line, and Space, pp. 9–30 (2006)

Neural adjustments to chromatic blur

MICHAEL A. WEBSTER *, YOKO MIZOKAMI,
LEEDJIA A. SVEC and SARAH L. ELLIOTT

Department of Psychology, University of Nevada, Reno, Nevada NV 89557, USA

Abstract—The perception of blur in images can be strongly affected by prior adaptation to blurry images or by spatial induction from blurred surrounds. These contextual effects may play a role in calibrating visual responses for the spatial structure of luminance variations in images. We asked whether similar adjustments might also calibrate the visual system for spatial variations in color. Observers adjusted the amplitude spectra of luminance or chromatic images until they appeared correctly focused, and repeated these measurements either before or after adaptation to blurred or sharpened images or in the presence of blurred or sharpened surrounds. Prior adaptation induced large and distinct changes in perceived focus for both luminance and chromatic patterns, suggesting that luminance and chromatic mechanisms are both able to adjust to changes in the level of blur. However, judgments of focus were more variable for color, and unlike luminance there was little effect of surrounding spatial context on perceived blur. In additional measurements we explored the effects of adaptation on threshold contrast sensitivity for luminance and color. Adaptation to filtered noise with a $1/f$ spectrum characteristic of natural images strongly and selectively elevated thresholds at low spatial frequencies for both luminance and color, thus transforming the chromatic contrast sensitivity function from lowpass to nearly bandpass. These threshold changes were found to reflect interactions between different spatial scales that bias sensitivity against the lowest spatial grain in the image, and may reflect adaptation to different stimulus attributes than the attributes underlying judgments of image focus. Our results suggest that spatial sensitivity for variations in color can be strongly shaped by adaptation to the spatial structure of the stimulus, but point to dissociations in these visual adjustments both between luminance and color and different measures of spatial sensitivity.

Keywords: Blur; adaptation; color; contrast sensitivity.

INTRODUCTION

Blur represents an important and conspicuous property of spatial stimuli, and is a feature that the visual system appears to continuously calibrate for. A focused edge has an amplitude spectrum in which contrast varies inversely with spatial frequency,

*To whom correspondence should be addressed. E-mail: mwebster@unr.nevada.edu

or as $1/f$. Attenuating the contrast at higher frequencies biases the spectrum and causes the edge to appear blurred, while boosting the amplitude at high frequencies instead causes the stimulus to appear too sharp. This $1/f$ scaling also characterizes the average amplitude spectrum of natural scenes, and many authors have suggested that visual coding has been shaped by this characteristic (Atick, 1992; Simoncelli and Olshausen, 2001). For example, the center-surround antagonism in retinal receptive fields acts to discount the low-frequency bias in natural images, so that the neural response is effectively whitened (Atick and Redlich, 1992; Srinivasan *et al.*, 1982). Similarly, in the spatial-frequency tuning of cortical cells, the average bandwidth increases roughly in proportion to the preferred frequency (or as $\sim f/1$), so that the responses to $1/f$ spectra are more nearly equated across spatial scale (Field, 1987). A vivid illustration of this match is that $1/f$ noise appears to have salient structure at many spatial scales, while white noise (which has a physically flat spectrum) instead appears composed only of high frequency structure (Field and Brady, 1997).

The visual system also recalibrates for image blur at much shorter time scales. An obvious example of this is the accommodative changes of the eye's optics in order to focus the retinal image. Yet neural responses may also adjust continuously to the degree of blur or sharpness in the retinal image, through the processes of adaptation. For example, Webster *et al.* (2002) found that after adapting to a blurry image, a physically focused image appeared over-sharpened. They suggested that such short-term adjustments may be important for fine tuning the match between spatial sensitivity and the spatial statistics of images, since this match could otherwise be corrupted by properties of the observer (e.g. refractive errors or changes with development) or, perhaps to a lesser extent, by properties of the environment (e.g. in hazy conditions). The fact that these adjustments occur rapidly and lead to strong perceptual aftereffects suggests that they may play an important role in spatial vision, and in turn, suggest that precisely matching sensitivity to the spatial statistics of scenes may be functionally important for spatial coding.

In this study we asked whether similar adjustments occur when the spatial statistics are defined by variations in color rather than luminance. That is, to what extent is the visual system able to calibrate spatial sensitivity based on the chromatic contrast in images? Information about color can support many spatial judgments, and indeed color vision may primarily serve as another means to form perception (De Valois, 2003; Mullen and Kingdom, 1991). However, the spatial properties of color coding are in some cases very different from luminance coding, and this has drawn attention to the question of whether and to what extent these properties are also optimized for the spatial structure of the environment. For many images the amplitude spectra for color variations falls near $1/f$ and thus parallels the structure of luminance variations (Burton and Moorhead, 1987; Párraga *et al.*, 1998; Ruderman *et al.*, 1998; Webster and Mollon, 1997). Moreover, independent components analysis of images yields opponent-color functions with spatial properties that share comparable features with the derived luminance components (Wachtler *et al.*, 2001). Yet retinal coding and psychophysical measures of contrast sensitivity at different scales are markedly different for color and luminance. While the luminance contrast sensitivity function (csf) is bandpass and extends to high spatial frequencies, sensitivity to color contrast shows little low-frequency attenuation and poorer spatial resolution (Mullen, 1985). The difference between the luminance and chromatic csf's has been accounted for by assuming that spatial color contrasts in images are noisier and thus cannot support discriminations at finer scales (Atick *et al.*, 1992), or by assuming that color is specialized for a particular subset of images with different structure. For example, Párraga *et al.* (2002) found that close-up images of colored objects like fruit tend to have steeper amplitude spectra for the chromatic contrast, and suggested that the low-pass csf for color may reflect an emphasis on encoding the low-frequency bias in such images.

The response of chromatic mechanisms to image blur remains unclear. Wuerger *et al.* (2001) found that the ability to discriminate changes in the magnitude of blur was very similar for luminance and chromatic patterns that were detected by the L and M cones, while very different for patterns that isolated S cones. This suggests that at least some types of color variation can support sensitive blur judgments. Yet the visual system may often 'ignore' this potential chromatic information in the perception of blur. For example, isoluminant stimuli are largely ineffective for driving the accommodative response (Switkes *et al.*, 1990; Wolfe and Owens, 1981), and in patterns with both luminance and chromatic contrast (i.e. stimuli characteristic of most natural scenes) the perceived blur may be strongly dominated by the luminance content of the scene. A striking example of this is illustrated in Wandell (1995, plate 7). Blurring only the light-dark variations in an image produced obvious changes in perceived image blur, yet when the same blur was applied only to the color variations the image remained perceptually well-focused.

We examined the responses to blur in chromatic stimuli by specifically asking whether the visual system can adapt to changes in the degree of blur or sharpening in chromatic images. Such adjustments might suggest that the spatial sensitivity for color is somehow matched to the spatial properties of the image, even if this match may not be important for the actual perception of blur. In previous studies we explored three different aspects of these adjustments for luminance-varying images. Webster and Miyahara (1997) measured how threshold contrast sensitivity is altered by adaptation to focused images or images with blurred or sharpened spectra, and found that adaptation produced pronounced and selective changes in the shape of the luminance csf. Webster *et al.* (2002) instead measured the actual perception of blur in stimuli after adapting to images that were blurred or sharpened, and again found strong after-effects. Finally, Webster *et al.* (2002) also observed strong changes in perceived blur when a focused image was presented within a blurred or sharpened surround. One goal of the present work was to test whether similar blur adaptation and blur induction effects occur for color-varying patterns, or whether these adjustments might reflect specialized processes within mechanisms

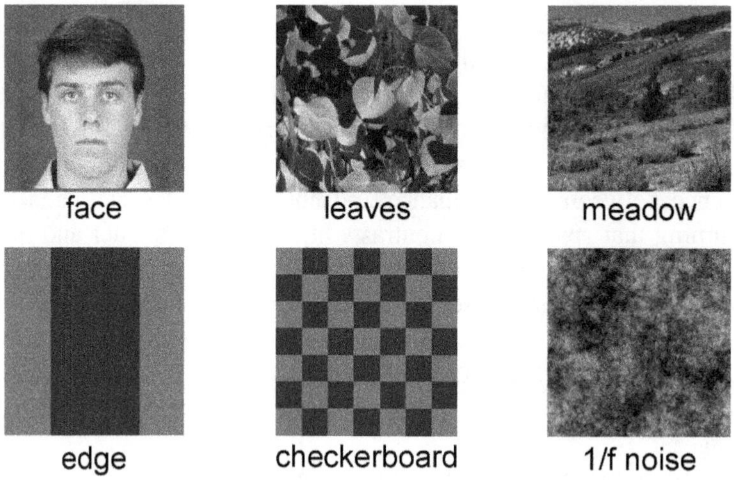

Figure 1. Images used to judge perceived focus when contrasts were defined by luminance or color variations.

sensitive to luminance. A second goal was to compare the pattern of adjustments for luminance and color to examine the extent to which the different blur adaptation and induction effects are functionally distinct.

METHODS

Stimuli consisted of images of natural objects or scenes, simple geometric patterns, or filtered noise, as described in Webster and Miyahara (1997) (Fig. 1). All images had an intensity resolution of 256 levels. For luminance images, these were used to define uniform gray-scale values corrected for the non-linearities of the display. For chromatic images, a look-up table was used to remap the luminance levels into chromatic contrasts. Chromaticities in the color images were varied along the two axes of the MacLeod-Boynton chromaticity diagram, which correspond to opposing signals in the long- and medium-wavelength sensitive cones (LM), or signals in the short-wavelength sensitive cones opposed by the combined signals in the L and M cones, (S-LM) (MacLeod and Boynton, 1979). These two axes are thought to be the principal axes underlying early post-receptoral color coding (Krauskopf *et al.*, 1982). The spatial capacities for the S axis are of further interest because the small number and sparse sampling by S cones severely restricts the spatial resolution of the S-cone pathways (Williams and Collier, 1983). For both axes we arbitrarily defined contrasts to equal 1.0 for the maximum complementary color excursions available on the monitor. The resulting stimuli had a mean chromaticity equivalent to Illuminant C (MacLeod-Boynton r, b coordinates of 0.6568, 0.01825) and maximum excursions of (0.7117, 0.01825) for the +L-M axis and (0.6568, 0.03388) for the +S axis. Isoluminance was defined photometrically based on calibrations of the display with a PR650 spectroradiometer.

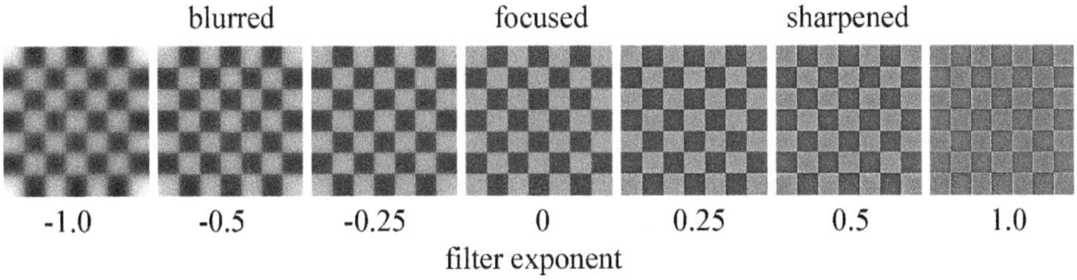

Figure 2. Examples of the image arrays, which varied the original spectrum ($s = 0$) in finely graded steps from moderately blurred ($s = -1$) to moderately sharpened ($s = +1$).

The images were filtered by multiplying the original amplitude spectrum by f^s, where f is the spatial frequency and s represented the change in the spectral slope of log amplitude *vs.* log frequency. Negative values of s steepened the original slope by reducing amplitude more at higher frequencies, and thus caused the image to appear blurred. Positive values instead flattened the slope by boosting the contrast at higher frequencies and thus sharpened the images. For most experiments we created an array of images by varying s from -1 to $+1$ in steps of 0.01, providing a series that varied from moderately blurred to moderately sharpened in finely graded steps (Fig. 2). Each image was rescaled after filtering so that it had the same mean luminance and rms contrast as the original image.

Responses to blur were assessed at suprathreshold by measuring the perceived blur of images and at threshold by measuring contrast sensitivity for grating patterns. The two tasks involved different procedures and displays and were tested for different subsets of observers. The observers included the authors and 5 additional subjects who were unaware of the specific aims of the experiment.

Suprathreshold blur judgments

For measurements of perceived blur, the images were shown on a Sony 20SE monitor controlled by a standard VGA card. Observers viewed the display binocularly in a dark room from a distance of 135 cm. At this distance, the 256×256 images subtended 4 deg centered within a 11.7 by 15.7 gray background, and spanned a range of horizontal or vertical frequencies from 0.25 to 32 c/deg. The mean luminance of the display was $\sim$10 cd/m^2. Perceived focus was measured with a forced-choice staircase procedure. On each trial a test image was shown for 500 ms, and the observer used a button box to respond whether the image appeared 'too blurred' or 'too sharp'. The slope (s) of subsequent images was varied in two randomly interleaved staircases, with the estimate of 'best focus' based on the mean of the final 6 reversals from each staircase. Settings were made either before or after adaptation to blurred or sharpened versions of the images. In adapting runs, subjects viewed a single image of a specified slope for 3 minutes. The image was randomly repositioned within a 6 deg field every 250 ms in order to prevent local differences in light adaptation. Test images were then shown for 500 ms and interleaved with 6 s periods of readaptation prior to each presentation. A uniform field was displayed for 500 ms between the adapting and testing images to prevent masking. In further experiments we also measured the effects on perceived blur of induction from a simultaneously presented surrounding field rather than adaptation from a prior stimulus. The stimulus configuration for these experiments is described below in the appropriate section of the results. Typically, 4 settings were made for each adaptation and induction condition, with the order counterbalanced for luminance and color and for different images.

Threshold contrast sensitivity

To measure contrast sensitivity functions following adaptation, stimuli were displayed on a Sony 20SE monitor controlled with a Cambridge Research Systems VSG graphics card, which allowed near threshold stimuli to be displayed with high resolution. Procedures and stimuli were similar to those used previously by Webster and Miyahara (1997). Adapting images in this case were all white noise filtered over a range of spectral slopes. These were adjusted for an rms luminance contrast of 0.35, or a nominal contrast of 1.0 for the chromatic patterns. They were shown at 8-bit resolution through the VSG framestore. The images subtended 5.6 deg and were delimited by a narrow black border from an 8 by 10.5 deg background of the same mean chromaticity and mean luminance (30 cd/m^2). At the 200 cm viewing distance the images contained frequencies ranging from 0.149 to 19.1 c/deg. Testing began with an initial 5-minute period of adaptation. To control for light adaptation in these experiments, observers viewed a rapid succession of 16 different noise images that all had the same spectral slope. A new image from the set was chosen at random every 300 ms. Following adaptation, a horizontal test grating was shown in the display field for 700 ms, with contrast ramped on and off during the first and last 150 ms with a Gaussian envelope ($\sigma = 50$ ms). The test and adapt stimuli were separated by a 500 ms uniform field, and each test presentation was preceded by a 6 s readaptation interval. In pre-adapt baseline settings, the same sequence was followed but with uniform fields in place of the adapting images. The gratings were displayed using the VSG waveform generator which provided a contrast resolution of 14 bits/gun. Subjects used a button box to adjust the grating contrast until it was at the detection threshold. Settings were made for test frequencies ranging from 0.25 to 16 c/deg in 1 octave steps. Results reported are the average of between 6 and 12 settings for each frequency.

RESULTS

Perceived focus of luminance and chromatic images

As a prelude to examining the effects of adaptation and induction for chromatic blur, we first compared the ability to simply judge the focus of images defined by luminance or chromatic contrast. Subjects were shown the 6 images in Fig. 1 and

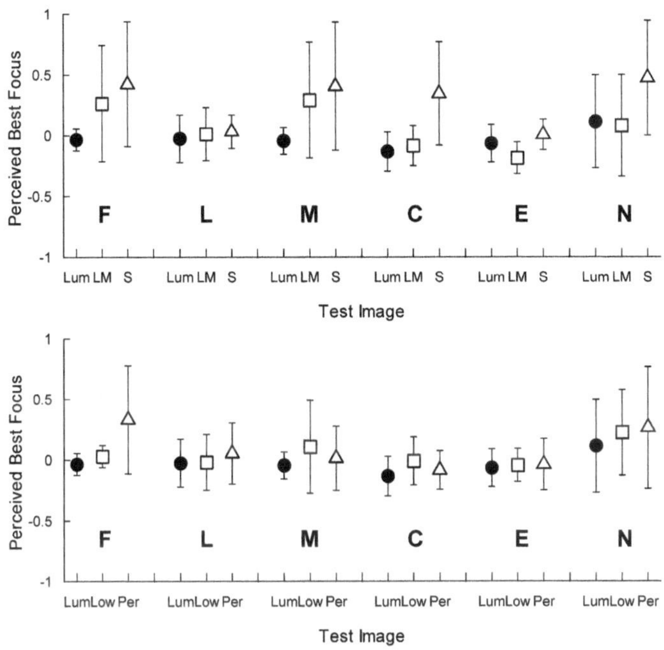

Figure 3. Slopes at which the images were judged to be best focused. Top panel: points plot the mean and standard deviations for 7 observers' settings when the same images were defined by luminance contrast (circles) or LM (squares) or S (triangles) chromatic contrast. Different triplets of settings correspond to the 6 images shown in Fig. 1. A setting of 0 corresponds to the original focused image, while a positive value means the original spectrum had to be sharpened in order for the image to appear focused. Bottom panel: settings for the original luminance images (circles) compared to settings when the luminance contrast was reduced by one-half (squares) or when the images were viewed at 7 deg in the periphery (triangles).

were asked to vary the spectral slope of each until they appeared physically focused. The mean settings for 7 observers are shown in Fig. 3. The top panel compares settings when the images varied in luminance (filled circles), in LM chromatic contrast (unfilled squares) or in S chromatic contrast (unfilled triangles). Values near zero indicate that the observers chose slopes close to the original, unfiltered spectra. For luminance, the chosen slopes were all close to the original images, and close to $1/f$ for the filtered noise. These results are similar to those reported previously by Tadmor and Tolhurst (1994) and Field and Brady (1997), and again show that subjects had accurate *a priori* knowledge of the image characteristics defining blur. Notably, settings for the luminance images were as consistent for the natural images as for the simple edges, and more consistent than for the filtered noise.

The settings for the chromatic images could differ from the settings for luminance-varying images in two ways. First, observers might be less sensitive to the slope of the amplitude spectra in the chromatic images, and thus their settings might be more variable. Second, the point of subjective best focus might differ for luminance-

Table 1.
Comparisons of the perceived focus settings for luminance and chromatic images, as plotted in Fig. 3

		Comparison	F	p
(a)		Luminance *vs.* LM		
	Mean	−0.03 *vs.* 0.064	1.36	NS
	Variance	0.049 *vs.* 0.154	3.13	<0.001
(b)		Luminance *vs.* S		
	Mean	−0.03 *vs.* 0.28	4.02	<0.001
	Variance	0.049 *vs.* 0.204	4.14	<0.001
(c)		High *vs.* Low Contrast		
	Mean	−0.03 *vs.* 0.052	1.50	NS
	Variance	0.049 *vs.* 0.076	1.55	NS
(d)		Foveal *vs.* Peripheral		
	Mean	−0.03 *vs.* 0.091	1.84	NS
	Variance	0.049 *vs.* 0.136	2.75	<0.001

F tests compare differences in the mean slope or in the variance of the settings for (a) luminance *vs.* LM varying images; (b) Luminance *vs.* S varying images; (c) high contrast *vs.* low contrast luminance images; or (d) luminance images presented in the fovea or the periphery. All comparisons are based on pooling the settings across the 6 different images and the 7 different observers.

varying and color-varying images. For example, if the original slope appeared blurred in the chromatic images, then observers might choose a shallower slope to set the focus. To examine this, we compared the mean and variance of the settings for the luminance and chromatic images (Table 1). This was done by pooling across the 6 different images and 7 observers. The variance in the settings was significantly higher for the LM and the S images compared to the luminance images. Moreover, the mean slope for the S images, but not the LM images, was significantly higher than for their luminance counterparts. As Fig. 3 suggests, these comparisons must be interpreted with caution since the differences between luminance and color were not consistent across the different images. (We do not know the bases for these differences.) Nevertheless, subjects in general were less consistent at judging the slopes in the chromatic images.

The differences between luminance and color could reflect properties of spatial coding that were specific to color, or more generally differences in the resolution or contrast sensitivity for the two types of stimuli. In the latter case we might be able to mimic the color results by displaying the luminance images at a lower contrast or under conditions that lower the resolution of the visual system. To assess this we repeated the measurements for the luminance patterns either at one-half the contrast or when they were viewed centered at 7 deg in the periphery. These comparisons are shown in the lower panel of Fig. 3. Reducing the contrast did not significantly affect the judgments, while with eccentric viewing variability increased while the mean settings remained similar to those for central fixation (Table 1). The latter is consistent with the compensation for blur in peripheral viewing reported by Galvin *et al.* (1997). Again, these differences are not conclusive, because they may not mimic the actual changes in effective contrast or resolution for the chromatic patterns. However, they leave open the possibility that the changes in subjective focus judgments for the S images (but not the LM images) are not accounted for simply by the differences in relative contrast or acuity for these images.

Adaptation to blur

We next tested whether the judgments of focus could be biased by prior adaptation. As discussed above, Webster *et al.* (2002) found that viewing blurred or sharpened images shifted the perceived focus of subsequently viewed images so that they appeared sharper or blurrier, respectively. The present experiments examined whether adaptation could similarly alter the perception of chromatic patterns. Because sensitivity to blur in the chromatic patterns was poorer (as noted above), we first trained subjects to learn the original focus by repeating the blur settings with feedback. The same judgments were then made without feedback after adapting to blurred or sharpened versions of the image.

Figure 4 shows the settings for 4 observers tested with luminance or chromatic versions of the checkerboard image. In all cases there is a consistent bias in the focus judgments following adaptation. Specifically, adaptation to the blurred checkerboard caused the original to appear too sharp, so that the image that appeared properly focused was now physically blurred, while prior adaptation to the sharpened image induced the opposite bias. Moreover, in most cases the magnitude of these shifts was comparable for luminance and color, and for both types of chromatic contrast. Thus despite the poorer sensitivity to blur in the chromatic patterns, for both luminance and color the judgments of blur appear similarly affected by adaptation.

The measurements in Fig. 4 were for adapting and test images that had the same type of contrast (i.e. both images were luminance or both were chromatic). Were observers adapting to the general property of blur *per se*, or to the blur along specific axes of color space? To test this, we made further measurements when the contrast defining the test and adapting stimuli fell along the same or different axes. This experiment also used a different image (the 'leaves' image of Fig. 1) to ensure that the blur aftereffects for color were not specific to simple binary patterns like the checkerboard image.

In this case subjects made settings after adapting to either a sharpened ($s = +0.5$) or blurred ($s = -0.5$) level of the adapting image. Again, these two stimuli should induce opposite shifts in the perceived focus of the test stimulus, and we therefore used the difference between the slopes chosen under these two conditions in order to compare the magnitude of the aftereffects. These differences are shown in the three bar charts in Fig. 5. The three panels plot the changes in the settings after adapting to images defined by luminance contrast (left panel), LM contrast (middle panel), or S contrast (right panel), while the three bars within each panel show how adaptation to blur along each axis affected a test pattern defined by the same or different axes. A value of zero means that the focus settings were equal after

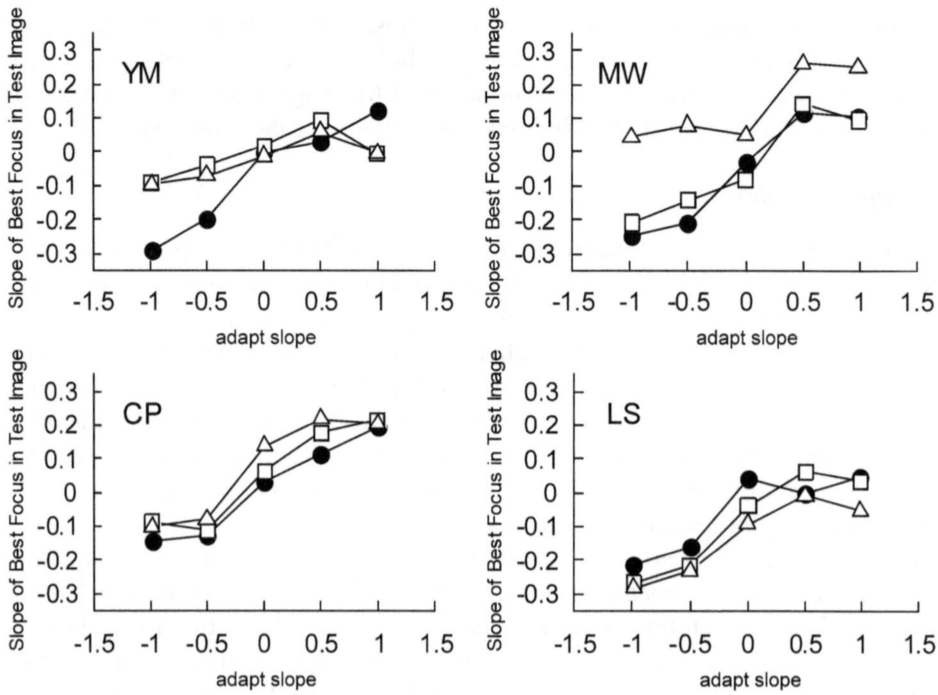

Figure 4. Changes in perceived blur following adaptation to blurred or sharpened images for luminance contrast (circles), LM chromatic contrast (squares) or S chromatic contrast (triangles). Each point plots the mean of 4 settings for the checkerboard pattern. The 4 panels show results for 4 different observers.

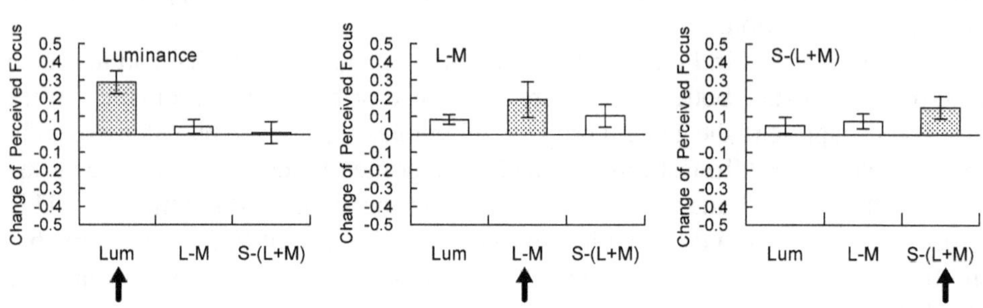

Figure 5. Changes in perceived blur when the adapting and test images both varied in luminance or along the same color axis (shaded bars), or when the adapt and test images varied along two different axes (unshaded bars). Bars show the difference in the blur settings after adaptation to a sharpened (+0.5) or blurred (−0.5) version of the 'leaves' image. Positive values indicate that the test image appeared more blurred after adapting to a sharp image than a blurred image. The three bars in each panel are for a luminance, LM, or S test image, each based on the mean settings for 5 observers (±1 standard deviation). The 3 panels show results for luminance (left), LM (right), or S adaptation.

blurred or sharpened adaptation, while a positive value indicates that the test image appeared more blurred after adapting to a sharpened image than a blurred image, consistent with the after-effects shown in Fig. 4. Once again there is clear adaptation for both luminance and color when the test and adapting images varied along the same axis (though in these settings it was generally stronger for luminance). Yet the biases are consistently weaker when the adapt and test axes differ. Thus the changes in blur involve changes in mechanisms that are selective for the luminance, LM, and S axes of color space. This selectivity also precludes the possibility that the adaptive shifts seen for color tests reflected judgments based on the luminance artifacts in the nominally chromatic patterns for, in that case, there should have been strong effects of luminance adaptation on the color tests. Instead, the color-selectivity indicates that the adaptation can adjust specifically to the blur carried by chromatic contrast.

Blur induction

As noted, Webster *et al.* (2002) also found that perceived blur is strongly influenced by the relative blur in the surrounding context. An example of these induction effects is illustrated in Fig. 6. The center segments of the bars are all square-wave edges, yet the luminance edge is perceptually sharpened by the blurred luminance surround while blurred by the sharpened surround. We used a similar stimulus arrangement to test for induction effects for chromatic contrast. The vertical edges of the test bars subtended 1 deg, and were shown within abutting surrounding edges that extended 3 deg above or below the test. To facilitate judging the test, a focused reference bar (without surrounding edges) was displayed 3 deg to the left, and observers adjusted the test edge until it appeared as sharp as the reference. Settings were made for surrounding edges that were either sharpened ($s = +0.5$) or blurred ($s = -0.5$), and we again compared the difference in the focus settings for the test edge under these two conditions.

Figure 7 plots the results for two observers. The three panels again correspond to inducing bars defined by luminance contrast (left), LM contrast (middle), or S contrast (right). For luminance, there is a strong simultaneous contrast interaction, consistent with the effects visible in the images in Fig. 6. However, when the center and surround edges were defined by either chromatic axis there was instead little induction. Thus, unlike the adaptation, for the conditions we tested color and luminance appeared to have different effects on the perception of relative spatial differences in blur. The weak induction for color again cannot be accounted for merely by the poorer acuity for color, because the induction effects for luminance actually become stronger when the images are viewed in the periphery (Webster *et al.*, 2001).

Figure 6. (See color plate IV) An illustration of blur induction for luminance or chromatic edges. Top: effect of luminance surrounds on luminance or S edges. The center row of luminance edges on the left are squarewave edges. However, these edges appear sharpened when the surround edges are blurred while blurred when the surround edges are sharpened. The same luminance surrounds induce an opposite aftereffect on S edges — the squarewave S edges at top right appear blurred (rather than sharpened) when surrounded by blurred luminance edges. Bottom: S surrounds instead have little effect on the luminance or S center squarewave, suggesting that there is little spatial blur induction from the S patterns.

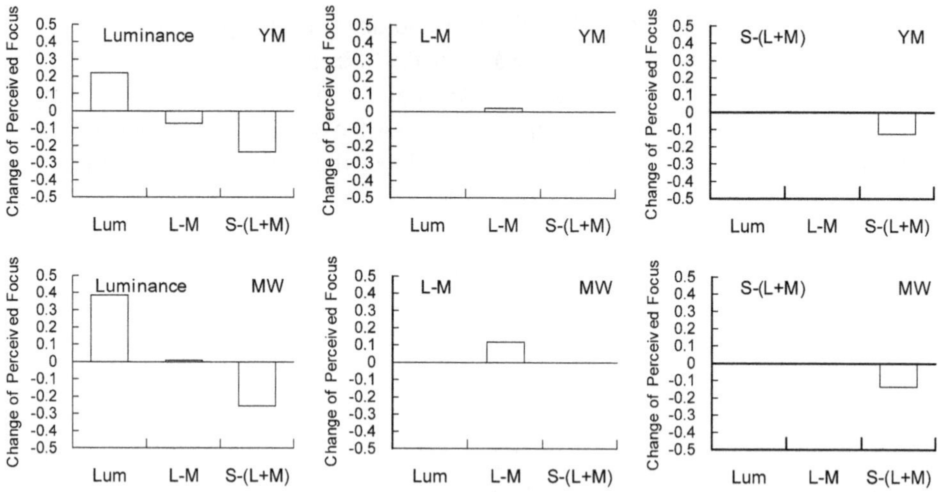

Figure 7. Blur induction for luminance or chromatic edges. Bars plot the change in perceived focus of a central edge in the presence of blurred (−0.5) or sharpened (+0.5) surround edges. A positive value indicates that the test edge appeared sharper when surrounded by the blurred edge. The three panels plot the results for luminance edges (left), LM edges (center) or S edges (right). The effects of luminance surrounds were tested for both luminance and color while, the color surrounds were tested only on the corresponding color test. The 2 rows show results for 2 observers.

The induction effects also revealed an asymmetric interaction between luminance and color. Specifically, surrounding luminance edges induced changes of opposite sign in the luminance and S chromatic test edges (Fig. 7, left column). That is, a blurred luminance surround caused the luminance test to appear sharper, yet the same surround caused an S edge to appear blurrier. The conditions yielding these opposite effects are illustrated by the images in Fig. 6. (In the converse case, shown in the lower image in Fig. 6, chromatic surrounds had no noticeable effect on a central luminance edge.) The luminance-induced blur in the S-cone pattern is reminiscent of the 'capture' of chromatic contrast by luminance edges (as in the blur illustration of Wandell, 1995), yet in this case the capture extended across a 1-deg gap at which there were no physical luminance edges.

Adaptation and threshold contrast sensitivity

In the next set of experiments we tested for adaptation to chromatic blur by examining how adaptation to blurred or sharpened images altered the shape of the contrast sensitivity function. Figure 8 shows measurements of contrast sensitivity as a function of spatial frequency before or after adapting to $1/f$ noise along the luminance axis or the LM or S chromatic axes. As Webster and Miyahara (1997) found previously for luminance-varying images, adaptation to the $1/f$ spectra characteristic of natural images induces a loss in contrast sensitivity at low and medium spatial frequencies, while sparing sensitivity at higher frequencies. Thus

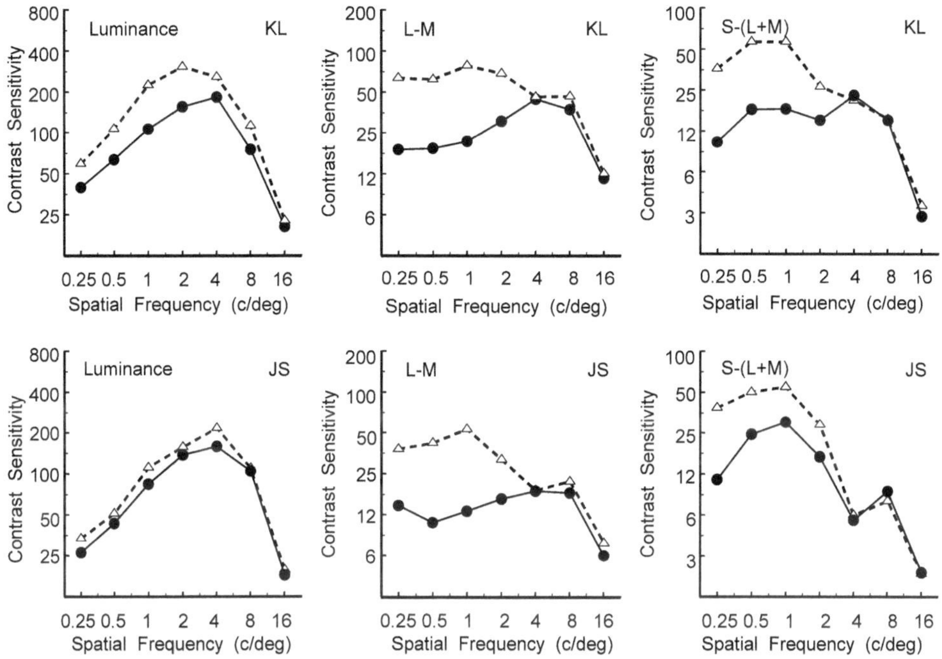

Figure 8. Contrast sensitivity functions measured before (triangles) or after (circles) adaptation to $1/f$ noise defined by luminance contrast (left), LM chromatic contrast (center), or S contrast (right). The 2 rows show results for 2 observers.

adaptation biases the overall sensitivity of the visual system toward higher spatial frequencies.

Under the present conditions these sensitivity changes turned out to be much stronger for chromatic contrast. Threshold elevations for both chromatic axes were pronounced at lower spatial frequencies, while again spared at the higher frequencies. Consequently, the chromatic csf switches from clearly lowpass before adaptation to weakly bandpass after adaptation. Figure 9 shows that the threshold changes for color also depend on the degree of blur in the adapting images. The different panels plot contrast sensitivity measured after adaptation to a range of LM noise images with amplitude spectral varying in slope from 0 (white noise) to −2.5 (strongly blurred; note that in this case slope values refer to the absolute slope and note the slope relative to $1/f$). As we found previously for luminance, white noise adaptation had relatively weak effects on sensitivity (Webster and Miyahara, 1997). But as the spectrum becomes more biased, there are progressively stronger and more selective changes in threshold at the lower frequencies, so that under adaptation to pronounced blur the chromatic csf is clearly bandpass. The shifts with the adapting spectrum reinforce the measurements of suprathreshold blur adaptation above, in showing that the visual system adaptively adjusts to changes in the spatial structure of the chromatic stimuli (though these adjustments may be much coarser than the

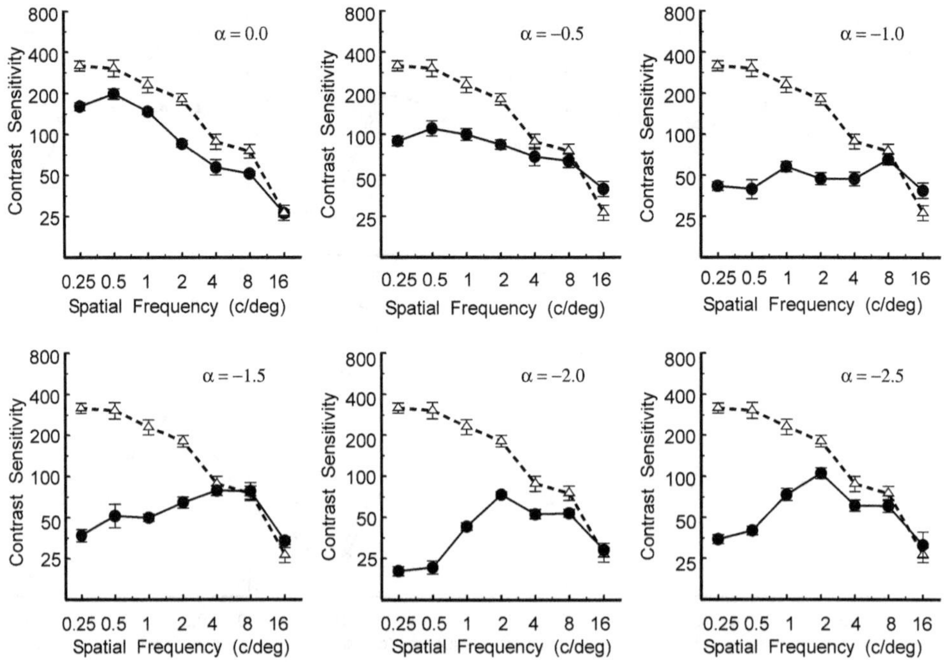

Figure 9. LM chromatic contrast sensitivity following adaptation to LM chromatic noise, for a single observer (MW) tested over a range of adapting slopes. Triangles plot the sensitivity before adaptation. Each panel plots the csf after adaptation to noise with the indicated spectral slope.

changes in surpathreshold blur, in the sense that much larger changes in adapting blur are required to produce a measurable change in the shape of the csf).

Again, for the conditions of our measurements the magnitude of these threshold changes was much larger for color than luminance. This is evident in Fig. 10, which plots the changes in contrast sensitivity (post-adapt/pre-adapt threshold) induced by the adaptation. The chromatic thresholds are taken from the plots in Fig. 9, while the luminance settings are from the study of Webster and Miyahara (1997) for the same observer and viewing conditions (their Fig. 10). For both luminance and color, the threshold changes clearly track the changes in the adapting spectra, but the selectivity and magnitude of these changes is more pronounced for color, with thresholds elevated by more than a log unit at the lowest frequencies.

The stronger adaptation effects for color are consistent with the higher intrinsic sensitivity of color mechanisms at low frequencies. That is, because the chromatic csf shows little low frequency attenuation, the effective contrast of natural spectra at low frequencies (i.e. the contrast after filtering by the csf) might be higher for the chromatic components than the luminance components. Thus we might expect more adaptation to the low frequencies for color. However, there are a number of complications with this simple interpretation. First, we made no attempt to equate the relative contrasts along the luminance and chromatic axes, and thus any differences could in part reflect differences in the overall contrast. To

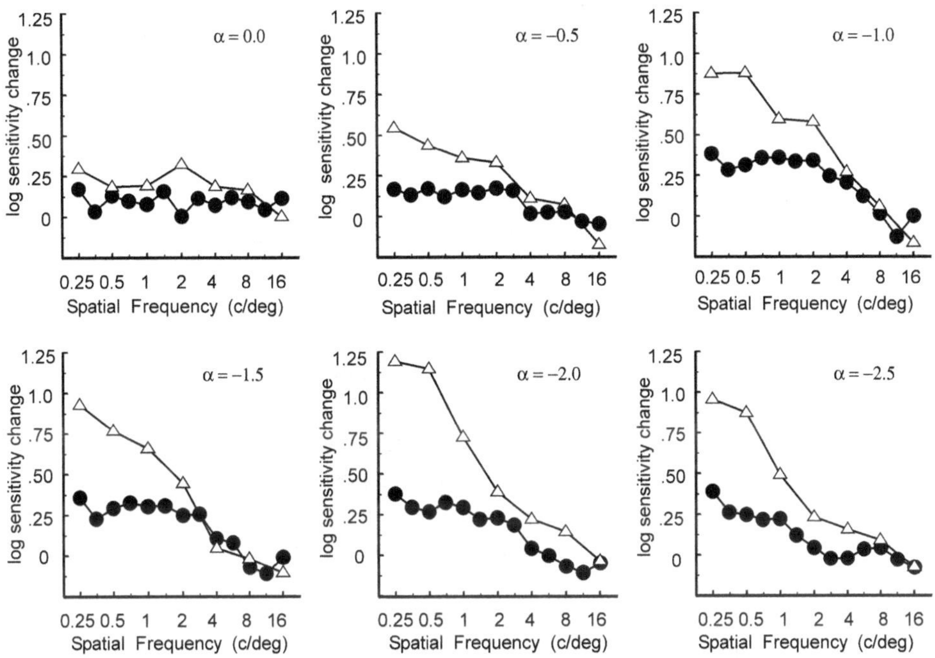

Figure 10. Change in log contrast sensitivity for LM gratings after adaptation to LM chromatic noise (triangles) or for luminance gratings after adaptation to luminance noise (circles). Chromatic results plot the change in pre- *vs.* post-adapt sensitivity for the settings in Fig. 9. Luminance results are for the same observer from the study of Webster and Miyahara (1997).

evaluate the relative adaptation states of color and luminance vision under natural viewing, it would be instructive to measure contrast sensitivity after adapting to actual images of natural scenes. A second complication is that at higher spatial frequencies the effects of optical chromatic aberrations become pronounced, so that these frequencies might be detected by the resulting luminance artifacts in the stimuli. We made no effort to correct for chromatic aberration, and thus it is possible that the adaptation for color is much less selective for spatial frequency than our results imply, because the thresholds at higher frequencies depend on detecting the encroaching luminance contrast. However, this does not alter the conclusion that sensitivity to chromatic variations in the stimulus may become more nearly bandpass under natural states of adaptation, nor that the shape of the chromatic csf can vary with the amount of blur in the image.

Threshold vs. suprathreshold adjustments to blur

The preceding results show that for both luminance and color, adaptation to blur can alter both the relative sensitivity to different spatial scales and the perceived blur in suprathreshold patterns. However, these two measures surprisingly point to different neutral states for the adaptation. Specifically, adaptation to a focused pattern does not alter perceived focus, as would be expected since this is the type of pattern we are already routinely adapted to and thus presumably should be already matched for. Yet exposure to the same focused pattern has large selective effects on the threshold csf, biasing sensitivity against lower frequencies, as if the visual system is *not* already matched to natural image spectra. In the final experiment we explored why adaptation to an '*a priori* unselective' stimulus like a $1/f$ spectrum leads to a selective change in thresholds. For these settings we restricted the stimuli to luminance images, since these provide a more sensitive measure of the adjustments and are less confounded by artifacts like chromatic aberration that might complicate the interpretation.

The failure of $1/f$ adaptation to influence thresholds at higher frequencies could result because of the paucity of physical contrast at those frequencies. By this account, the selective changes in the csf result simply because the intrinsic sensitivity of the visual system does not compensate for the low-frequency bias in the physical stimulus. However, an alternative possibility is that the adaptation cannot be predicted simply from the amplitude spectrum of the stimulus, because the visual system does not adapt independently to the contrast at different scales. This latter possibility is suggested by previous studies of adaptation to simple compound gratings. These studies have shown that the threshold changes following adaptation to a grating (e.g. of frequency $3f$) are reduced when the same adapting grating is presented in the presence of a lower fundamental frequency (e.g. of frequency f) (Klein and Stromeyer, 1980; Nachmias *et al.*, 1973; Tolhurst, 1972; Tolhurst and Barfield, 1978). That is, the lower frequency components in the stimulus appear to block adaptation to the higher frequency components.

We tested whether similar interactions occur for natural image spectra. To examine this, we again adapted to $1/f$ noise images but this time filtered out frequencies below a given cutoff. The cutoff frequencies ranged from 0.5 to 16 c/deg. The filtering removed all frequencies below the cutoff and did not rescale contrast, so that frequencies above the cutoff remained at the same amplitude as in the original unfiltered image. Observers again adapted to a random sequence of noise images with the same spectra, and as before set thresholds for a range of frequencies before or after adaptation. We also increased the viewing distance so that the stimuli subtended 4 deg, so that in this case frequencies in the images ranged from 0.25 to 32 c/deg.

Figure 11 plots the change in sensitivity to a fixed test frequency as a function of the cutoff frequency of the adapting stimuli. Thus the individual curves do not show the effects of individual adapting spectra as in previous figures, but rather directly show how the sensitivity at a given test frequency varied with changes in the adapting spectrum. The two curves plot the results for two observers. Lowering the cutoff frequency always resulted in an increase in the physical contrast of the adapting stimuli. Yet the largest threshold changes consistently occurred when the test frequency corresponded to the cutoff frequency — or in other words to the lowest frequency component in the adapting stimulus. In particular, there are clear threshold changes at the highest frequencies (8 and 16 c/deg) when the

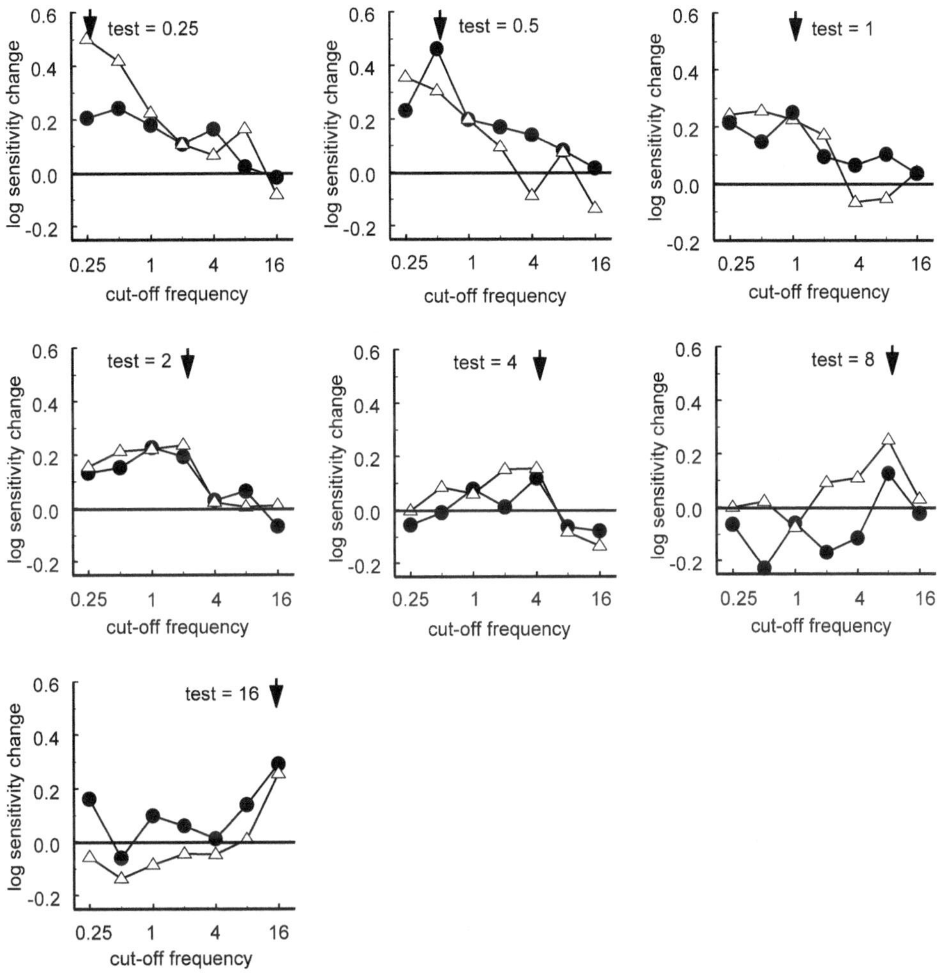

Figure 11. Changes in thresholds for test gratings as a function of the cutoff frequency of the adapting spectra. Each panel plots the results for the single test frequency indicated as a function of the lowest frequency in the adapting image. The 2 curves plot the results for 2 different observers.

adaptation was restricted to those frequencies, even though sensitivity at these same frequencies remained largely unaffected when the adapting images were broadened in bandwidth. These results are thus consistent with the adaptation effects reported previously for compound gratings. They suggest that the biases induced in contrast sensitivity arise, not because visual sensitivity is otherwise mismatched to natural image spectra, but because the effects of adaptation at different spatial scales are not independent. In particular, the contrast changes appear to reflect adaptation to the largest scale or 'size' in the adapting stimulus.

Webster *et al.* (2002) similarly concluded that the effects of adaptation on blur perception could not be accounted for by the overall amplitude spectrum. They found that how adaptation transferred across different images depended on the actual physical focus of the images and not on the average contrast at different spatial scales (which can vary across physically focused images because of differences in the density of structure at different scales (Field and Brady, 1997)). Thus, like judgments of image blur (Field and Brady, 1997; Tadmor and Tolhurst, 1994), the adaptation is controlled by the actual blur in the image and not by the global spectrum.

This leaves unanswered the question of why adaptation-induced changes in the csf are not manifest in the judgments of image focus. One possibility is that the pattern of the sensitivity changes is different at high test contrasts (Snowden and Hammett, 1992), such as those examined in our blur task. However, Webster and Miyahara (1997) found similar selective changes in perceived contrast in suprathreshold gratings of moderate contrast. This suggests that the difference may instead be due to the nature of the test stimuli. The visual cues to blur are surprisingly poorly defined, and as the preceding paragraph suggests, may depend on attributes that are not readily captured by the amplitude spectrum. Thus sensitivity to blur and to sine wave gratings may tap very different processes. Consistent with this, Wuerger *et al.* (2001) found that the differences in the luminance and chromatic csf's could not predict performance in their blur discrimination task.

DISCUSSION

Color vision provides a useful context for exploring the match between spatial coding and the visual environment, for it provides the visual system with additional sources of information about the world, and thus the chance to test whether design principles developed in one context have general validity (Atick, 1992; Simoncelli and Olshausen, 2001). There are many clear cases where the encoding of color seems strongly driven by the properties of the environment, as for example in the variations in spectral sensitivity with environmental niche (Lythgoe, 1979). Recently, MacLeod and von der Twer (2003) explored the match between post-receptoral color coding and the distribution of colors in the natural environment. They showed that contrast coding for color can be closely predicted from natural color distributions and from the principle that the visual system is designed to minimize the overall error in the representation of different levels in these distributions. This match makes sense, for clearly a fundamental function of our color vision is to provide us with information about color.

However, chromatic variations in images also carry information about the spatial properties of objects. In this study we examined the extent to which spatial sensitivity based on color reflect the characteristic spatial structure of natural images. The role of color in this case is less certain. On the one hand, color is clearly a useful cue for spatial vision, and supports spatial discriminations that are nearly as fine as those supported by luminance (Krauskopf and Farell, 1991; Webster *et al.*, 1990). Yet on the other hand, it is clear that some aspects of spatial vision, such as visual acuity, are markedly degraded for chromatic contrast. Moreover, there are a number of visual judgments that become impoverished at isoluminance, in part because the visual system fails to encode information that is potentially available in the chromatic content of the stimulus (Mullen and Kingdom, 1991). As noted in the Introduction, this may specifically occur in the case of blur, since color can support fine blur discriminations (Wuerger *et al.*, 2001) yet does not appear important for accommodation (Switkes *et al.*, 1990; Wolfe and Owens, 1981) or for judgments of blur in images that also vary in luminance (Wandell, 1995).

Regardless of their potential functional role, the present results suggest that the spatial properties of chromatic mechanisms are shaped by the chromatic spatial structure of images. The primary evidence for this is the changes we observed in perception and sensitivity to color patterns when observers adapted to images with altered spectra. Changes in perceived blur and threshold sensitivity varied in systematic ways as the adapting image spectra were progressively blurred, and the adaptation affected perceived blur selectively for luminance and chromatic contrast. Our results thus parallel the evidence for spatial- and color-selective adaptation reported previously from measures of grating adaptation (Bradley *et al.*, 1988). The qualitative similarity between the adaptation effects for luminance and color suggest that at least some aspects of the mechanisms encoding blur are functionally similar for luminance and chromatic contrast, as also suggested by the work of Wuerger *et al.* (2001).

Yet we also observed important differences in blur perception for color. First, observers were generally much less sensitive to the absolute blur in the chromatic patterns, and for the S stimuli tended to perceive the chromatic images as blurrier than their luminance counterparts. Secondly, we found little evidence for blur induction for the chromatic patterns, suggesting that color cannot be used to adjust for differences in relative blur across the visual field. The blur induction for luminance may be related to contrast gain control, in which the perceived contrast of a central region is reduced in the presence of a higher contrast surround (Chubb *et al.*, 1989). However, strong contrast induction is also observed at isoluminance, and these contrast changes show some selectivity for the spatial properties of the stimuli (D'Zmura and Singer, 1999). Thus it is surprising that it is specifically the feature of blur that fails at isoluminance in induction.

Again, these differences are unlikely to result simply from the lower acuity for color, and may reflect characteristics of spatial vision that are somehow unique to color. What are these characteristics? One possibility is that they reflect the representation of borders or edges in chromatic images. Chromatic patterns — and particularly S-cone mediated patterns — are poor at mediating the perception of well-defined spatial transitions (Boynton *et al.*, 1977). In fact, one common method for empirically defining isoluminance is to vary the relative luminance of two colors until they form a 'minimally distinct border' (Boynton, 1978). This does not reflect poor resolution alone, for color also exhibits an active filling-in process which may be held in check by luminance edges (Pinna *et al.*, 2001). For example, adding luminance contrast often facilitates the ability to discriminate a difference in color between two regions (Eskew *et al.*, 1991; Montag, 1997; Switkes *et al.*, 1988), for without it the regions tend to 'bleed' together, as in the classic observation of the limited hues distinguishable in the spectrum (Smeulders *et al.*, 1994). This capture of color by luminance may underlie the failure to perceive chromatic blur when luminance components of the image are in focus (Wandell, 1995) and may account for the blurring of S-cone patterns induced by blurred luminance surrounds (Figs 6 and 7). Thus it may be the failure of color to support the perception of edges, rather than a failure to encode spatial structure, that accounts for the differences we observed.

Our results also highlight the different effects of adaptation on threshold sensitivity and suprathreshold blur perception. Observers were highly sensitive to blur in most luminance images and rapidly adapted to modest changes in the blur or sharpness of the adapting images. In contrast, the changes in detection thresholds showed a comparatively weak dependence on the specific adapting slope and strong sensitivity changes even to the $1/f$ patterns to which the visual system is normally exposed and thus presumably matched. Conversely, sensitivity and adaptation to blur in the chromatic patterns were weaker even though the threshold changes were substantially larger than for luminance. Thus it was generally not possible to predict the changes in blur perception from the ways in which adaptation did or did not alter the threshold contrast sensitivity function. This is consistent with our finding that the changes in the csf with adaptation cannot be accounted for by independent gain changes at different spatial scales. Instead, they appear to reflect an adjustment to the lowest spatial 'grain' in the image. The csf is routinely used to characterize the spatial sensitivity of the visual system for luminance and chromatic stimuli. However, our results show that the shape of the csf may change substantially under states of adaptation that are likely to characterize natural viewing conditions, even to the extent that the lowpass spatial sensitivity that is considered a hallmark of color vision can give way to a bandpass function more typical of luminance coding. Perhaps more importantly however, these differences in threshold sensitivity do not readily translate into differences in other perhaps more ecologically important spatial judgments, such as the perception of blur. This suggests that in relating visual coding to natural images it may be important to choose a dependent measure that is itself natural and functionally important.

Acknowledgement

This research was supported by National Eye Institute Grant EY-10834.

REFERENCES

Atick, J. J. (1992). Could information theory provide an ecological theory of sensory processing? *Network* **3**, 213–251.

Atick, J. J. and Redlich, A. N. (1992). What does the retina know about natural scenes? *Neural Computation* **4**, 196–210.

Atick, J. J., Li, Z. and Redlich, A. N. (1992). Understanding retinal color coding from first principles, *Neural Computation* **4**, 559–572.

Boynton, R. M. (1978). Ten years of research with the minimally distinct border, in: *Visual Psychophysics: Psychophysics and Physiology*, Armington, J. C., Krauskopf, J. and Wooton, B. (Eds). Academic Press, New York.

Boynton, R. M., Hayhoe, M. M. and MacLeod, D. I. A. (1977). The gap effect: chromatic and achromatic visual discrimination as affected by field separation, *Optica Acta* **24**, 159–177.

Bradley, A., Switkes, E. and De Valois, K. K. (1988). Orientation and spatial frequency selectivity of adaptation to color and luminance gratings, *Vision Research* **28**, 841–856.

Burton, G. J. and Moorhead, I. R. (1987). Color and spatial structure in natural scenes, *Applied Optics* **26**, 157–170.

Chubb, C., Sperling, G. and Solomon, J. A. (1989). Texture interactions determine perceived contrast, *Proc. Nat. Acad. Sci. USA* **86**, 9631–9635.

De Valois, K. K. (2003). The role of color in spatial vision, in: *The Visual Neurosciences*, Chalupa, L. M. and Werner, J. S. (Eds), Vol. 2, pp. 924–935. MIT Press, Cambridge.

D'Zmura, M. and Singer, B. (1999). Contrast gain control, in: *Color Vision: From Genes to Perception*, Sharpe, L. T. and Gegenfurtner, K. R. (Eds), pp. 369–385. Cambridge University Press, Cambridge.

Eskew, R. T., Stromeyer, C. F., III., Picotte, C. J. and Kronauer, R. E. (1991). Detection uncertainty and the facilitation of chromatic detection by luminance contours, *J. Optic. Soc. Amer. A* **8**, 394–403.

Field, D. J. (1987). Relations between the statistics of natural images and the response properties of cortical cells, *J. Optic. Soc. Amer. A* **4**, 2379–2394.

Field, D. and Brady, N. (1997). Visual sensitivity, blur, and the sources of variability in the amplitude spectra of natural images, *Vision Research* **37**, 3367–3383.

Galvin, S. J., O'Shea, R. P., Squire, A. M. and Govan, D. G. (1997). Sharpness overconstancy in peripheral vision, *Vision Research* **37**, 2035–2039.

Klein, S. and Stromeyer, C. F., III. (1980). On inhibition between spatial frequency channels: adaptation to complex gratings, *Vision Research* **20**, 459–466.

Krauskopf, J. and Farell, B. (1991). Vernier acuity: Effects of chromatic content, blur and contrast, *Vision Research* **31**, 735–749.

Krauskopf, J., Williams, D. R. and Heeley, D. W. (1982). Cardinal directions of color space, *Vision Research* **22**, 1123–1131.

Lythgoe, J. N. (1979). *The Ecology of Vision*. Clarendon Press, Oxford.

MacLeod, D. I. A. and Boynton, R. M. (1979). Chromaticity diagram showing cone excitation by stimuli of equal luminance, *J. Optic. Soc. Amer.* **69**, 1183–1186.

MacLeod, D. I. A. and von der Twer, T. (2003). The pleistochrome: optimal opponent codes for natural colours, in: *Colour Perception: Mind and the Physical World*, Mausfeld, R. and Heyer, D. (Eds), pp. 155–184. Oxford University Press, Oxford.

Montag, E. D. (1997). Influence of boundary information on the perception of color, *J. Optic. Soc. Amer. A* **14**, 997–1006.

Mullen, K. T. (1985). The contrast sensitivity of human colour vision to red-green and blue-yellow chromatic gratings, *J. Physiol.* **359**, 381–409.

Mullen, K. T. and Kingdom, F. A. A. (1991). Colour contrast in form perception, in: *Vision and Visual Dysfunction 6: The Perception of Colour*, Gouras, P. (Ed.), pp. 198–217. Macmillan, London.

Nachmias, J., Sansbury, R., Vassilev, A. and Weber, A. (1973). Adaptation to square-wave gratings: in search of the elusive third harmonic, *Vision Research* **13**, 1335–1342.

Párraga, C. A., Brelstaff, G., Troscianko, T. and Moorehead, I. R. (1998). Color and luminance information in natural scenes, *J. Optic. Soc. Amer. A* **15**, 563–569.

Párraga, C. A., Troscianko, T. and Tolhurst, D. J. (2002). Spatiochromatic properties of natural images and human vision, *Current Biology* **12**, 483–487.

Pinna, B., Brelstaff, G. and Spillmann, L. (2001). Surface color from boundaries: a new 'watercolor' illusion, *Vision Research* **41**, 2669–2776.

Ruderman, D. L., Cronin, T. W. and Chiao, C.-C. (1998). Statistics of cone responses to natural images: implications for visual coding, *J. Optic. Soc. Amer. A* **15**, 2036–2045.

Simoncelli, E. P. and Olshausen, B. A. (2001). Natural image statistics and neural representation, *Ann. Rev. Neurosci.* **24**, 1193–1216.

Smeulders, N., Campbell, F. W. and Andrews, P. R. (1994). The role of delineation and spatial frequency in the perception of colours in the spectrum, *Vision Research* **34**, 927–936.

Snowden, R. J. and Hammett, S. T. (1992). Subtractive and divisive adaptation in the human visual system, *Nature* **355**, 248–250.

Srinivasan, M. V., Laughlin, S. B. and Dubs, A. (1982). Predictive coding: a fresh view of inhibition in the retina, *Proc. Roy. Soc. London B* **216**, 427–459.

Switkes, E., Bradley, A. and De Valois, K. K. (1988). Contrast dependence and mechanisms of masking interactions among chromatic and luminance gratings, *J. Optic. Soc. Amer. A* **5**, 1149–1162.

Switkes, E., Bradley, A. and Schor, C. (1990). Readily visible changes in color contrast are insufficient to stimulate accommodation, *Vision Research* **30**, 1367–1376.

Tadmor, Y. and Tolhurst, D. J. (1994). Discrimination of changes in the second-order statistics of natural and synthetic images, *Vision Research* **34**, 541–554.

Tolhurst, D. J. (1972). Adaptation to square-wave gratings: inhibition between spatial frequency channels in the human visual system, *J. Physiol.* **226**, 231–248.

Tolhurst, D. J. and Barfield, L. P. (1978). Interactions between spatial frequency channels, *Vision Research* **18**, 951–958.

Wachtler, T., Lee, T.-W. and Sejnowski, T. J. (2001). Chromatic structure of natural scenes, *J. Optic. Soc. Amer. A* **18**, 65–77.

Wandell, B. A. (1995). *Foundations of Vision.* Sinauer, Sunderland, MA.

Webster, M. A. and Miyahara, E. (1997). Contrast adaptation and the spatial structure of natural images, *J. Optic. Soc. Amer. A* **14**, 2355–2366.

Webster, M. A. and Mollon, J. D. (1997). Adaptation and the color statistics of natural images, *Vision Research* **37**, 3283–3298.

Webster, M. A., De Valois, K. K. and Switkes, E. (1990). Orientation and spatial-frequency discrimination for luminance and chromatic gratings, *J. Optic. Soc. Amer. A* **7**, 1034–1049.

Webster, S. M., Webster, M. A., Taylor, J., Jaikumar, J. and Verma, R. (2001). Simultaneous blur contrast, *Proc. SPIE* **4299**, 414–422.

Webster, M. A., Georgeson, M. A. and Webster, S. M. (2002). Neural adjustments to image blur, *Nature Neurosci.* **5**, 839–840.

Williams, D. R. and Collier, R. J. (1983). Consequences of spatial sampling by a human photoreceptor mosaic, *Science* **221**, 385–387.

Wolfe, J. M. and Owens, D. A. (1981). Is accommodation colorblind? Focusing chromatic contours, *Perception* **10**, 53–62.

Wuerger, S. M., Owens, H. and Westland, S. (2001). Blur tolerance for luminance and chromatic stimuli, *J. Optic. Soc. Amer. A* **18**, 1231–1239.

Color, Line, and Space, pp. 31–44 (2006)

Brightness contrast inhibits color induction: evidence for a new kind of color theory

JAMES GORDON [1,*] and ROBERT SHAPLEY [2]

[1] *Psychology Department, Hunter College of CUNY, 695 Park Avenue, New York, NY 10021, USA*
[2] *Center for Neural Science, New York University, 4 Washington Place, New York, NY 10003, USA*

Abstract—A gray region can be made to look colored by a colored surround. This phenomenon, chromatic induction, depends on color differences around the boundary of the region. We performed experiments on chromatic induction with small, initially achromatic, targets on nine different colored surrounds ranging in color from blue to red. Using scaling of saturation as our measure of perceived color strength, we found that chromatic induction is at its maximum when the brightness contrast at the boundary between target and surroundings is minimal. This implies that the neural mechanism in the cerebral cortex that mediates the appearance of brightness at a boundary inhibits the activity of chromatic mechanisms at that same boundary. Observers matched the apparent brightness and luminance of each of the colored surrounds. For surround colors where brightness and luminance matches differ, brightness contrast, not luminance contrast, controls chromatic induction. These new findings, taken together with other evidence, require a new theory of color appearance that includes mutually inhibitory interactions between color and brightness mechanisms that are sensing color and brightness contrast at visual boundaries.

Keywords: Color induction; brightness contrast; color contrast; saturation; mutual inhibition; edges; Kirschmann's 3rd Law.

INTRODUCTION

The perception of the color of a surface depends strongly on the color difference across the boundary of the surface, on visual edge contrast (Krauskopf, 1963; Yarbus, 1967; Valberg, 1974; Grossberg and Mingolla, 1985; Krauskopf *et al.*, 1986; Ratliff, 1992; Shevell and Wei, 1998). One of the most striking visual phenomena illustrating the power of edges to influence the visual perception of surfaces they enclose is chromatic induction: the visual perception of surface hue on an achromatic surface caused by the hue around its boundary. Even though

*To whom correspondence should be addressed. E-mail: jim@cns.nyu.edu

one is aware that the interior surface has not changed physically, it changes hue perceptually. There are some results that suggest that much of the hue assigned to a surface by the brain is a result of edge contrast rather than local reflectance. In stabilized vision, hue fills in long distances from an unstabilized boundary (Krauskopf, 1963; Yarbus, 1967), and this can even be seen with voluntary fixation in the periphery of the visual field (Krauskopf, 1963).

Mysteriously, chromatic induction can be variable in strength, and sometimes seems weak, especially in unconvincing textbook demonstrations. This variability can be accounted for in part by our present findings that magnitude of brightness contrast at an edge reduces chromatic induction from that boundary. Figure 1 is a demonstration of what we have found: this figure shows the induction of hue into gray targets by a green background. In this figure, the large surrounding regions are uniform in brightness and chromaticity — green to the left and gray to the right. The surrounds are adjusted to be approximately the same perceived brightness. On each surround there is a column of square-shaped targets that increase in brightness from bottom to top. At each height in the figure, the square in the gray surround and the square on the green surround are physically identical. The maximal induction of redness occurs in the targets in the middle of the figure on the green surround. These are the squares whose brightnesses are nearly equal to that of the green surround (see Fig. 1, legend). In perceptual experiments using many different background hues, we measured that maximal chromatic induction occurs always at or around the point of equal brightness between target and surround, whatever the hue of the inducing background. When the target is brighter or darker than the surrounding colored region, the induced hue is less. This implies that there are inhibitory interactions in the visual system between achromatic and chromatic signals evoked at surface boundaries.

Our findings bear on a long-standing dispute about the fundamentals of spatial effects on color perception. The laws of chromatic induction were early findings in psychophysics. Kirschmann's 3rd Law (Kirschmann, 1891) stated that chromatic induction was greatest when brightness contrast was minimal. Subsequently the 3rd Law was disputed, particularly by Kinney (1962), who stated that chromatic

Figure 1. (See color plate V) Demonstration of dependence of chromatic induction on brightness contrast. There are nine square targets on each rectangular surround; each horizontal pair of targets is identical in spectral energy distribution and brightness. On an achromatic surround each target appears colorless (right panel) and target brightness increases from bottom to top. On the green surround (left panel), the targets near the middle of the picture take on a pink hue complementary to the hue of the surround.

The two surrounds are approximately matched for brightness, and the fifth target in the middle of the vertical array is equal in brightness (this may vary somewhat across observers and across displays) to the brightness of the gray surround — so the small square target disappears on the right hand side because the small square and the background are the same brightness (that is, there is zero contrast). For most subjects the fifth target in the left (green) panel is perceived as the most saturated pink of the nine targets; it is the target with least brightness contrast. Saturation decreases for targets both below and above this target.

induction was proportional to the luminance ratio between inducing and induced areas: the higher the luminance of the induced area, the less the induction. Jameson and Hurvich (1959) in one particular version of their opponent color theory considered a mechanism for chromatic induction consistent with Kinney's results (1962). This mechanism therefore was of necessity not sensitive to the magnitude of the luminance contrast between target (induced region) and surrounding (inducing) region. Specifically, the theory of Jameson and Hurvich calculated apparent color from the color and brightness signals evoked from the interior of regions, rather than those evoked by boundary contrast (see Discussion below for a full consideration of opponent color theory). Others have examined the question of the rules of chromatic induction with several different experiments but did not reach a definitive conclusion (Jameson and Hurvich, 1959; Bergstrom *et al.*, 1978). We believe that their failure to obtain clear results is related to the same methodological problems that caused Kinney to dispute the Kirschmann result. We will examine this controversy in detail in the Discussion after presentation of our results. The main qualitative empirical difference between Kirschmann and Kinney is the following: Kirschmann's 3rd Law predicts that chromatic induction from a surrounding region into a darker test (induced) region will be less than into a test region that is equal in brightness to the inducing region. Kinney's work implies chromatic induction will be greater into the darker test (induced) region. Our experiments confirm decisively that chromatic induction is less into a dark gray than into a gray target that is of equal brightness with its chromatic surround, for a large set of surrounding hues. Readers can see this result directly already in the demonstration of Fig. 1 where the amount of perceived hue in the target is diminished for targets darker than the chromatic background. We will return to this issue in the Discussion.

The results of our experiments require a revision of the classical opponent mechanisms theory of color perception to include the mutually inhibitory effect of brightness contrast and chromatic contrast on each other. The classical formulation does not account for the inhibitory effect of strong negative brightness contrast on chromatic induction. We offer a preliminary version of a revised theory of color appearance in the Discussion.

METHODS

Participants

Five observers, three males and two females, ranging in age from 22 to 55 participated in this experiment. Color vision was assessed with Ishihara Plates and the Farnsworth-Munsell 100 hue test — all observers were color normal.

Apparatus and procedures

Circular 1.6 deg foveal achromatic stimuli (CIE coordinates $x = 0.33$, $y = 0.33$), ranging in luminance from 3 to 60 cd/m^2, were presented for 5 seconds surrounded contiguously by various chromatic annular surrounds. All stimuli were presented using a Silicon Graphics Indigo Computer and 19″ Silicon Graphics Monitor (256 gray levels, 60 Hz frame rate, 1280×1024 pixels) viewed at a distance of 1 m. Stimuli were calibrated with a Photo Research Spectrascan 703a spectroradiometer and a look-up table used for linearization. The nine surround colors with their CIE x, y values, and luminances (in cd/m^2) were blue/red (0.399, 0.208, 22.7), blue (0.152, 0.068, 6.4), blue/green (0.164, 0.114, 9.8), green/blue (0.205, 0.286, 38.2), green (0.235, 0.401, 34.5), green/yellow (0.284, 0.594, 38.2), yellow (0.458, 0.466, 35.6), red/yellow (0.558, 0.392, 27.7), and red (0.627, 0.338, 18.8). To maintain neutral adaptation, the 5 observers viewed an achromatic field (CIE $x = 0.33$, $y = 0.33$; 24 cd/m^2) between each trial. Observers used saturation scaling (100% indicated a target that appeared completely saturated, 0% a target that appeared achromatic) to specify the amount of chromatic induction. Scaling values for four repeats of each presentation were averaged. This technique, introduced by Jameson and Hurvich (1959) has been shown to be a very rapid and reliable method for determining the quantity of hue (Gordon *et al.*, 1994). Even though absolute saturation values are variable across observers, they are very consistent in the relative changes that observers show across conditions. In conjunction with hue scaling, saturation scaling yields color distance metrics such as wavelength discrimination that are indistinguishable from those derived using more traditional methods.

In separate experiments we measured the luminance of test fields needed to produce equibrightness and equiluminance values for each observer. We used heterochromatic flicker photometry (HFP) on the same SGI monitor as that on which the induction experiments were displayed, to measure an observer's individual equiluminant values for each of the chromatic surrounds. Each observer adjusted the luminance of a variable achromatic target, alternated at 15 Hz with each chromatic stimulus, to minimize flicker. The procedure was repeated three times for each chromatic surround with different starting points for each trial, and the average of the three settings was taken as the equiluminant value. We will call these luminance values F-match.

The observers also rated the brightness of each of the achromatic test stimuli as brighter or dimmer than the chromatic surrounds in order to find their equal-brightness values. The brightness matches were performed on stimuli presented for three seconds. Each of the 12 achromatic test stimuli was presented in a random order within a given chromatic surround (100 trials for each of the nine surrounds). The average of repeated comparisons (see Fig. 2 for examples) were fitted with psychometric functions and the means of the fitted functions used to estimate the equal-brightness luminances of the test for each surround. We will call these luminance values B-match.

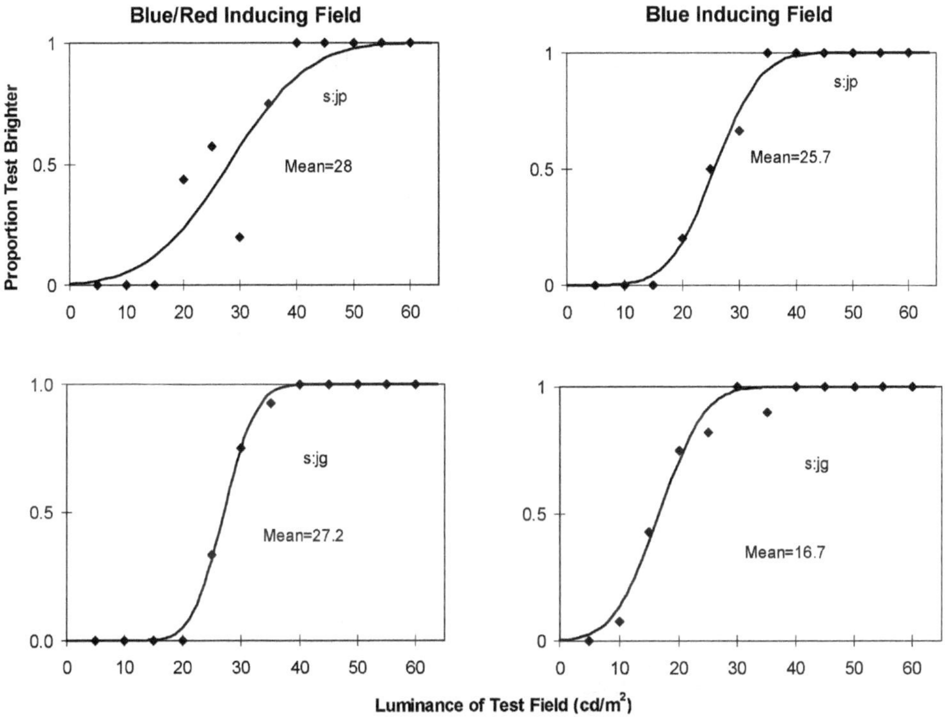

Figure 2. Proportion of times each gray test field was judged brighter than the surrounding chromatic stimulus. Data are shown for two observers, for two surrounds. The fitted curve is a cumulative normal function; the mean of this function is taken as the value at which the test spot's brightness equals that of the chromatic surround.

RESULTS

In the experiments, observers rated apparent color strength or saturation for circular targets of different intensities on colored backgrounds (as described in the Methods section). The targets appeared as different shades of white or gray or black when placed against a neutral background, but some of the 'gray' targets were noticeably colored (with an approximately complementary hue) when surrounded by color. The colored surround's luminance was held fixed. The graphs in Fig. 3 plot the apparent saturation of the target's (induced) color as a function of the luminance of the target, for two observers on two different colored surrounds. These are representative data sets. In all cases the scaled saturation peaked at an intermediate target luminance, as is evident in Fig. 3 (and also in the demonstration in Fig. 1).

The question is: how does the target's brightness and/or luminance compare to its surround, when the scaled saturation is maximal? To answer this question, we measured the luminance of the test, for each surround, at which induced chromatic saturation reached its peak (see Methods). We will call this value MaxSat. Then we determined the luminance of the test required to make it equal in brightness (B-match) to each colored surround. Finally (using flicker photometry) we found the luminance of the test that made it equiliminant for each observer (F-match) to each colored surround. MaxSat, B-match, and F-match means are shown in Fig. 4 for each of the nine chromatic surrounds.

For all observers, MaxSat was approximately equal to B-match. Thus, chromatic induction was maximal (MaxSat) when the achromatic central target had a brightness (B-match) that was approximately equal to the brightness of its chromatic surround. Chromatic induction was weaker when the central target was brighter *and also when it was dimmer* than the inducing colored surround. It is worth noting that, in these experiments, the luminance of the target was the same for B-match and F-match (see Methods) for most colored backgrounds we used. This result seemed

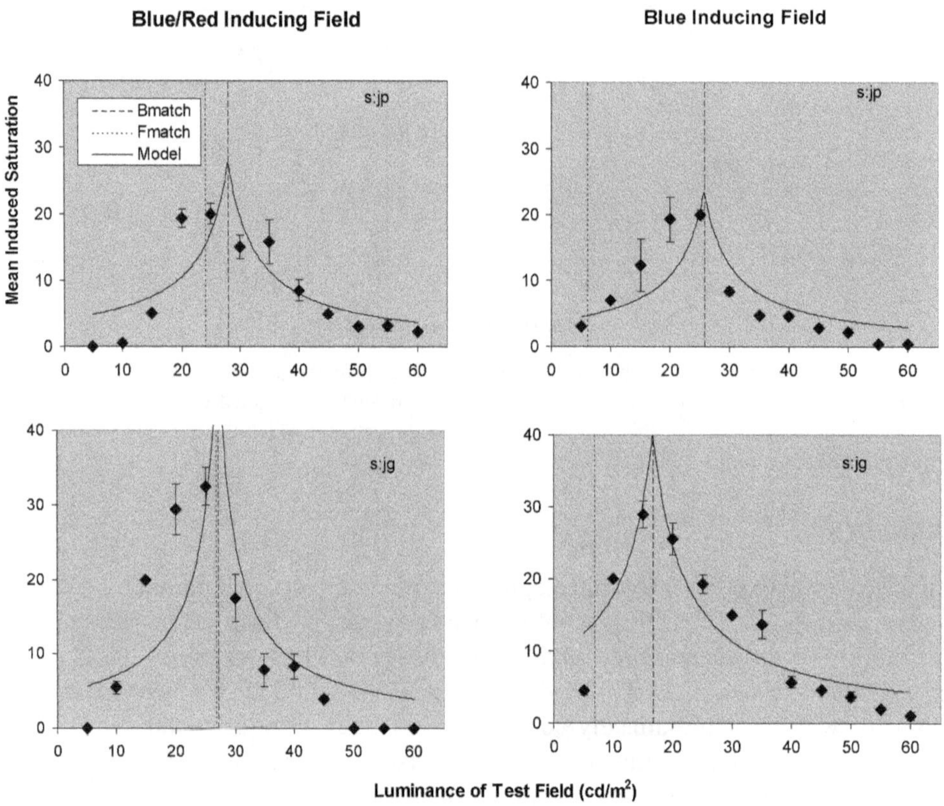

Figure 3. The apparent saturation of the target's (induced) color as a function of the target's luminance, for two subjects observing targets on two different colored surrounds. These are representative data sets of saturation scaling (see Methods section) by 2 out of our population of 5 subjects, with 2 of the 9 surround colors. The main feature of the data is that in all cases the apparent saturation peaked at an intermediate target luminance, near the luminance that was equal in brightness to that of the chromatic surround.

The points (means of four trials) with error bars (± one SEM) are fitted with continuous curves derived from our new proposed nonlinear mutual inhibition theory that is explained in the text. The vertical fine dashed and coarse dashed lines indicate the subjective equiluminance (F-match) and equal brightness (B-match) values of the target luminance, respectively.

a little surprising to us at first (*cf.* Wagner and Boynton, 1972; Burns *et al.*, 1982) but in fact similar results have been reported previously (Ayama and Ikeda, 1998). Ayama and Ikeda found, as did we, that while all observers required increased luminance for brightness matches to blue lights, some observers did not require this increase in luminance for brightness matches to red lights. For those colors where B-match = F-match, we could not decide whether it was equal brightness or equal luminance that was necessary for maximum color induction. However, for blue and blue-green surrounds, B-match and F-match were quite far apart. For blue surrounds, maximum induction was nearer to the B-match than to the F-match (Fig. 3, right panels). Therefore, based on the data from chromatic induction on blue and blue-green surrounds, we conclude that it is brightness not luminance that must be equated for maximal chromatic induction.

It is also of interest to note that while the values for F-match shown in Fig. 4 differ across the colored surrounds by more than a factor of four, the values for B-match across the colored backgrounds are within a factor of two. That means that the apparent brightnesses of the colored inducing fields used in our experiments were within a factor of two of each other.

We present the relative differences of our measured maximum saturation values, across the range of inducing colors, in Fig. 5. The figure displays {(MaxSat − B-match)/MaxSat}, the relative difference between the luminance of the target for peak color induction, MaxSat, and the luminance at which it was equal in brightness

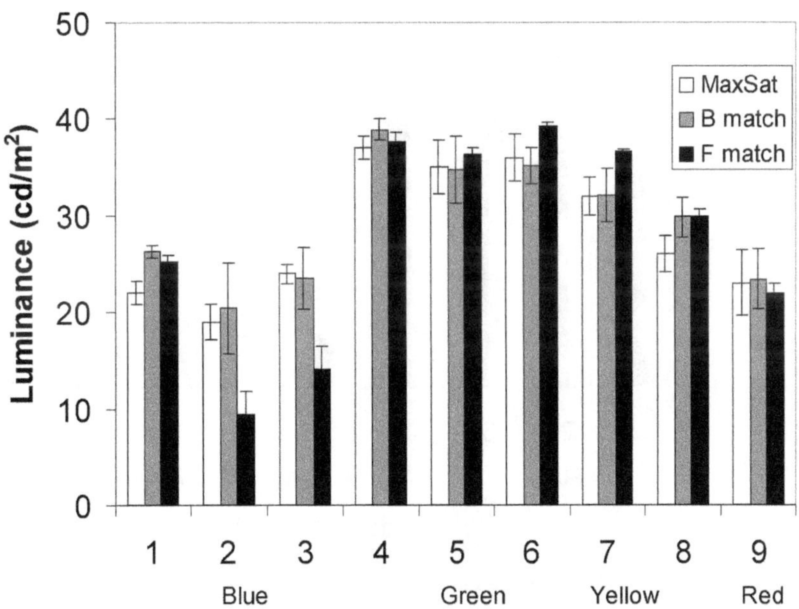

Figure 4. Luminances of the target fields for each of the inducing fields. The points show equiluminance (F match), equibrightness (B match) and maximum scaled saturation (MaxSat). Points are means across the five observers, error bars are ± one SEM. For all background colors the luminance for MaxSat is like that for B match.

to the colored surround (B-match): this ratio is equivalent to the brightness contrast at the point of maximum color induction. For all colors we used, maximal induction is at minimal brightness contrast. The amount of brightness contrast needed to reduce chromatic induction to half of its maximum was computed to be an average of 36.5% across all observers and inducing colors. Thus there must be a substantial amount of brightness contrast present to suppress chromatic induction, approximately 30–50× the threshold for brightness detection. This is consistent with the results of Miyahara *et al.* (2001) who found no effect on magnitude of induction for small changes in luminance around the isoluminant point.

Figure 5 also shows {(MaxSat − F-match)/MaxSat}, the relative difference between the luminance of the target for MaxSat and the value determined by flicker photometry, F-match. It is clear that in the blue region of the spectrum, where F-match and B-match differ from one another, there is a large relative difference between the test required to achieve maximum saturation and that required for equiluminance.

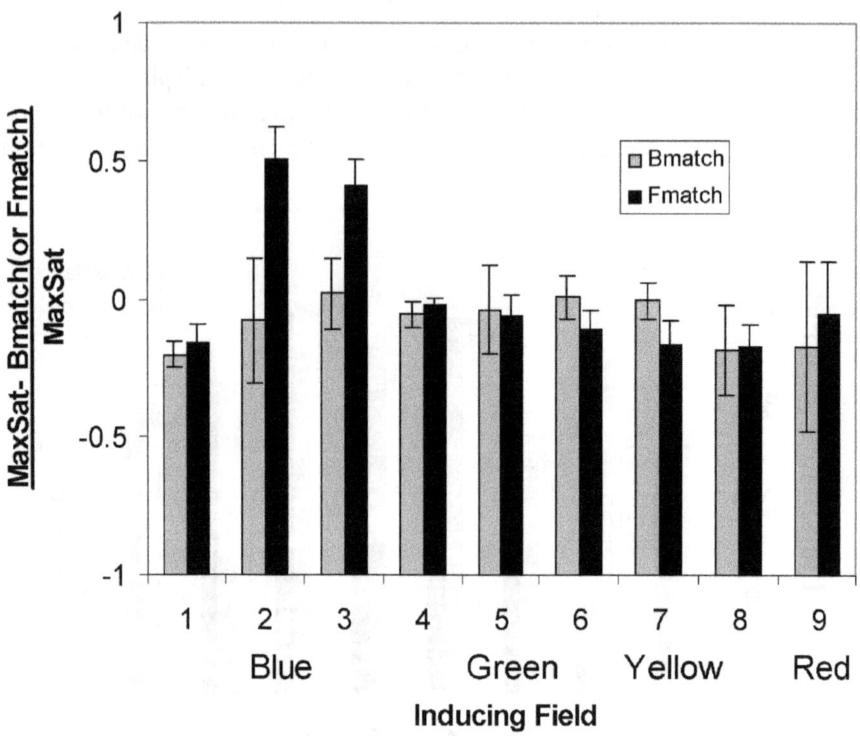

Figure 5. The difference between the luminance of the target for peak color induction MaxSat and the luminance at which it was equal in subjective brightness to the colored surround (B match) or equiluminant to the colored surround (F match), normalized by MaxSat: this is equivalent to the contrast at the point of maximum color induction. For all nine colors we used, maximal induction is at minimal contrast when targets were adjusted by B match but not by F match. Points are means across the five observers, error bars are ± one SEM.

DISCUSSION

Our data prove that chromatic induction is greatest when there is no brightness contrast between a test area and its surround, in support of Kirschmann's 3rd Law (Kirschmann, 1891). Brightness contrast, whether of positive or negative sign, suppresses chromatic induction. Brightness suppresses color just as color suppresses brightness (Alpern, 1964), at inducing boundaries. Based on the results with the blue and blue/green surrounds (Figs 3, 4, and 5), one can conclude that it is not the luminance (F-match) of a region but its brightness (B-match) that is crucial in the effect on chromatic induction. The dependence on apparent brightness carries some implications about the site in the visual system at which chromatic induction occurs. It is generally believed that luminance signals depend on the activity of the M-magnocellular system, and that signals for luminance are elaborated first in the retina and then passed to magnocellular LGN and thence to V1 (Shapley, 1990; Lee *et al.*, 1990). Signals about brightness and hue are generated in the cerebral cortex in extrastriate visual areas (DeValois and DeValois, 1993). Our evidence suggests that neural interactions in color must take place at a central nervous system site at which brightness is computed, presumably at higher levels of the visual system in extrastriate visual areas of the cerebral cortex.

As suggested above, there was a direct experimental test of Kirschmann's 3rd Law by Kinney (1962), the results of which were interpreted to mean that chromatic induction grows stronger and stronger the darker is the test target. However, our experimental results are clear: chromatic induction is strongest when the test target and inducing surround are nearly the same brightness, as can be observed for instance in Fig. 1. It is important to try to understand why Kinney (1962) obtained different results. Kinney (1962) reported that induction grew stronger on relatively darker test targets in experiments in which she varied the inducing surround's luminance and held the target's luminance fixed. This is the converse of what we did in our experiments, where the target luminance varied on a chromatic surround of fixed luminance. The demonstration in Fig. 1 parallels the design of our experiments, with a fixed inducing surround and varying target luminance. To test whether or not it matters whether the target or the inducing surround varies in luminance, we created the demonstration in Fig. 6. This figure uses the experimental design of Kinney (1962). The colorimetric purity of the seven green surrounding fields is the same but they are at seven different luminances, ascending from top to bottom in luminance. This was done by keeping the ratio of red/green primaries fixed in the color-inducing surrounding regions, and varying their luminances. Under these conditions, most people observe that chromatic induction is greatest when the subjective brightness of the target and the colored surround are nearly equal, and that color induction is reduced as surround brightness increases above the point of equality. (It is also perceived that color induction is diminished when the color-inducing surround is darker than the target, but this is not a point of contention between the classical studies; everyone observes this latter effect.)

So the explanation for disagreements about chromatic induction is not which brightness, that of the target or that of the surround, is varied. Rather we think it has to do with how the amount of hue is measured as brightness varies. When Kinney (1962) observed more color induction at backgrounds brighter than equal-brightness, backgrounds that caused the targets to appear quite dark, this result could have occurred because she used a matching technique. To match the appearance of the darkened targets, her observers may have had to add quite a lot of color to the dark matching squares, more than they used to match the targets in the equal brightness case. This is because dark targets of a given colorimetric purity appear less saturated with color than targets of a higher brightness. The matching technique would say there is more color in the darker targets. Nevertheless, the appearance of these dark targets, targets on a higher than equal brightness background, is of much lower color saturation than equal brightness targets as one can see in Figs 1 and 6, and in our data (Fig. 2 for instance).

While this difficulty in measuring color appearance with a matching technique can account for Kinney's data, it does not fully explain the results of Bergstrom and colleagues (Bergstrom and Derefeldt, 1975; Bergstrom *et al.*, 1978). They used either saturation scaling or hue cancellation, but not matching, and varied target luminance as we did, and surround luminance as Kinney (1962) did. Under some conditions they obtained results like ours (maximum chromatic induction at equal brightness) and under others they did not. They pointed out that the outcome was quite sensitive to several variables, including pre-adaptation. We believe this may be the key point explaining the variability of Bergstrom's results, and the consistency of our results. The way we ran our experiments with a gray field, with a brightness approximately equal to that of the colored surround region, as a pre-adaptation field, and with relatively long observation times, prevented fluctuations in pre-adaptation from having a big effect on our results. Bergstrom *et al.* (1978) also worked at a lower mean luminance than we did and it is possible that color induction is weaker or more variable at dimmer light levels. One or a combination of these factors could

Figure 6. (See color plate V) Color induction with surrounds that vary in luminance. This demonstration is the converse of that shown in Fig. 1. Here the targets are all equal to one another in luminance and all have the same spectral energy distribution. They are neutral in color on an achromatic surround, as in the right hand side of the figure. Changes in the appearance of the targets are caused by the different surrounds. The seven green surrounds on the left were designed to have the same shaped spectral energy distribution (constant colorimetric purity) but to increase in brightness from top to bottom of the figure. Each gray surround on the right was subjectively matched in brightness to the green surround next to it.

The gray target on the middle surround on the left is nearly matched in brightness with its green surround, as is the corresponding gray target on gray surround on the right. The target on the middle background on the left, on the surround it nearly matches in brightness, appears to most observers the most colored (a dark pink). The crucial comparison is with the target just below it, placed on a somewhat brighter green surround. To most observers this lower target appears darker and less saturated than the physically matching target just above it. Note that the matches may vary somewhat from observer to observer and from display to display.

have caused Bergstrom *et al.* not to obtain the clear confirmation of Kirschmann's 3rd Law that we found.

Our results support a theoretical picture of highly interactive cortical mechanisms for hue and brightness perception. Hue and brightness signals are strongest at boundaries between regions (Ratliff, 1985; Cornsweet, 1970). Evidence for the importance of boundaries comes from the filling-in of color in stabilized images (Krauskopf, 1963; Yarbus, 1967), and from the phenomenon of Gauzkontrast: the enhancement of chromatic contrast in a visual pattern when it is viewed through gauze or coarse cloth such that edges are effectively masked (Berliner, 1949, p. 31). At edges, we suppose that there is mutual inhibition between neural mechanisms for hue and brightness in a kind of 'winner-take-all' network. Perception is ruled by the strongest signal for boundary identification. This is consistent with masking experiments (De Valois and Switkes, 1983; Switkes *et al.*, 1988), with previous results on suppression of brightness contrast by color contrast (Alpern, 1964), and with experiments on darkness induction (Shinomori *et al.,* 1997).

This edge-based color theory is also consistent with the conceptual framework of opponent-mechanisms theory for color vision (proposed by Hering, 1920) and elaborated later (Jameson and Hurvich, 1961; Eskew *et al.*, 1991). However, to account for our results one must include four new ideas: (1) that the opponent mechanisms are not independent but *interact strongly* in influencing color perception; (2) that neuronal signals about chromatic induction are *mainly evoked at edges of regions*; (3) that the responses of the putative color-induction mechanisms are *functions of the edge contrast*; and (4) that *interactions* between color mechanisms (including a black-white mechanism) *take place at the neural representations of these edges.* For instance, our results require that the magnitude of brightness contrast at a visual edge evokes an achromatic signal that suppresses color contrast signals evoked by that same edge. Results on chromatic masking of achromatic patterns (Alpern, 1964; De Valois and Switkes, 1983; Switkes *et al.*, 1988) suggest there is a corresponding chromatic inhibitory signal that suppresses brightness contrast. More generally, we hypothesize that all the multiple chromatic and achromatic mechanisms that are excited at an edge engage in a mutually inhibitory interaction. Note that, at very low contrasts near threshold, some interactions may be facilitatory (D'Zmura and Lennie, 1986; De Valois and Switkes, 1983; Switkes *et al.*, 1988). Chromatic induction is a special case of the general case of colored regions surrounded by other colored regions; the interactions that have become evident in chromatic induction are presumably active in every color judgment (*cf.* Hering, 1920).

Formulations of color theory have not so far taken such contrast interactions into account in a quantitative way (Jameson and Hurvich, 1959; Eskew *et al.*, 1991). Jameson and Hurvich (1964) do consider Kirschmann's 3rd law briefly and indicate that it may be correct, and that it may be that it is the induction of Blackness or Whiteness that causes the decrease in the saturation of the induced colors. But they did not incorporate this idea into a quantitative model, and in the 1964 paper they refer in a misleading fashion to their earlier work (Jameson and Hurvich,

1959) in which the mathematical model and some of the data actually contradict Kirschmann's third law. Thus, our results necessitate a new color theory that includes explicitly the contribution of nonlinear mutual inhibition between color and brightness contrasts at edges, a theory that begins to quantify the observations we have offered in this paper. Here is one initial working hypothesis for such a new color theory: that perceived color saturation in a chromatic induction experiment is the ratio of stimulus color contrast to the magnitude of stimulus brightness contrast when brightness contrast exceeds a criterion amount:

$$\%\text{Induced Saturation} = \frac{\text{Max}\%\text{Saturation}}{1 + \beta(\text{BrightnessContrast})}. \quad (1)$$

Brightness contrast is the absolute value of the difference between the luminance of each test target and the luminance of the test target matched in brightness to the surround, normalized by the luminance of the test target matched in brightness to the surround, thus {BrightnessContrast = |{(Lum(target)−(B-match))/(B-match)}|. It is assumed that Max%Saturation is proportional to edge contrast in the most responsive color mechanism at the target-surround boundary. Equation (1) accounts qualitatively for the data we have observed as evidenced by the qualitative fit of the solid curves and the points in Fig. 3 (see legend). The use of contrasts in equation (1) is based on the idea that neural signals in the visual system will be approximately proportional to contrast magnitude. So equation (1) is equivalent to the assumption that brightness and color neurons (or neural networks) mutually inhibit each other via a kind of nonlinear shunting inhibition. Other nonlinear mutual inhibitory interactions between brightness and color might be postulated: Equation (1) is a simple first approximation that accounts for the experimental data, at least qualitatively. Doubtless the theory could be refined to give better quantitative fit to the data, but at present we know of no alternative theory that accounts for these data even qualitatively. Note that the contrast as we have defined it is Weber Contrast (*cf.* Shapley and Enroth-Cugell, 1984). We also tried using Michelson (or Rayleigh) Contrast and it was clear that Weber Contrast yields better fits to our data.

Perception of the color of a reflecting surface depends upon discounting the color of the illumination; this is the primary purpose of mechanisms for color constancy (D'Zmura and Lennie, 1986; Blackwell and Buchsbaum, 1988; Brainard *et al.*, 1993). Chromatic induction is one major mechanism for color constancy. It tends to maintain the perception of a colored surface at its 'correct' color when the surface and its surroundings are illuminated by a colored light source that is peaked towards one end or the other of the visible spectrum. Our findings indicate that color constancy will fail more in scenes with high brightness contrast in them than in low contrast scenes because chromatic induction will be less in the high contrast scenes. This may have practical application in commercial design and in fine arts.

It is important to realize that chromatic induction and edge-based signals about color may not be the only color signals that are important in color perception. Recent studies of the neurophysiology of color vision (Shapley and Hawken, 2002;

Johnson *et al.*, 2004), coupled with classical psychophysics of color detection (Mullen, 1985), indicate that there is at least one other neuronal mechanism for color perception, one that is sensitive to local color signals and not to edge contrast. The edge-based system we have studied in this paper must in some way combine with the local region-based color system to generate the full rich experience of color.

Acknowledgements

We thank David Israel, Elizabeth Johnson, and Julie Pastagia for their participation and help in this study, and Israel Abramov for comments and discussion. This work was supported by NIH Grant EY01472.

REFERENCES

Alpern, M. (1964). Relation between brightness and color contrast, *J. Opt. Soc. Amer.* **54**, 1491–1492.

Ayama, M. and Ikeda, M. (1998). Brightness-to-luminance ratio of colored light in the entire chromaticity diagram, *Color Research and Application* **23**, 274–287.

Bergstrom, S. S. and Derefeldt, G. (1975). Effects of surround/test field luminance ratio on induced color, *Scand. J. Psychol.* **16**, 311–318.

Bergstrom, S. S., Derefeldt, G. and Holmgren, S. (1978). Chromatic induction as a function of luminance relations, *Scand. J. Psychol.* **19**, 265–276.

Berliner, A. (1949). *Lectures on Visual Psychology*. Professional Press, Chicago.

Blackwell, K. T. and Buchsbaum, G. (1988). Quantitative studies of color constancy, *J. Opt. Soc. Amer. A* **5**, 1772–1780.

Brainard, D. H, Wandell, B. A. and Chichilnisky, E. J. (1993). Color constancy: From physics to appearance, *Current Directions in Psychological Science* **2**, 165–170.

Burns, S. A., Smith, V. C., Pokorny, J. and Elsner, A. E. (1982). Brightness of equal-luminance lights, *J. Opt. Soc. Amer.* **72**, 1225–31.

Cornsweet, T. (1970). *Visual Perception*. Academic Press, New York.

De Valois, R. L. and De Valois, K. K. (1993). A multi-stage color model, *Vision Research* **33**, 1053–1065.

De Valois, K. K. and Switkes, E. J. (1983). Simultaneous masking interactions between chromatic and luminance gratings, *J. Opt. Soc. Amer.* **73**, 11–18.

D'Zmura, M. and Lennie, P. (1986). Mechanisms of color constancy, *J. Opt. Soc. Amer. A* **3**, 1662–1672.

Eskew, R. T., Jr., Stromeyer, C. F., 3rd, Picotte, C. J. and Kronauer, R. E. (1991). Detection uncertainty and the facilitation of chromatic detection by luminance contours, *J. Opt. Soc. Amer. A* **8**, 394–403.

Gordon, J., Abramov, I. and Chan, H. (1994). Describing color appearance: hue and saturation scaling, *Perception and Psychophysics* **35**, 2999–3010.

Grossberg, S. and Mingolla, E. (1985). Neural dynamics of form perception: boundary completion, illusory figures, and neon color spreading, *Psychol. Rev.* **92**, 173–211.

Hering, E. (1920). *Grundzüge der Lehre vom Lichtsinn*. Springer-Verlag, Berlin. *Outlines of a Theory of the Light Sense*, Translated by Hurvich, L. M. and Jameson, D. (1964). Harvard University Press, Cambridge, MA.

Jameson, D. and Hurvich, L. M. (1959). Perceived color and its dependence on focal, surrounding, and preceding stimulus variables, *J. Opt. Soc. Amer.* **49**, 890–898.

Jameson, D. and Hurvich, L. M. (1961). Opponent chromatic induction: experimental evaluation and theoretical account, *J. Opt. Soc. Amer.* **51**, 46–53.

Jameson, D. and Hurvich, L. M. (1964). Theory of brightness and color contrast in human vision, *Vision Research* **4**, 135–154.

Johnson, E. N., Hawken, M. J. and Shapley, R. (2004). Cone inputs in macaque primary visual cortex, *J. Neurophysiol.* **91**, 2501–2514.

Kinney, J. A. S. (1962). Factors affecting induced color, *Vision Research* **2**, 503–525.

Kirschmann, A. (1891). Ueber die quantitativen Verhältnisse des simultanen Helligkeits- und Farbencontrastes, *Philosophische Studien* **6**, 417–491.

Krauskopf, J. (1963). Effect of retinal image stabilization on the appearance of heterochromatic targets, *J. Opt. Soc. Amer.* **53**, 741–744.

Krauskopf, J., Zaidi, Q. and Mandler, M. B. (1986). Mechanisms of simultaneous color induction, *J. Opt. Soc. Amer. A* **3**, 1752–1757.

Lee, B. B., Pokorny, J., Smith, V. C., Martin, P. R. and Valberg, A. (1990). Luminance and chromatic modulation sensitivity of macaque ganglion cells and human observers, *J. Opt. Soc. Amer. A* **7**, 2223–2236.

Miyahara, E., Smith, V. C. and Pokorny, J. (2001). The consequences of opponent rectification: the effect of surround size and luminance on color appearance, *Vision Research* **41**, 859–871.

Mullen, K. T. (1985). The contrast sensitivity of human colour vision to red-green and blue-yellow chromatic gratings, *J. Physiol.* **359**, 685–688.

Ratliff, F. (1985). The influence of contour on contrast: from cave painting to Cambridge Psychology, *Trans. Amer. Phil. Soc.* **75**, 1–19.

Ratliff, F. (1992). *Paul Signac and Color in Neo-Impressionism.* Rockefeller University Press, New York.

Shapley, R. (1990). Visual sensitivity and parallel retinocortical channels, *Annual Review of Psychology* **41**, 635–658.

Shapley, R. and Enroth-Cugell, C. (1984). Visual adaptation and retinal gain controls, in: *Progress in Retinal Research*, Osborne, N. and Chader, G. (Eds), Vol. 3, pp. 263–346. Pergamon Press, Oxford.

Shapley, R. and Hawken, M. J. (2002). Neural mechanisms for color perception in the primary visual cortex, *Curr. Opin. Neurobiol.* **12**, 426–432.

Shevell, S. K. and Wei, J. (1998). Chromatic induction: border contrast or adaptation to surrounding light, *Vision Research* **38**, 1561–1566.

Shinomori, K., Schefrin, B. E. and Werner, J. S. (1997). Spectral mechanisms of spatially induced blackness: data and quantitative model, *J. Opt. Soc. Amer. A* **14**, 372–387.

Switkes, E., Bradley, A. and De Valois, K. K. (1988). Contrast dependence and mechanisms of masking interactions among chromatic and luminance gratings, *J. Opt. Soc. Amer. A* **5**, 1149–1162.

Valberg, A. (1974). Color induction: dependence on luminance, purity, and dominant or complementary wavelength of inducing stimuli, *J. Opt. Soc. Amer.* **64**, 1531–1540.

Wagner, G. and Boynton, R. M. (1972). Comparison of four methods of heterochromatic photometry, *J. Opt. Soc. Amer.* **62**, 1508–1515.

Yarbus, D. L. (1967). *Eye Movements and Vision.* Plenum Press, New York.

Color, Line, and Space, pp. 45–57 (2006)

Colour contrast influences perceived shape in combined shading and texture patterns

FREDERICK A. A. KINGDOM [1,*], KAREN WONG [1],
ALI YOONESSI [1] and GOKHAN MALKOC [2]

[1] *McGill Vision Research Unit, 687 Pine Av. W., Room. H4-14, Montreal, PQ, H3A 1A1, Canada*
[2] *Department of Psychology, Dogus University, Acıbadem-Kadıkoy, 34722, Istanbul, Turkey*

Abstract—The 'colour-shading effect' describes the phenomenon whereby a chromatic pattern influences perceived shape-from-shading in a luminance pattern. Specifically, the depth corrugations perceived in sinusoidal luminance gratings can be enhanced by spatially non-aligned, and suppressed by spatially aligned sinusoidal chromatic gratings. Here we examine whether colour contrast can influence perceived shape in patterns that combine shape-from-shading with shape-from-texture. Stimuli consisted of sinusoidal modulations of texture (defined by orientation), luminance and colour. When the texture and luminance modulations were suitably combined, one obtained a vivid impression of a corrugated depth surface. The addition of a colour grating to the texture-luminance combination was found to enhance the impression of depth when out-of-phase with the luminance modulation, and suppress the impression of depth when in-phase with the luminance modulation. The degree of depth enhancement and depth suppression was approximately constant across texture amplitude when measured linearly. In the absence of the luminance grating however, the colour grating had no phase-dependent affect on perceived depth. These results show that colour contrast modulates the contribution of shading to perceived shape in combined shading and texture patterns.

Keywords: Colour; texture; shape-from-shading.

INTRODUCTION

The role that colour vision plays in the analysis of form is of considerable interest to vision scientists (e.g. see review by Regan, 2000). A popular technique for studying this topic is to measure performance on a given form task using isoluminant (colour-only) and isochromatic (luminance-only) stimuli. However, much can be learnt about the role of colour vision in form perception by studying how colour and

*To whom correspondence should be addressed. E-mail: fred.kingdom@mcgill.ca

luminance interact in stimuli that emulate the spatio-temporal relationships between colour and luminance found in the natural visual world (Kingdom, 2003).

The 'colour-shading effect' (Kingdom, 2003; Kingdom *et al.*, 2005) is a recent example of this approach. When a chromatic grating is added to a differently-oriented luminance grating, one obtains an impression of a corrugated surface – an instance of 'shape-from-shading'. We term this 'depth-enhancement'. However, when a second chromatic grating of the same orientation and spatial phase as the luminance grating is added, the impression of depth is reduced or eliminated. We term this 'depth-suppression'. These observations show that shape-from-shading, hitherto studied primarily in the achromatic domain (Lehky and Sejnowski, 1988; Ramachandran, 1988; Attick *et al.*, 1996; Sun and Perona, 1997), can be profoundly influenced by colour contrast. Kingdom (2003) argued that the depth-enhancing and depth-suppressing capabilities of colour contrast observed in colour/luminance patterns reveal that the visual system has certain built-in assumptions. These are: (1) chromatic variations, and those luminance variations that are spatially aligned with them, arise from changes in surface reflectance; (2) pure, or near-pure luminance variations, arise from spatially non-uniform illumination, such as shading and shadows. Although the physical relationships that underpin these assumptions have been appreciated by vision scientists for some time (Rubin and Richards, 1982; Cavanagh, 1991; Mullen and Kingdom, 1991; Olmos and Kingdom, 2004), the colour-shading effect is the first evidence that these assumptions are built into the fabric of the human visual system.

In natural scenes there are often multiple cues to surface shape. Does the modulatory influence of colour contrast on the perception of surface shape occur with cues other than shading, such as texture, and does colour contrast influence the contribution of shading to surface curvature when it is present alongside other cues? It is possible that the influence of colour contrast on shape-from-shading is reduced, or even eliminated, when surface information other than colour is present, because in such circumstances the surface-*versus*-illumination interpretative role of colour contrast becomes redundant. The aim of the present study is to determine whether colour contrast influences perceived shape in patterns that combine shape-from-shading with shape-from-texture.

Orientation-defined textures are a powerful perceptual cue to surface shape (Li and Zaidi, 2000), and this is to be expected given how informative are variations in local orientation for surface shape analysis (Stevens, 1988; Knill, 2001). Moreover, orientation-defined textures have been shown to combine synergistically with shading to create strong impressions of depth (Mamassian and Landy, 2001). For these reasons we have chosen orientation-defined textures as our second cue to surface shape.

In our two previous studies of the colour-shading effect (Kingdom, 2003; Kingdom *et al.*, 2005), subjects adjusted the amplitude of a random-element, disparity-modulated stereo-grating to match the perceived depth of the corrugations in the plaid patterns. One criticism of this method is that binocular viewing the patterns may have obscured subtle depth differences because stereo-vision would signal 'flat'. For this reason we have also measured the perceived depth in our test stimuli using a new method, in which subjects viewed the stimuli monocularly and adjusted the height of a bar to indicate the amount of depth they perceived. Details of these procedures are provided below.

METHODS

Subjects

The four authors were test subjects. All had normal or corrected-to-normal visual acuity and normal colour vision. For the stereo-matching procedure, AY, FK and KW were employed. For the bar-adjustment procedure, GM, AY and KW were employed.

Stimuli

Generation and display. The stimuli were generated by a VSG2/5 graphics card (Cambridge Research Systems) and displayed on a Sony Trinitron F500 flat-screen monitor. The R (red), G (green) and B (blue) gun outputs of the monitor were gamma-corrected after calibration with an Optical photometer (Cambridge Research Systems). The spectral emission functions of the R, G and B phosphors were measured using a PR 640 spectral radiometer (Photo Research), with the monitor screen filled with red, green or blue at maximum luminance. The CIE coordinates of the monitors' phosphors were R: $x = 0.624$, $y = 0.341$; G: $x = 0.293$, $y = 0.609$; B: $x = 0.148$, $y = 0.075$.

Test stimuli. Example stimuli are shown in Fig. 1. The stimuli were constructed from three types of sinusoidally-modulated gratings: an orientation-modulated (OM) texture grating, a luminance modulated (LM) grating and a colour modulated (HM) grating (H stands for hue; we prefer this to CM, which is widely used to denote 'contrast-modulated'). All grating components had a spatial frequency of 0.75 cpd and an orientation of $-45°$ (left-oblique). The stimuli were presented in a circular, hard-edged window of diameter 4.5 deg. The OM and LM gratings were 'black-white', and were produced by modulating all three RGB phosphors in phase. The OM and LM gratings were presented together on one page of video memory, the HM on another. The two video pages were alternated at the monitor's frame-rate of 180 Hz, and thus the stimuli were refreshed at 90 Hz.

The OM grating comprised 600 Gabor patches with luminance spatial frequency 8.0 cpd, bandwidth at half-height of 1 octave, phase 90 deg (making them odd-symmetric, and hence d.c. balanced) and contrast 25%. The Gabors were randomly

Figure 1. (See colour plate VI) Example stimuli. Left hand panels = zero texture amplitude; right hand panels = 45 deg texture amplitude. Top pair = no colour conditions; middle pair = colour in-phase conditions; bottom pair = colour out-of-phase conditions.

positioned with the constraint that no two Gabors could be less than 2 standard deviations apart. The orientation of the Gabors was modulated sinusoidally around a mean orientation of 45 deg (right-oblique) and in the direction of 45 deg, producing texture bars at −45 deg (left-oblique). The phase of orientation modulation was randomized on each trial. The amplitude of orientation modulation determined by how much the orientation of the Gabors changed throughout one cycle of orientation modulation, and this was an independent variable with values 0.0, 5.63, 11.25, 22.5 and 45 deg. Examples of the 0 deg (left column) and 45 deg (right column) amplitude conditions are shown in Fig. 1.

The LM grating was combined with the OM grating multiplicatively. Multiplying the OM by the LM grating simulated real shading, which changes the average luminance but not contrast of local luminance variations. The phase of the LM grating was always set to 0 deg with respect to the orientation grating, and as can be seen in Fig. 1b, this phase relationship produced an impression of a corrugated surface illuminated from above-right. Note that with the OM grating, the d.c. orientations (45 deg) lie at the apparent peaks and troughs of the depth corrugation. The contrast of the LM grating was an independent variable, and values of 0.0, 0.037, 0.075, 0.15 and 0.3 were employed.

The HM grating was 'red-green', and was designed to isolate the post-receptoral chromatic mechanism that differences the L (long-wavelength-sensitive) and M (middle-wavelength-sensitive) cones. However, this choice of HM grating was arbitrary. Kingdom *et al.* (2005) showed that the colour direction of both the depth-enhancing and depth-suppressing chromatic gratings in the colour-shading effect was unimportant: it was their colour *contrast* that mattered. For this reason the details of the generation of the HM grating will not be given here, but the interested reader can consult the aforementioned article. The HM grating was fixed throughout at a (relatively high) L-M cone contrast of 0.08. There were three conditions: no colour grating, colour in-phase, and colour out-of-phase, as illustrated in Fig. 1, top to bottom respectively. For the in-phase condition (Figs 1c, d), the HM grating was added in phase to the LM grating, but with the polarity (0 or 180 deg) randomised, producing either a 'bright-red/dark-green', or 'dark-red/bright-green' grating. For the out-of-phase condition (Figs 1e, f), the HM grating was randomly allocated either 90 or 270 deg phase with respect to the LM grating.

Match stimuli. Two methods were used to estimate the apparent depth of the corrugations. The first method employed a random-Gabor stereo-grating with left-oblique corrugations of the same spatial frequency and orientation as the test stimuli. The matching stereo-pairs were positioned below the test stimuli on either side of the centre of the monitor screen. Stereo-fusion was achieved by viewing the display through a modified 8-mirror Wheatstone stereoscope. Viewing distance along the light path through the stereoscope was 105 cm. The details of the construction of the stereo-grating have been given elsewhere (Kingdom *et al.*, 2005) and will not be repeated here.

The second method used to estimate the perceived depth of the corrugations employed a vertical bar positioned below the test stimulus. The bar was 2 deg in height by 12 min in width, and was a shade of grey 0.12 in contrast darker than the background. The bar could be 'filled' with a shade of grey of 0.44 in contrast up to a height chosen by the subject. Viewing distance was also 105 cm.

Procedures

Isoluminance setting. Although our stimuli were not isoluminant, we wanted to minimize the likelihood that our HM grating introduced a luminance component to the stimulus. To this end we used an HM grating that when presented on its own was isoluminant according to the criterion of minimum perceived motion. To produce such a grating, we set an HM grating with an L-M cone contrast of 0.025 to drift at about 1.0 Hz. Subjects pressed a key on the CB3 response box (Cambridge Research Systems) to add or subtract an equal amount to both the L and -M cone contrasts until perceived motion was at a minimum. The resulting ratio of (absolute) L to M cone contrast weightings necessary to make the HM grating isoluminant was for GM 1.1, AY 1.67, KW 1.14 and FK 1.78.

Matching perceived depth using the adjustable stereo-grating. Subjects used the keys on the response box to adjust the amplitude of the depth corrugations in the stereo-grating until they matched the apparent depth of the corrugations in the test stimulus. There was no time limit. Subjects were encouraged to let their eyes roam freely over the stimuli to avoid any fading from prolonged fixation. During each session all conditions were presented in random order, and for all subjects except KW there were eight repeat sessions and therefore eight measurements per condition. For KW there were between four and twelve sessions/repeat measures, as her data formed part of a series of pilot experiments which explored the parameter space of the stimuli.

Matching perceived depth using the adjustable bar. Subjects used the keys on the response box to adjust the height of the bar to match the perceived depth of the corrugation in the test stimulus. No instructions were given as to which parts of the stimulus should be attended; subjects were simply told to set the bar height to indicate how much the bars 'stood out'. The whole display was viewed monocularly by the dominant eye, with the non-dominant eye covered by a patch. For all subjects there were eight sessions/repeat measures using this method.

RESULTS

To repeat: there were three colour conditions (no-colour, colour in-phase, colour out-of-phase), five texture amplitudes (0.0, 5.63, 11.25, 22.5 and 45 deg) and five shading contrasts (0.0, 0.037, 0.075, 0.15 and 0.3), making a total of $3 \times 5 \times 5 = 75$ conditions. Figures 2 and 3 show the mean settings for the adjustable stereo-grating and adjustable bar methods respectively, with the no-colour, in-phase and out-of-phase conditions shown in separate graphs. In all graphs, the magnitude of

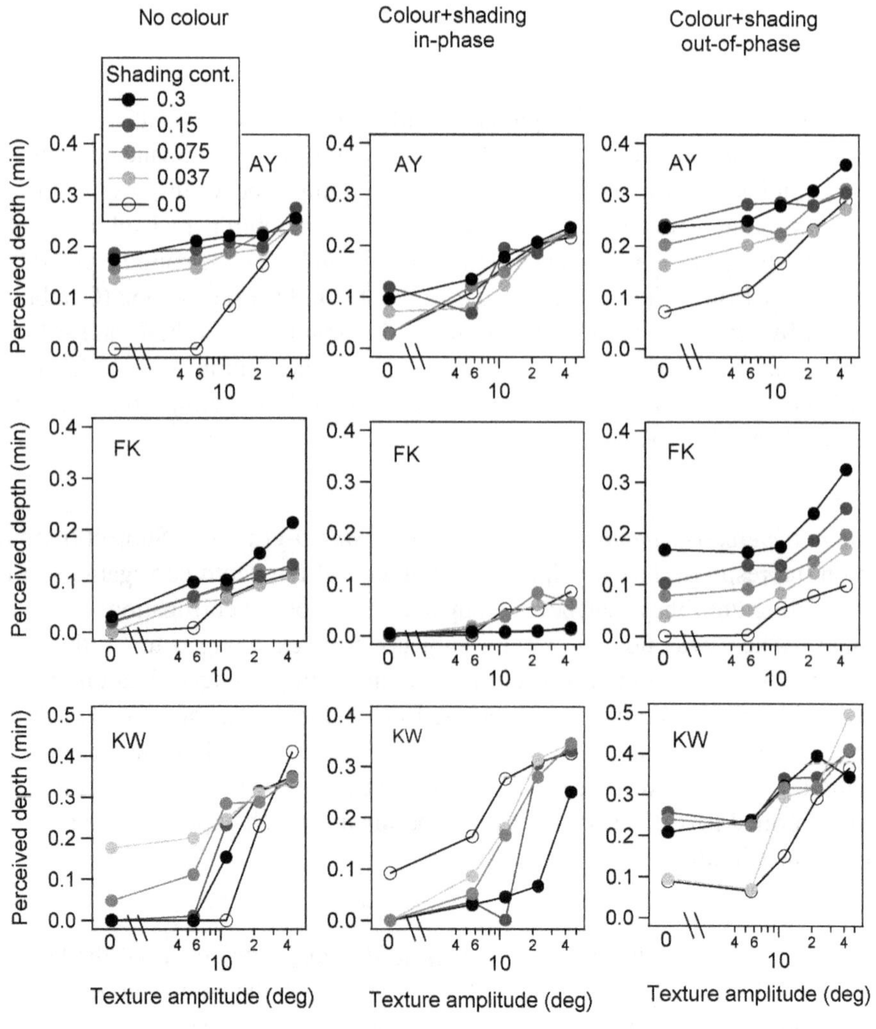

Figure 2. Results obtained using the stereo-grating matching method. In each graph perceived depth is plotted against the amplitude of texture modulation. Symbols with different shades of grey are for different luminance contrasts of the 'shading' grating. Left column = no-colour condition, middle column = colour in-phase condition, right column = colour out-of-phase condition. Error bars have been omitted to avoid cluttering.

perceived depth is plotted against texture amplitude, with separate plots for each shading contrast.

First note that in general an increase in either texture amplitude or shading contrast increased perceived depth. The effects of shading contrast are, however, largely absent in the in-phase conditions; this is most likely a floor effect resulting from the influence of in-phase colour contrast, which appears to reduce perceived depth to such an extent that little room is left for the effects of shading contrast. The effects of shading contrast are also anomalous in KW's no-colour condition, where, if anything, an increase in shading contrast reduced perceived depth, a result for which we have no explanation. KW's results are in general more erratic than the

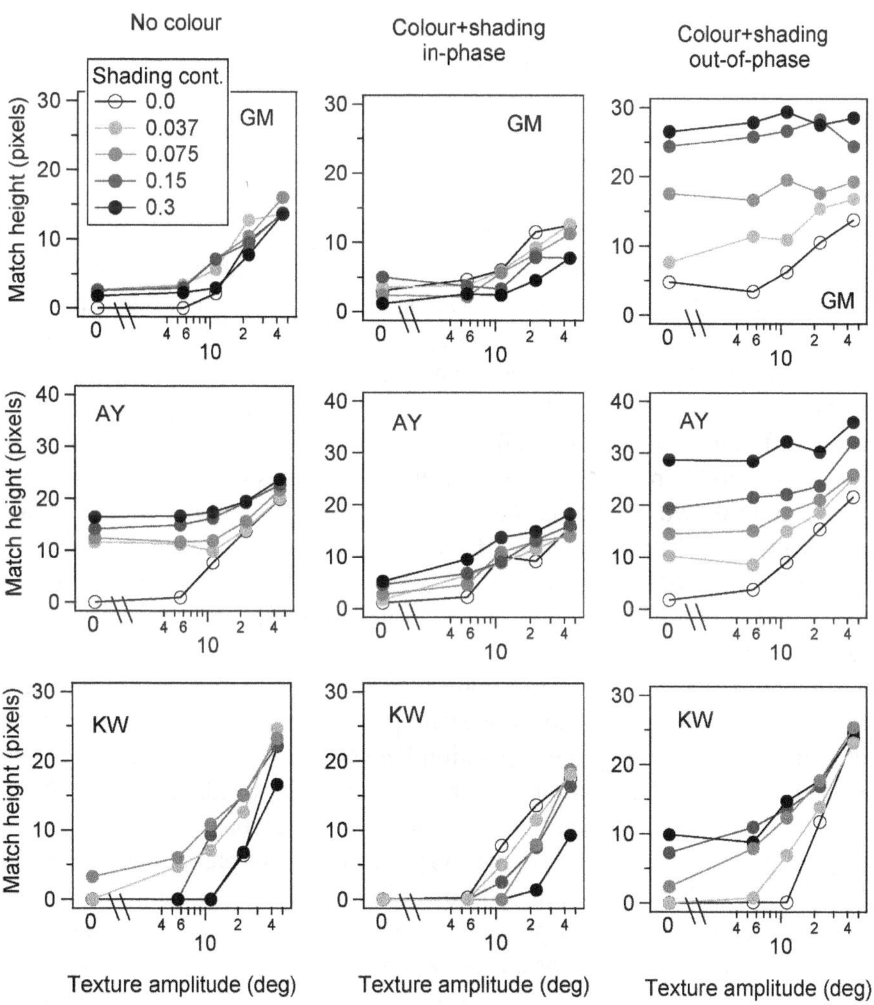

Figure 3. Results obtained using the adjustable bar method. The ordinates in each plot give perceived depth in terms of bar height. Otherwise, the plots and symbols are the same as in Fig. 2.

other subjects, but in spite of this the overall influence of colour contrast on her results is consistent with the other subjects.

Consider now the effects of colour contrast. When compared to the no-colour condition, adding colour contrast in-phase with the shading reduced perceived depth, whereas adding colour contrast out-of-phase with the shading increased perceived depth. These effects appear to be robust across subjects and method. A particularly noteworthy aspect of the data is that adding colour contrast out-of-phase with the shading causes the different shading contrast conditions to spread out vertically. In other words out-of-phase colour contrast results in shading contrast having a more differentiated impact on perceived depth.

One of the aims of this study was to determine the extent to which the colour-shading effect operated when cues to surface shape other than shading, specifically orientation-based texture variation, were present. In order to put the data into a manageable form that gives an indication of the impact of colour contrast on perceived depth in the context of texture amplitude, we performed the following analysis. For each shading contrast we subtracted the raw no-colour settings from both the colour in-phase and colour out-of-phase settings. We also performed the same subtractions with logarithmically transformed values, in order to determine whether there was a proportionate, rather than absolute change in the impact of colour with texture amplitude. We then averaged the subtracted results across (a) all positive shading contrasts and (b) subjects. The same analysis was performed separately for the no shading (zero shading contrast) conditions, though only using raw values, as there were insufficient data points with positive values to justify the logarithmic analysis. The results are shown in Fig. 4. The two graphs on the left show the differences in raw values for the conditions in which shading was absent, the two graphs in the middle the differences in log values for the non-zero shading contrasts, and the two graphs on the right the differences in raw values for the no shading contrasts. The top graphs are for the stereo matching procedure, the bottom graphs for the adjustable bar procedure. Squares are for the differences between the no colour and in-phase conditions, circles for the differences between the no-colour and out-of-phase conditions.

Four features of Fig. 4 are worth noting. First, when shading is present, colour contrast has a pronounced and consistent phase-dependent effect on perceived depth. This is evidenced by the separation between the in-phase and out-of-phase plots in the four left-hand graphs, with no overlap in standard errors. Second, colour contrast has no significant *phase-dependent* impact on perceived depth when shading is absent, as indicated by the closeness of the in-phase and out-of-phase plots in the two right hand graphs. Although in the lower right hand graph there is an apparently significant colour phase effect at the highest texture amplitude, this is a lone data point and arguably best ascribed to chance. Third, the impact of colour contrast when shading is present is more-or-less constant across texture amplitude for the raw (linear) values, but decreases slightly for the log transformed data, i.e. decreases slightly when computed proportionately. Fourth, both measurement

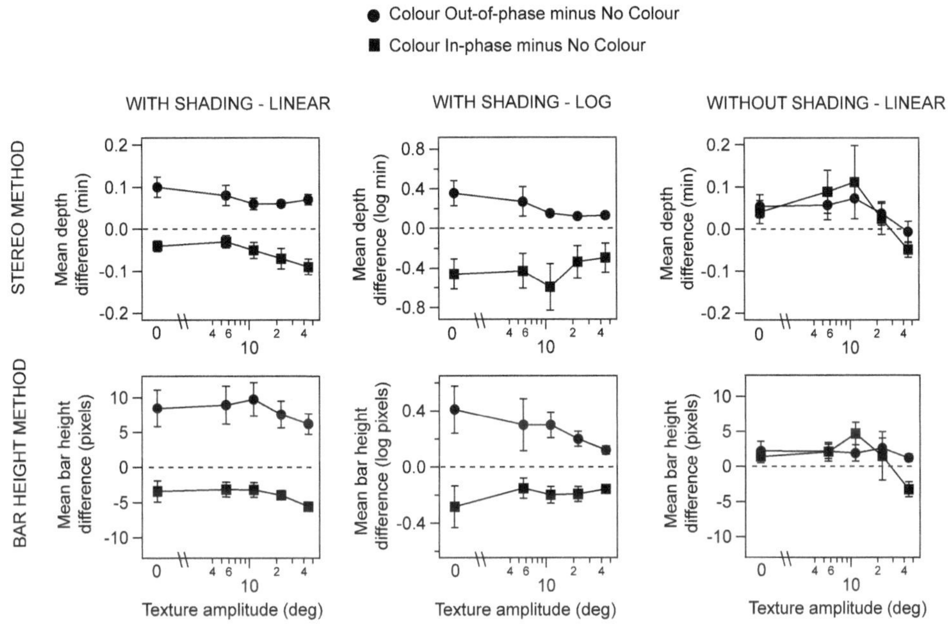

Figure 4. Left and middle panels show the mean differences, calculated across shading contrast and subject, between the no-colour and with-colour contrast conditions. The right-hand panels show the mean differences between the no-colour and with-colour contrast conditions for just zero shading contrast. The top graphs are for the stereo-grating matching procedure, the bottom graphs for the adjustable bar procedure. Filled squares are the mean differences between the no-colour and in-phase conditions, filled circles the mean differences between the no-colour and out-of-phase conditions.

techniques produce more-or-less similar results, as evidenced by the similarity between the top and bottom sets of graphs.

DISCUSSION

The results of the present study can be summarised as follows.

(1) Colour contrast has a phase-dependent influence on perceived depth in patterns that combine shape-from-shading with shape-from-texture.

(2) The influence of colour contrast on perceived depth is similar across texture amplitude when computed linearly, but diminishes slightly with texture amplitude when computed proportionately.

(3) Colour contrast has no phase-dependent influence on perceived depth in the absence of shading.

That colour contrast has a phase-dependent influence on perceived shape confirms previous findings using patterns that contain only colour and luminance contrast (Kingdom, 2003; Kingdom *et al.*, 2005), and extends the findings to the situation where texture is an additional cue to surface shape. In the experiments here, we only

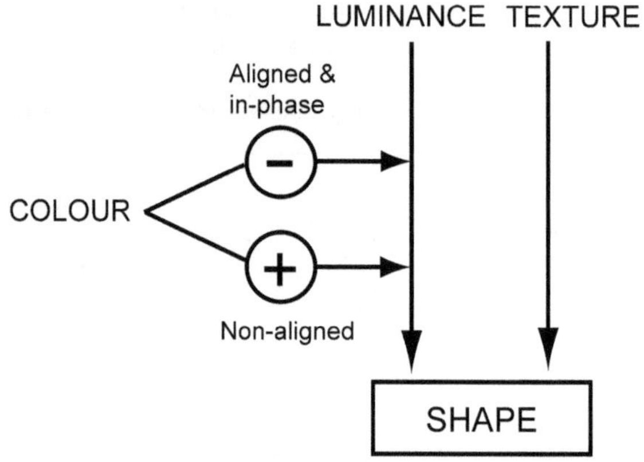

Figure 5. Schematic model of the conclusion of the study.

employed in-phase (0 or 180 deg) or out-of-phase (90 or 270 deg) relations between the colour and luminance gratings. Based on the results of Kingdom (2003), we would expect that other phase relationships would produce results intermediate between maximum depth-suppression and maximum depth-enhancement, with no depth-suppression or depth-enhancement (i.e. no effect of colour contrast) for phase differences of 45, 135, 225 and 315 deg. That colour contrast has no phase-dependent influence on pure shape-from-texture, i.e. when no shading is present, is a new, though not unexpected finding. On ecological grounds it is hard to see why chromatic variations should influence perceived shape-from-texture, because textural variation is invariably a surface rather than illumination property.

We found that *in the presence of shading* the magnitudes of both colour-driven depth-enhancement and colour-driven depth-suppression were more-or-less constant across texture amplitude when computed linearly, and diminished slightly with texture amplitude when computed proportionately. These findings are broadly compatible with the idea that perceived depth involves the weighted sum of shading and texture cues, with the shading cue weighted according to the robustness of the shading interpretation (see Landy *et al.*, 1995). In this scheme colour contrast promotes, or inhibits the shading interpretation based on its spatial relationship with the luminance pattern. The essential relationship between colour, shading and texture revealed by the present study is illustrated in Fig. 5. The figure demonstrates that chromatic variations impact on the contribution of luminance variations to perceived shape in combined shape-from-shading and shape-from-texture patterns, but do not impact on shape-from-texture itself.

Although the principal aim of our study was to investigate the impact of colour contrast on perceived depth in shading/texture patterns, our results support the well-attested observation that orientation variations elicit strong impressions of surface curvature (Stevens, 1988; Li and Zaidi, 2000; Mamassian and Landy, 2001). To what do these impressions of surface curvature correspond, in terms of geometry? The visual system is presumably making the assumption that the

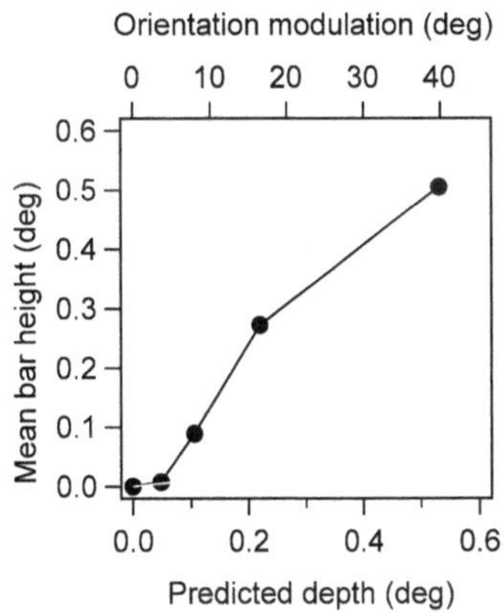

Figure 6. Actual versus predicted settings of bar height for the texture-only conditions, according to a model that assumes that the orientation variations in the stimuli arise from the projection of straight lines lying on a sinusoidally folded surface, slanted 53 deg away from the observer. See text for details.

orientation variations in the stimuli arise from the projection of straight contours that lie across an undulating surface, whose folds are oriented obliquely and are slanted away from the observer. It can be shown that if the surface is folded sinusoidally, the local orientations, i.e. tangents of the contours, are near-sinusoidal, with an amplitude of orientation modulation a_θ given by $\tan^{-1}[a_s 2\pi f \sin(s_\theta)]$, where a_s is the amplitude of surface modulation, f its spatial frequency, and s_θ the slant of the surface away from the fronto-parallel (upright) plane. The effect of s_θ assumes parallel projection. By inverting this equation we obtain the amplitude of surface modulation that would give rise to a given amplitude of orientation modulation, i.e. $a_s = \tan(a_\theta)/2\pi f \sin(s_\theta)$. If we now assume that our subjects' settings reflected the perceived depth, i.e. distance between the peaks and troughs of the surface modulation, then we can determine how well these settings are predicted by (twice) the values of a_s calculated from the above equation for any surface slant s_θ. Figure 6 shows the average subject's settings obtained from the no-colour, zero-shading-contrast conditions using the bar-height adjustment method. These are plotted, in degrees, against $2a_s$, also in degrees, for a slant of 53 deg — the value which produced the closest absolute match between the predicted and actual settings. The settings are reasonably well predicted. However, although the near-linear relationship between predicted and actual settings revealed in Fig. 6 is consistent with the model described above, the 53 deg estimate of perceived slant should be treated with caution, as there are many stimulus factors that are likely to affect the 'gain' of perceived depth, such as the use of randomly positioned micropatterns rather than parallel contours, and these will affect the estimate of perceived slant.

The unique and positive role that colour vision appears to play in the perception of shape-from-shading in our mixed shading and texture patterns is pertinent given that colour vision is traditionally considered the poor cousin of luminance vision in its capacity to provide information about the third dimension, i.e. depth. For example, stereoscopic depth judgements of isoluminant stimuli are generally worse than those with isochromatic stimuli (Lu and Fender, 1972; Gregory, 1977; Livingstone and Hubel, 1987; Kingdom and Simmons, 2000). The results here demonstrate that colour vision *in combination* with luminance and texture can significantly enhance impressions of depth. This positive role of colour vision in the perception of depth complements other positive roles in the analysis of image-structure, such as in the detection of fruit and flowers in foliage (Mollon, 1989; Sumner and Mollon, 2000; Domini and Lucas, 2001), the identification of shadows (Kingdom, Beauce and Hunter, 2004) and the memory of scenes (Gegenfurtner and Rieger, 2000).

Acknowledgements

This research was supported by Canadian Institute of Health Research grant #MOP-11554 given to F. K.

REFERENCES

Attick, J. J., Griffin, P. A. and Redlich, A. N. (1996). Statistical approach to shape from shading: reconstruction of three-dimensional face surfaces from single two-dimensional images, *Neural Computation* **8**, 1321–1340.

Cavanagh, P. (1991). Vision at equiluminance, in: *Vision and Visual Dysfunction Volume V: Limits of Vision*, Kulikowski, J. J., Murray, I. J. and Walsh, V. (Eds), pp. 234–250. CRC Press, Boca Raton, FL.

Domini, N. J. and Lucas, P. W. (2001). Ecological importance of trichromatic vision to primates, *Nature* **410**, 363–365.

Gegenfurtner, K. R and Rieger, J. (2000). Sensory and cognitive contributions of color to the recognition of natural scenes, *Current Biology* **10**, 805–808.

Gregory, R. L. (1977). Vision with isoluminant colour contrast: 1. A projection technique and observations, *Perception* **6**, 113–119.

Kingdom, F. A. A. (2003). Colour brings relief to human vision, *Nature Neuroscience* **6**, 641–644.

Kingdom, F. A. A. and Simmons, D. R. (2000). The relationship between colour vision and stereoscopic depth perception (Invited Paper), *J. Soc. 3D Broadcasting and Imaging* **1**, 10–19.

Kingdom, F. A. A., Beauce, C. and Hunter, L. (2004). Colour vision brings clarity to shadows, *Perception* **33**, 907–914.

Kingdom, F. A. A., Rangwala, S. and Hammamji, K. (2005). Chromatic properties of the colour-shading effect, *Vision Research* **45**, 1425–1437.

Knill (2001). Contour into texture: information content of surface contours and texture flow, *J. Opt. Soc. Amer. A* **18**, 12–35.

Lehky, S. R. and Sejnowski, T. J. (1988). Network model of shape-from-shading: neural function arises from both receptive and projective fields, *Nature* **333**, 452–454.

Landy, M. S., Maloney, L. T., Johnston, E. B. and Young, M. (1995). Measurement and modelling of depth cue combination: in defense of weak fusion, *Vision Research* **35**, 389–412.

Li, A. and Zaidi, Q. (2000). Perception of three-dimensional shape from texture is based on patterns of local orientation, *Vision Research* **40**, 217–242.

Livingstone, M. S. and Hubel, D. H. (1987). Psychophysical evidence for separate channels for the perception of form, color, movement, and depth, *J. Neurosci.* **7**, 3416–3468.

Lu, C. and Fender, D. H. (1972). The interaction of color and luminance in stereoscopic vision, *Investigat. Ophthalmol.* **11**, 482–489.

Mamassian, P. and Landy, M. S. (2001). Interaction of prior visual constraints, *Vision Research* **41**, 2653–2668.

Mollon, J. D. (1989). 'Tho' she kneel'd in that place where they grew ...' The uses and origins of primate colour vision, *J. Exper. Biol.* **146**, 21–38.

Mollon, J. (2000). 'Cherries among the leaves': The evolutionary origins of color vision, in: *Color Perception: Philosophical, Psychological, Artistic and Computational Perspectives*, Steven, D. (Ed.), pp. 10–30. Oxford University Press, New York.

Mullen, K. T. and Kingdom, F. A. A. (1991). Colour contrast in form perception, in: *The Perception of Colour*, Gouras, P. (Ed.), pp. 198–217. Volume 6 in *Vision and Visual Dysfunction*, Cronly-Dillon, J. (Ed.). Macmillan, Oxford, UK.

Olmos, A. and Kingdom, F. A. A. (2004). A biologically-inspired algorithm for separating reflectance and shading images, *Perception* (in press).

Ramachandran, V. S. (1988). Perception of shape form shading, *Nature* **331**, 163–166.

Regan, D. (2000). *Human Perception of Objects*, Sinauer Associates, Sunderland, Massachusetts.

Rubin, J. M. and Richards, W. A. (1982). Color vision and image intensities: When are changes material? *Biol. Cybern.* **45**, 215–226.

Stevens, K. A. (1988). The line of curvature constraint and the interpretation of 3D shape from parallel surface contours, in: *Natural Computation*, Richards, W. (Ed.). MIT Press, Cambridge, MA.

Sumner, P. and Mollon, J. D. (2000). Catarrhine photopigments are optimised for detecting targets against a foliage background, *J. Exp. Biol.* **23**, 1963–1986.

Sun, J. and Perona, P. (1997). Shading and stereo in early perception of shape and reflectance, *Perception* **26**, 519–529.

Color, Line, and Space, pp. 59–70 (2006)

Fechner-Benham subjective colors do not induce McCollough after-effects

VINCENT A. BILLOCK [1,*], THOMAS DITZINGER [2,3], J. A. SCOTT KELSO [2] and BRIAN H. TSOU [4]

[1] *General Dynamics, Inc., Suite 200, 5200 Springfield Pike, Dayton, OH 45431, USA*
[2] *Center for Complex Systems, Florida Atlantic University, 777 Glades Road, Boca Raton, FL 33431, USA*
[3] *Current address: Springer-Verlag, Tiergartenstr. 17, 69121 Heidelberg, Germany*
[4] *US Air Force Research Laboratory, Wright Patterson Air Force Base, OH 45433, USA*

Abstract—Fechner-Benham subjective color is widely believed to be governed by local interactions in early (probably retinal) mechanisms. Here we report three lines of phenomenological evidence that suggest otherwise: subjective colors seen in spatially extended stimuli (a) are dependent on global aspects of the stimuli; (b) can become multistable in position; and (c) even after being stabilized do not support the creation of McCollough's colored after-effects — a cortically based phenomenon generally thought to be more central than Fechner-Benham color. These phenomena suggest a central locus that controls perception of subjective color, characterized by pattern dependent interactions among cortical mechanisms that draw their inputs from peripheral units.

Keywords: Color; Fechner-Benham subjective color; McCollough effect; multistability.

1. INTRODUCTION

Among the most arresting of all visual illusions are the subjective (Fechner-Benham) colors that appear on rotating or flickering achromatic patterns (e.g. Fig. 1a). Many studies suggest that subjective colors arise from lateral inhibitory interactions in color opponent neurons, perhaps as early as the color opponent midget ganglion cells of the retina. If subjective colors do arise so early in visual processing, it should be possible to use them as psychoanatomical probes (Julesz, 1971; Teller, 1984; Blake, 1995) of phenomena presumed to arise at later stages of processing (e.g. the McCollough oriented color after-effects; McCollough, 1965). While attempting to generate a McCollough after-effect by adaptation to Fechner-

*To whom correspondence should be addressed. E-mail: Vince.Billock@wpafb.af.mil

Benham subjective color, we found three independent lines of phenomenological evidence that Fechner-Benham subjective colors are produced or controlled at a more central locus than is often thought.

2. BACKGROUND

Subjective color is generally produced by temporally modulating a pattern with a fixed phase relationship to a flickering background. Dichoptic presentation of pattern and background result in a seemingly normal Benham stimulus (e.g. Fig. 1a), but no color is perceived, leading many to infer that subjective color is produced prior to binocular integration in striate cortex (Hess, 1952; Gregory, 1968; Campenhausen, 1968, 1973; Karvellas, 1969; Jarvis, 1977). Jarvis (1977) reports psychophysical attributes of subjective colors that are consistent with retinogeniculate receptive field properties: (1) flickering spots are as effective as stripes at inducing subjective color; (2) there is an optimal width (as gauged by the purity of induced color) of the target that is consistent with the center sizes of retinal ganglion cells; (3) saturation of subjective colors improves asymptotically as the width of the gap between target stripes is increased, as would be expected if an inhibitory surround were involved. This is consistent with the finding that Benham colors only appear on the edges of broad inducing stripes (Hasegawa, 1971; Both and Campenhausen, 1978). Some other studies that support a possible retinal basis for subjective colors include Zrenner (1983), Courtney and Buchsbaum (1991), Campenhausen *et al.* (1992), Grunfeld and Spitzer (1995) and Rotgold and Spitzer (1997). Evidence for later processing is addressed in the Discussion.

A color phenomenon likely to lie later in the processing hierarchy is the McCollough (1965) effect. If one adapts to a repetitive colored pattern (e.g. a red/black vertical grating) for several minutes, complementary illusory colors are seen in an achromatic version of the adaptation pattern. The effects are even stronger if adaptation to the red/black vertical grating is paired with adaptation to a green/black horizontal grating. These McCollough after-effects last much longer than simple adaptation phenomena (on the order of days or weeks; Riggs *et al.*, 1974) and are believed to be produced cortically (Stromeyer, 1978; Barlow, 1990; Dodwell and Humphrey, 1990). Suggestions that the McCollough effect might have an early locus in visual cortex (Shute, 1979) were prompted by reports of oriented color contrast sensitive cells in V1 (Michael, 1978a, b) that could be adapted by oriented chromatic patterns. However, studies of McCollough after-effects in very specific conjunctions of linked stimuli (Stromeyer, 1978; Harris, 1980; Barlow, 1990) and in complex geometries (Dodwell and O'Shea, 1987; Dodwell and Humphrey, 1990) suggested a still later locus. Although it is possible to produce a weak after-effect with locally orthogonal, globally random contingencies, the McCollough effect is stronger for globally orthogonal opponent pairs. For example, strong McCollough effects are produced by concentric circles and fan-shaped patterns, which are complementary pairs in Lie geometry, arguing for a rather sophisticated geometric analysis (Dodwell and O'Shea, 1987), with potential loci in either V1 or extrastriate cortex. In V1, because of a logarithmic retinocortical mapping, both concentric circles and fans plot to V1 as simple mutually orthogonal gratings (Schwartz, 1977, 1994), simplifying the required neural interactions. In the extra-striate cortex, Gallant *et al.* (1993) found V4 cells selective to specific geometries, including fan shapes or to concentric circles. Two imaging studies favor an extrastriate locus. Barnes *et al.* (1999) found perception of McCollough after-effects correlated with activity in left anterior fusiform gyrus, ventrolateral frontal cortex and left insula. Similarly, Humphrey *et al.* (1999) found a McCollough correlate in the lingal and fusiform gyri, although they cite evidence implicating V1 as well. So, if the McCollough effect is produced more centrally than Fechner-Benham subjective colors (and lies in the same processing stream; Blake, 1995), then it should be possible to produce a McCollough effect from adaptation to subjective colored patterns. We sought to test this proposition.

3. PRELIMINARY OBSERVATIONS: PHENOMENOLOGY OF EXTENDED SUBJECTIVE COLOR GRATINGS

The McCollough (1965) effect is easily produced by adaptation to oriented color gratings. We attempted to create oriented illusory colored gratings by using rotating cylinders, each with a single Benham-like repetitive stimulus pattern (e.g. Fig. 2a, wrapped around a cylinder, sized so that the top and bottom met). The alternative was to use an equivalent temporal modulation on a CRT to generate this illusory grating; however, modulating a light source in this way can introduce physical color artifacts (Jarvis, 1977, 1993). In this way, we expected to produce red-gray, green-gray, or blue-gray square wave gratings; however, the colors seen in these extended gratings were unexpectedly poor compared to those seen in non-repetitive designs (e.g. Fig. 1), viewed in the same illumination. Often the colors disappeared altogether. Since subjective colors are believed to be generated at a retinogeniculate locus characterized by center/surround interactions (Jarvis, 1977; Zrenner, 1983; Campenhausen *et al.*, 1992), we attempted to strengthen the colors by altering the ratio of the widths of the black and white bars of the grating (e.g. Fig. 2b). Our reasoning was based on Jarvis' (1977) observation that subjective color saturation improves asymptotically as the gap between target stripes is increased, and on the center/surround rationale of Westheimer's (1967) spatial sensitization paradigm. We found that decreasing the ratio of stripe width to gap improved the colors, but surprisingly, the colors appeared only on about half of the physically identical stripes. Moreover, the effect was multistable: colored stripes often exchanged positions with achromatic ones. This effect occurred for the entire range of black:white stripe width ratios tested (4 : 6 to 1 : 9), and was unaffected by the absolute size of the stripes on the retina. Finally, we found that illusory colors were both strengthened and made uniform on all the stripes by addition of one or more black reference lines to the pattern (e.g. Fig. 2c). All of these effects

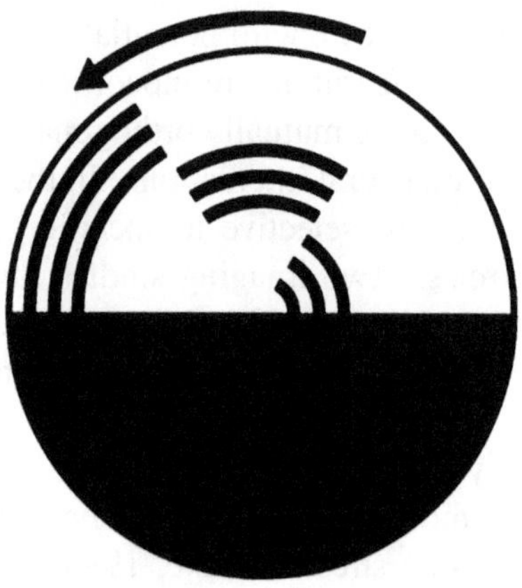

Figure 1. When this Benham's disk (Benham, 1894) is spun at moderate speeds (about 4–10 Hz), the arcs form colored circles on a gray background. Most observers report that (for the direction of spin shown) the exterior arcs appear red or reddish brown. The middle arcs appear green and the interior arcs appear blue or violet. The order of subjective colors reverses if the disk is rotated in the opposite direction.

were seen in a variety of illuminants, ranging from fluorescent to incandescent to indirect sunlight, and were reported by naive observers as well as by the authors. Now, armed with a suitable adapting grating, we sought to test whether McCollough after-effect colors could be produced by adaptation to subjective color.

4. EXPERIMENT 1. USING EXTENDED BENHAM GRATINGS AS MCCOLLOUGH EFFECT INDUCERS

4.1. Methods: Production of subjectively colored extended gratings

For these measurements, three subjects were employed. All of the subjects had normal color vision, as tested with pseudoisochromatic test plates. In addition, two of the subjects had previously been tested using more rigorous means, including Rayleigh matches and the Farnsworth F_2 tritan test plate (Naval Submarine Medical Research Laboratory, Groton, CT). Experiments were in accord with our institution's human use guidelines. Two observers were authors, the third was a colleague skilled in psychophysical measurements but naive to the rationale for the experiments. Each stimulus (a slice of which is shown in Fig. 2c) was printed on a sheet of white paper and wrapped around a spinning cylinder driven by a variable speed motor under the observers' control. An incandescent illuminant approximating CIE Illuminant A was used because subjective colors are more saturated under this illuminant than under other available illuminants (Verriest and Seki, 1964). Luminance of the white and black portions of the Benham stimulus were approximately 565 and 49 cd/m^2 respectively. Depending on the direction of rotation, the gray bars of this grating turn either red or blue, allowing us to study McCollough after-effect induction in both the red/green and blue/yellow systems with the same physical pattern.

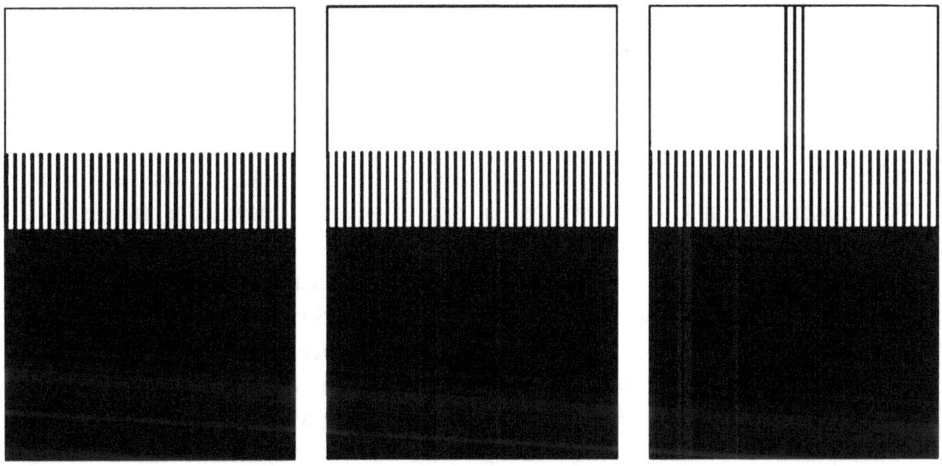

(a) (b) (c)

Figure 2. (a) We attempted to produce an extended grating of Fechner-Benham illusory color by using just one set of spinning arcs at a time. The one depicted here would be expected to produce either a red-and-light-gray grating or a blue-and-light-gray grating depending on direction of rotation (we also used stimuli that produce illusory green). The stimulus pattern shown is wrapped around a cylinder (so that the top and the bottom of the pattern meet) and spun, using a variable speed motor under the subject's control. For this extended pattern, little red or blue is seen at any rotation rate (the perceived pattern is mostly of darker gray stripes on a lighter gray background). (b) If the black stripes are made narrower than the white spaces between, color appears on some but not all the rotating stripes, and the positions of the colored stripes are multi-stable. (c) If a reference pattern of unmodulated black stripes is added, the colors become stronger and much more uniform. The resulting pattern consists of, for example, alternating stripes of broader light gray and narrower illusory red, with one or more black reference stripes down the middle, and should be (but is not) an effective stimulus for inducing a McCollough after-effect.

Subjects free viewed the stimulus binocularly, at a distance of 0.46 m. The subjective grating subtended 20° and contained 60 cycles, yielding an effective spatial frequency of 3 c/deg. The duty cycle (width ratio of arcs to background intervals) was fixed at 3.5 : 6.5 (the most moderate duty cycle that produced optimal color). Subjects adapted to the vertical grating for 15 min (a duration that Allan and Siegel (1997) found produced good after-effects in single gratings, and which Webster *et al.* (1988) used to rule out McCollough effects for contrast colors). Subjects were instructed to keep viewing distance and head orientation constant and were encouraged to blink and make horizontal eye movements to prevent discomfort, Troxler fading, or retinal after-images. Subjects were also instructed to adjust rotation rate to compensate for any gradual shifts in the saturation of the subjective colors (Le Rohellec and Vienot, 1995). After a few minutes of rest to allow retinal after-effects to dissipate, subjects viewed achromatic versions of the inducing pattern and were asked to comment on the presence of color.

4.2. Results

None of the observers reported a McCollough effect after adapting to either the red or blue subjective color grating. No after-effect colors were observed in the achromatic (stationary) version of the adapting pattern, nor were they seen in enhanced contrast, reduced contrast or reversed contrast versions of that pattern generated on CRTs. We are aware of several differences (necessitated by the requirements for producing subjective color) between our stimulus and the standard stimuli used in McCollough effect experiments. Subjective colors are desaturated and appear darker than their gray backgrounds — both attributes that might be expected to work against, but not to eliminate production of a McCollough effect (Stromeyer, 1978). (Note however that Shute (1979) reported that desaturated stimuli are quite effective in producing McCollough effects.) Also, our manipulation of duty cycle was unusual, as was our use of single colored gratings rather than orthogonal adapting pairs — some experimenters have found that this produces weaker after-effects (Stromeyer, 1978). Based on the limited literature on these conditions (e.g. Stromeyer, 1978; Allan and Siegel, 1997), 15 min should have been a sufficient time to induce an effect, if one were inducible at all. However, without a control experiment, we could not be sure if the combination of these conditions could have prevented the production of a McCollough after-effect.

5. EXPERIMENT 2: MCCOLLOUGH PHYSICAL COLOR CONTROL EXPERIMENT

5.1. Methods

Since the subjective color stimuli combine several attributes that might reduce their potency as McCollough effect inducers, as a control we created a physically colored stimulus of similar appearance to the subjective gratings and tested its ability to induce the McCollough after-effect. We matched the subjective red grating because it is more saturated than subjective blue (Jarvis, 1977) and hence has the better chance of producing an after-effect (see Allan and Siegel (1997) for a discussion of how perceptual salience improves aftereffects). Two of the three subjects who participated in the original experiment were available for this control experiment. Colorimetric matches for chromaticity and achromatic contrast were made on a photometrically and colorimetrically calibrated VisionWorks display system (Vision Research Graphics, Inc., Durham, NH). The subjective color apparatus from the previous experiment was positioned next to the Vision Works system. For the McCollough adaptation pattern, we created a monitor-based physically colored grating with red bars, and gray background in 3.5 : 6.5 arc to background width ratio. The red bars were set to the chromaticity and luminance of the colorimetric match to the subjective colors. We used a gray for the background with the same chromaticity as our CIE A white illuminant (0.448, 0.407) and matched it for achromatic contrast with the gray seen between the stripes in the Benham stimulus. The resulting grating was a duplicate of the subjective color grating in spatial frequency, orientation and duty cycle, and was judged a good likeness in chromaticity and achromatic contrast in side-by-side comparisons. Subjects adapted to these gratings for 15 min with the same conditions used for the previous experiment. After a few minutes of rest to allow retinal aftereffects to dissipate, subjects viewed achromatic versions of the inducing pattern and were asked to comment on the presence of color. The test grating was re-inspected 4 h and 24 h later (subjects were asked to refrain from ingesting caffeine, both prior to adaptation and during the 24 h test period; Shute, 1979). The experiment was later repeated using an equal energy gray in the adaptation and test gratings.

5.2. Results

Unlike adaptation to the subjective color grating, 15 min of adaptation to the matched physical stimulus produced a vivid greenish McCollough after-effect color in a 3.5 : 6.5 achromatic reference pattern (with the same achromatic contrast as the inducing pattern). The after-effect persisted for at least 24 h for both subjects. The after-effect was specific to the inducing orientation; no color after-image was produced by inspection of the test grating at an orthogonal orientation. Using an equal energy gray (0.333, 0.333) instead of CIE A in the adaptation and test grating produced similar results, both in appearance of the McCollough aftereffect colors and in their duration. We conclude that the failure of Fechner-Benham subjective colors to induce McCollough effects is not due to the physical stimulus conditions.

6. DISCUSSION

6.1. Summary

(1) We attempted to produce extended illusory colored gratings by mounting a repetitive stimulus pattern (Fig. 2a) on a rotating cylinder, but the colors were poor and fleeting. (2) These colors were strengthened by reducing the width of the black stripes relative to the background (e.g. Fig. 2b). However, this produced a grating in which only some of the physically identical stripes were subjectively colored and the locations of these colored stripes were unstable. (3) Addition of an achromatic reference pattern (e.g. Fig. 2c) — even a single black stripe — strengthened, stabilized and made uniform the entire subjectively colored grating. (4) These global dependencies and multi-stabilities are inconsistent with models of subjective color that posit highly local processing within retinogeniculate units. (5) The vivid illusory color gratings produced by Fig. 2c-type stimuli were incapable of inducing a McCollough effect, even after lengthy adaptation periods, while physical colored gratings colorimetrically matched to the subjective stimuli produced vivid long lasting McCollough effects. (6) These results run counter to our assumptions of early local processing in subjective color phenomena and later global processing in

the McCollough effect. Below, we rethink these assumptions and try to place our findings in perspective.

6.2. Relationship of our findings with other studies of induced color as stimuli

One kind of subjective color — after-image color — induces McCollough effects (Murch and Hirsch, 1972; Humphrey *et al.*, 1989; Webster *et al.*, 1988). Keith White (unpublished, personal communication) finds that Bidwell colors (after-image colors seen in lieu of the masked physical colors that produced them) induce McCollough effects (and also confirms the ineffectiveness of Fechner-Benham colors in producing McCollough effects). Conversely, Webster *et al.* (1987, 1988) found that although McCollough colors produce color contrast, the illusory colors created by color contrast (colored shadows) do not induce McCollough effects. The production of contrast colors by McCollough effects is explicable, because lateral inhibition can be found on many levels of visual processing. However, since color contrast mechanisms appear as early as V1, it is odd that color contrast does not drive a McCollough effect. Webster *et al.* (1988) note that contrast effects in colored shadows may require huge receptive fields that occur late in visual processing.

6.3. Other indications of multi-stability and long-range interactions in subjective color

A few previous studies found indications of multi-stability for non-Benham-like subjective colors, e.g. the multi-stable subjective colors generated within a perimeter defined by the apparent movement of many icons (Mayzner, 1975). Similarly, Williams *et al.* (1991) inferred cortical multi-stability in Brewster's subjective colors, from the desaturation of these colors, relative to that predicted by cone aliasing models. It has long been appreciated that temporal modulation from an adjacent area affects subjective color (Fry, 1933). Some such interactions are surprisingly long in range, extending over about 40° of visual angle (Campenhausen and Schramme, 1995), very similiar to the 45° range found for color spreading in the watercolor illusion (Pinna *et al.*, 2001). At least two modern studies report long range/non-retinal interactions in subjective color. Schramme and Supply (1997) studied subjective colors in proximity to physical/subjective colors, both of which modify perception of subjective color. They found the same interactions occur when the test color is induced in one eye and the inducing color is produced in the other. Distant color induction between subjective colored features has also been reported by Le Rohellec and Vienot (1997).

6.4. Psychoanatomy of Fechner-Benham subjective color

Section 2.2 summarized some properties of subjective color suggestive of retino-geniculate mechanisms with center-surround color opponent (Type I) receptive fields. Type I receptive fields respond to both chromatic and achromatic stimuli

(De Valois and Pease, 1971; Ingling and Martinez, 1983), making the multiplexed signal potentially ambiguous to cortical units receiving it (Billock, 1989). In the space/time domain, these receptive fields behave as if center and surround responses add for chromatic stimuli but subtract for achromatic stimuli. Zrenner (1983) posits that the phase differences of the arcs and backgrounds of Benham's disk act to change the phase relationship of stimulated P cell centers and surrounds by advancing or retarding the response of the center relative to the surround. This allows a luminance signal to mimic the phase characteristics of a color signal. This retinal encoding stage may be a necessary, but not a sufficient condition for the production of illusory color. The multi-stability and pattern dependency found by us and previous studies may reflect a cortical color stage. The earliest possible neural locus for pattern dependency in subjective colors would be cortical units sensitive to the spatial form of chromatic patterns, i.e. the double opponent cells of V1 and V2, which come in several geometries (Michael, 1978a, b; 1979). A theory of how these cells extract chromatic information from P cells is available (Billock, 1991), but this local and linear model cannot account for either global pattern effects or multi-stability. A non-linearily coupled network of such units might suffice.

6.5. Multiple stages complicate psychoanatomy

The above studies suggest that interacting cortical mechanisms are involved in the interpretation and/or modification of subjective colors established by phase dependent lateral interactions in early vision. We note that Broese and O'Shea (1995) posit two stages to the production of McCollough colors: after-effect colors that lie on the edges of test stimuli, and a subsequent filling-in process. They suggest that edge colors are associated with local features (and hence likely earlier processing), while the filling-in color is promoted by global organization (and hence likely later processing). This raises the possibility that the inability to produce a McCollough effect from Fechner-Benham subjective colors could be because an essential cortical stage of Fechner-Benham color coding lies after an essential early stage in the production of McCollough effects.

Acknowledgements

This research was supported by NIMH grants, by an NRC/AFOSR Research award and by the Deutsche Forschungsgemeinschaft. We thank Nigel W. Daw for a stimulating discussion on receptive fields and illusory colors, and Paul Havig, Robert Hestle and William McLean for technical assistance.

REFERENCES

Allan, L. G. and Siegel, S. (1997). Contingent color aftereffects: reassessing old conclusions, *Perception and Psychophysics* **59**, 129–141.

Barnes, J., Howard, R. J., Senior, C., Brammer, M., Bullmore, E. T., Simmons, A. and David, A. S. (1999). The functional anatomy of the McCollough contingent color after-effect, *Neuroreport* **10**, 195–199.

Barlow, H. B. (1990). A theory about the functional role and synaptic mechanism of visual after-effects, in: *Vision: Coding and Efficiency*, C. Blakemore (Ed.), pp. 363–375. Cambridge University Press, Cambridge, UK.

Benham, C. E. (1984). An artificial spectrum top, *Nature* **51**, 200.

Billock, V. A. (1989). Benham's top: an information theoretic approach, *Optics News* **15**, 90.

Billock, V. A. (1991). The relationship between simple and double opponent cells, *Vision Research* **31**, 33–42.

Blake, R. (1995). Psychoanatomical strategies for studying human visual perception, in: *Early Vision and Beyond*, Papathomas, T. V., *et al.* (Eds), pp. 17–25. MIT Press, Cambridge, MA.

Both, R. and Campenhausen, C. von (1978). Sensitivity of a sensory process to short time delays: a study in pattern induced flicker colors (PIFCs), *Biological Cybernetics* **30**, 63–74.

Broerse, J. and O'Shea, R. P. (1995). Local and global factors in spatially-contingent coloured aftereffects, *Vision Research* **35**, 207–226.

Campenhausen, C. von (1968). Über den ursprungsort von musterinduzierten flickerfarben im visuellen system des menschen, *Zeitschrift für vergleichende Physiologie* **61**, 355–360.

Campenhausen, C. von (1973). Detection of short time delays between photic stimuli by means of pattern induced flicker colors (PIFCs), *Vision Research* **13**, 2261–2272.

Campenhausen, C. von, Hofstetter, K., Schramme, J. and Tritsch, M. F. (1992). Color induction via non-opponent lateral interactions in the human retina, *Vision Research* **32**, 913–923.

Campenhausen, C. von and Schramme, J. (1995). 100 years of Benham's top in colour science, *Perception* **24**, 695–717.

Courtney, S. M. and Buchsbaum, G. (1991). Temporal differences between color pathways in the retina as a possible origin of subjective colors, *Vision Research* **31**, 1541–1548.

Daw, N. W. (1972). Color-coded cells in goldfish, cat and rhesus monkey, *Investigative Ophthal. Vis. Sci.* **11**, 411–416.

De Valois, R. L. and Pease, P. L. (1971). Contours and contrast: Responses of monkey lateral geniculate cells to luminance and color figures, *Science* **171**, 694–696.

Dodwell, P. C. (1991). Transformations after Gibson: biological and ecological constraints on vision, in: *Pattern Recognition in Man and Machine*, Watt, R. J. (Ed.), pp. 19–29. CRC Press, Boca Raton, FL.

Dodwell, P. C. and O'Shea, R. P. (1987). Global factors generate the McCollough effect, *Vision Research* **27**, 569–580.

Dodwell, P. C. and Humphrey, G. K. (1990). A functional theory of the McCollough effect, *Psychol. Rev.* **97**, 78–89.

Fry, G. A. (1933). Color phenomena from adjacent retinal areas for different temporal patterns of intermittent white light, *Amer. J. Psychol.* **45**, 714–721.

Gallant, J. L., Braun, J. and van Essen, D. C. (1993). Selectivity for polar, hyperbolic and cartesian gratings in macaque visual cortex, *Science* **259**, 100–103.

Gregory, A. H. (1968). Interaction of visual stimuli, *Nature* **219**, 734–735.

Grunfeld, E. D. and Spitzer, H. (1995). Spatiotemporal model for subjective colors based on colour coded ganglion cells, *Vision Research* **35**, 275–283.

Harris, C. S. (1980). Insight or out of sight? Two examples of perceptual plasticity in the human adult, in: *Visual Coding and Adaptability*, Harris, C. S. (Ed.), pp. 95–149. Lawrence Erlbaum, Hillsdale, NJ.

Hasegawa, T. (1971). Effect of spatiotemporal interactions on Fechner-Benham subjective color, *Acta Chromatica* **2**, 49–52.

Hess, E. H. (1952). Subjective color: retinal vs. central origin, *Amer. J. Psychol.* **65**, 278–280.

Humphrey, G. K., Dodwell, P. C. and Emerson, V. F. (1989). Pattern-contingent color aftereffects on noninduced patterns, *Perception and Psychophysics* **38**, 343–353.

Humphrey, G. K., James, T. J., Gati, J. S. Menon, R. S. and Goodale, M. A. (1999). Perception of the McCollough effect correlates with activity in extrastriate cortex: a functional magnetic resonance imaging study, *Psychol. Sci.* **10**, 444–448.

Ingling, C. R. Jr. and Martinez, E. (1983). The relationship between spectral sensitivity and spatial sensitivity for the primate r-g X-channel, *Vision Research* **23**, 1495–1500.

Jarvis, J. R. (1977). On Fechner-Benham subjective colour, *Vision Research* **17**, 445–451.

Jarvis, J. R. (1993). Subjective colour from achromatic flicker, *Vision Research* **33**, 1131–1132.

Julesz, B. (1971). *Foundations of Cyclopean Perception.* University of Chicago Press, Chicago, IL.

Karvellas, P. (1969). *A Quantitative Study of the Subjective Color Phenomena.* Ann Arbor, University Microfilms.

Le Rohellec, J. and Vienot, F. (1995). Contribution of two colour opponent mechanisms to Fechner-Benham subjective colors, in: *Colour Vision Deficiencies XII*, Drum, B. (Ed.), pp. 251–258. Kluwer, Dordrecht, The Netherlands.

Le Rohellec, J. and Vienot, F. (1997). Distant colour induction on Fechner-Benham top, in: *Colour Vision Deficiencies XIII*, Cavonius, C. R. (Ed.), pp. 459–464. Kluwer, Dordrecht, The Netherlands.

Mayzner, M. S. (1975). Studies of visual information processing in man, in: *Information Processing and Cognition*, Solso, R. L. (Ed.), pp. 31–54. Lawrence Erlbaum. Hillsdale, NJ.

McCollough, C. (1965). Color adaptation of edge-detectors in the human visual system, *Science* **149**, 1115–1116.

Michael, C. (1978a). Color vision mechanisms in monkey striate cortex: simple cells with dual opponent-color receptive fields, *J. Neurophysiol.* **41**, 1233–1249.

Michael, C. (1978b). Color sensitive complex cells in monkey striate cortex, *J. Neurophysiol.* **41**, 1250–1266.

Michael, C. (1979). Color sensitive hypercomplex cells in monkey striate cortex, *J. Neurophysiol.* **42**, 726–744.

Murch, G. M. and Hirsh, J. (1972). The McCollough effect created by complementary afterimages, *Amer. J. Psychol.* **85**, 241–247.

Pinna, B., Brelstaff, G. and Spillmann, L. (2001) Surface color from boundaries: a new 'watercolor' illusion, *Vision Research* **41**, 2669–2676.

Riggs, L. A., White, K. D. and Eimas, P. D. (1974). Establishment and decay of orientation-contingent aftereffects of color, *Perception and Psychophysics* **16**, 535–542.

Rotgold, G. and Spitzer, H. (1997). Role of remote adaptation in perceived subjective color, *J. Optical Soc.Amer. A* **14**, 1223–1230.

Schramme, J. and Supply, N. (1997). Cortical modification of pattern induced flicker colors, *Investigative Ophthal. Vis. Sci.* **38** (suppl.), 900.

Schwartz, E. L. (1977). Spatial mapping in the primate sensory projection: analytic structure and relevance to perception, *Biolog. Cybernet.* **25**, 181–194.

Schwartz, E. L. (1994). Topographic mapping in primate visual cortex: history, anatomy and computation, in: *Visual Science and Engineering*, Kelly, D. H. (Ed.), pp. 293–359. Marcel Dekker, New York.

Shute, C. C. D. (1979). *The McCollough Effect.* Cambridge University Press, Cambridge, UK.

Stromeyer, C. S. (1978). Form-color aftereffects in human vision, in: *Handbook of Sensory Physiology, Vol. 8, Perception*, Held, R., Leibowitz, H. and Teuber, H. L. (Eds), pp. 97–142. Springer-Verlag, Berlin.

Teller, D. Y. (1984). Locus questions in visual science, in: *Visual Coding and Adaptability*, Harris, C. S. (Ed.), pp. 151–176. Lawrence Erlbaum, Hillsdale, NJ.

Verriest, G. and Seki, R. (1964). Les chromaticites des couleurs subjectives suscitees par la rotation du disque de Fechner-Benham, *Revue d'Optique Theorique et Instrumentale* **43**, 53–63.

Webster, W. R., Crassini, B. and Willenberg, K. (1987). Simultaneous color contrast from McCollough effects is spatially contingent, *Perception and Psychophysics* **41**, 402–408.

Webster, W. R., Day, R. H. and Willenberg, K. (1988). Orientation-contingent color aftereffects are determined by real color, not induced color, *Perception and Psychophysics* **44**, 43–49.

Westheimer, G. (1967). Spatial interaction in human cone vision, *J. Physiol.* **190**, 139–154.

Williams, D. R., Sekiguchi, N., Haake, W., Brainard, D. and Packer, W. (1991). The cost of trichromacy for spatial vision, in: *From Pigments to Perception*, Valberg, A. and Lee, B. B. (Eds), pp. 11–22. Plenum, New York.

Zrenner, E. (1983). *Neurophysiological Aspects of Color Vision in Primates*. Springer-Verlag, Berlin.

Color, Line, and Space, pp. 71–90 (2006)

Cone contrast computations: physical versus perceived background and colour constancy

A. DAUGIRDIENE [1], I. J. MURRAY [2], H. VAITKEVICIUS [3] and J. KULIKOWSKI [2,*]

[1] *Department of Psychology and Didactics, Vilnius Pedagogical University, Studentu 39, LT-08106 Vilnius, Lithuania*
[2] *Visual Sciences Lab, Moffatt Building, PO Box 88, Faculty of Life Sciences, University of Manchester, Manchester M60 1QD, UK*
[3] *Department of General Psychology, Faculty of Philosophy, Vilnius University, Didlaukio 47, 2057 Vilnius, Lithuania*

Abstract—The light reflected from an object depends both on the object's surface and on the illuminant. Visual systems attempt to resolve this intrinsic ambiguity by comparing the light reflected from the object with respect to the background by computing the difference between the object-background light sampled by three cones. The cone-contrasts for the sample-background stimulus under the test illumination should correspond to the cone-contrasts for samples matched in appearance under the standard background (C). The validity of this cone-contrast rule (Whittle, 2003) and its possible link with stability of perceived colour was studied here using six test illuminants.

A successive asymmetric colour-matching task with 40 simulated Munsell samples (value 7, chroma 4) on a neutral background (N7) was used. The subject adjusted the sample under standard illuminant C to match the colour appearance of the test sample under one of the test illuminants. Brunswik Ratio (BR) was used as an index of stability of colour appearance.

When computed with respect to the reference illuminant C, the cone contrast rule was violated (particularly for S-cone-contrast). However, if a new reference point based on the perceived colour of the neutral background under the test illumination was used, the cone contrast rule applied. That is, when cone contrasts of the matching samples are computed with respect to this perceived background, they correspond to cone contrasts of the test stimuli. This represents a form of discounting the illuminant for the purpose of determining an object's cone-contrast against the background, which does not vary with background illumination. These cone contrasts, however, do not provide any information about the colour appearance of objects under particular viewing conditions, unless calibrated against a standard by allowing subjects to learn particular colours.

Keywords: Cone contrast; colour appearance; colour matching; colour constancy.

*Correspondence and reprint requests should be addressed to.
E-mail: janus.Kulikowski@manchester.ac.uk

INTRODUCTION

One of the most important aims of all visual systems is to extract invariant features from components of the retinal image. In the colour domain, objects have different reflecting characteristics, but light reflected from surfaces also depends on the ambient illumination. Therefore, to be biologically useful, vision must utilise an algorithm to retrieve information about the colour of objects in a scene by estimating surface reflectance from reflected light that depends also on ambient illumination characteristics. This is called colour constancy. Yet colour constancy may have different meanings, which correspond to different experimental tasks (for reviews see Hurlbert 1999; Foster 2003).

If emphasis is put on the colour itself, the observer estimates colour appearance when, for example, it is important to judge the ripeness of fruits. Conversely, the colour may be a factor when differencing objects against each other and against their background. Then the components of a scene may have shifts of colours throughout the day as the illumination varies, but the ratio of these shifts and, consequently, the scene remains constant.

The traditional approach to constancy of colours is that vision can discount the illuminant after adaptation (Judd, 1940). However, complete adaptation rarely occurs or is desirable and the adaptation process is highly complex and almost certainly composed of a multi-stage process culminating in multiple chromatic channels (see review by Webster, 2003). Among the factors hypothesized as leading to an estimate of the illuminant is the visual system's ability either to compute the mean colour of the scene (Buchsbaum and Gottschalk, 1983; Land, 1986; Linnell and Foster, 2002) — the grey-world hypothesis — or to adopt the colour of the brightest part in the scene as indicative of the characteristics of the illuminant (e.g. Land and McCann, 1971; McCann *et al.*, 1976). Another hypothesis is that the illuminant is estimated by statistical analysis of the scene (Goltz and MacLeod, 2002).

However, Kraft and Brainard (1999) and Kraft *et al.* (2002) showed that none of the classical theories could be entirely correct because environments can be created for which the theories unambiguously predict zero constancy and nonetheless relatively good constancy is observed.

Craven and Foster (1992) and Foster and Nascimento (1994) simplified the problem by dealing with 'relational colour constancy' of objects. In their paradigm, two scenes are compared in quick succession and observers report whether they see either an illuminant or material change (see Foster, 2003). This relational constancy holds true for both artificially simulated patterns (Nascimento and Foster, 1997) and natural scenes (Nascimento *et al.*, 2002). These authors demonstrated that cone contrasts, or cone-excitation ratios, are the decisive parameter in relational colour constancy. In this case, colour appearance of individual components is irrelevant and only *a very* weak link with true colour 'appearance' constancy is reported (Nascimento *et al.*, 2005).

Moreover, the issue of whether the relative activation of the three cone types can account for perceived colour has a venerable and somewhat neglected history. The ideas can be traced from von Kries (1905) through Ives (1912) and Le Grand (1948) to Chichilnisky and Wandell (1995) to name only a few contributors. Whittle (2003) offers an excellent review. What he calls the cone contrast rule emphasises the differencing mechanisms in the retina, which suggest that perceived colour depends, initially at least, on the three cone contrasts rather than on the quantum catch of each class of cone. The question we address here is: how strong is the relationship between the cone contrast rule and perceived colour? We argue that, though providing important raw data, cone contrasts bear little relationship to colour appearance or to the mechanisms that stabilize our colour perception (see Discussion).

Following our previous studies (Daugirdiene *et al.*, 2000, 2002; Stanikunas *et al.*, 2004), our first aim was to test the possible role of object/background chromatic contrast in determining colour constancy. We investigated colour appearance of single samples on a neutral background when briefly illuminated by a test illuminant. These experimental conditions were chosen particularly because they induce significant shifts in colour appearance under different illuminants. Presentation time was chosen to be brief since the constancy index may improve with adaptation (e.g. Daugirdiene *et al.*, 2000, 2002). Conversely, the border contrast of sample/background stimuli should be the same. Such conditions allow the intricacies of the relationship between colour appearance and cone-contrasts to be tested.

Our second aim was to test whether the physical cone-contrast in the matching sample/background stimulus is relevant to colour constancy, bearing in mind the reported differences between physical and so-called perceived cone contrasts (e.g. Switkes and Crognale, 1999). We show that it is crucial to take account of the colour appearance of the neutral background when it is illuminated by the test illuminant. It turns out that the cone contrast rule holds if, and only if, cone-contrasts are computed from the difference between matching sample and the perceived, not the physical background.

METHODS

Apparatus and stimuli

Forty Munsell hues, value7/, chroma/4 and one neutral N7 were simulated on a Barco monitor (model CC1D 121) driven by a VSG card; both were switched on at least 1 h before the experimental session. Calibration was carried out using a Photo Research (SpectraScan PR650) spectrophotometer (4 nm).

The stimulus consisted of a centrally located square subtending 2° superimposed on a neutral 20° background. Both the sample and background were simulated as illuminated either by one of the six test illuminants or the reference illuminant C (20 $\mathrm{cd\,m^{-2}}$). The reflectance functions of all samples and backgrounds under all illuminants were computed using Parkkinen functions (Parkkinen *et al.*, 1989).

Illuminant C (averaged daylight of colour temperature 6774 K) was adopted as the reference and six test illuminants were used as described in Table 1. Two test illuminants were chosen close to the CIE daylight or Planckian locus; A (black-

Table 1.
Chromaticity coordinates of the reference and six illuminants

CIE 1976	C	A	S	R	G	Y	V
u′	0.201	0.256	0.174	0.262	0.172	0.194	0.212
v′	0.461	0.524	0.392	0.453	0.467	0.518	0.377

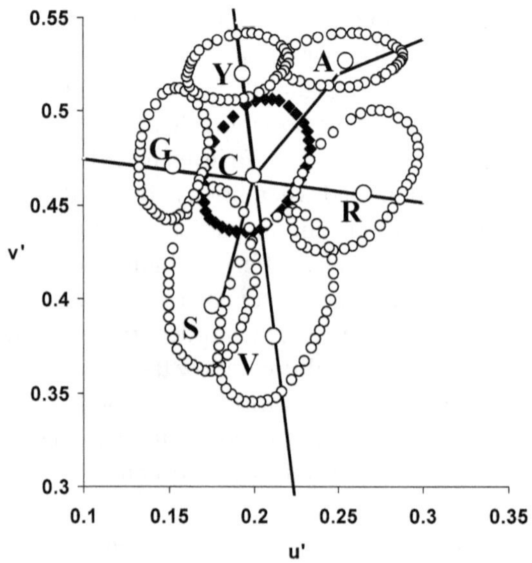

Figure 1. The CIE 1976 chromaticity coordinates (u′, v′) of the colour stimuli used. Illuminant C was used as a reference illuminating a neutral background (N7, see the central point) and chromaticity of 40 reference samples are indicated by filled circles around it. Test stimuli consist of the same 40 samples (small empty circles) on the same neutral background under test illuminants (Planckian: A, S and cardinal G, R, Y, V, see large empty circles). Thick lines are: cardinal–tritanopic (vertical) and cardinal red-green (horizontal) and Planckian (oblique curve). Note that the cardinal illuminants are chosen to make the loci of the test samples tangent to the locus of the reference samples. This means, for example, that the cardinal green sample under the test illuminant G reflects the same light as the cardinal red sample under C.

body radiation at temperature 2856 K, similar to light of a tungsten bulb) and S (infinite-black body temperature, light of extremely blue sky). The remaining four illuminants were chosen to be cardinal, i.e. lying on the cardinal axes; red-green R, G and yellow-violet Y, V. Figure 1 illustrates the CIE 1976 chromaticity coordinates (u′, v′) of the 40 Munsell samples (under C — central ellipse) and under all test illuminants A, S, R, G, Y and V. The points within each ellipse specify lights reflected from the neutral background (N7) illuminated by C and six test illuminants respectively (large empty circles). The chromaticity coordinates of illuminant A, C, S are fixed, but the chromaticity of the cardinal illuminants (G, R, Y, V) were chosen in such a way that the loci of Munsell samples were contiguous with each other. That is, the reflectance of some samples under a test illuminant has the chromaticity coordinates of the opposite colour samples: they would be so perceived if the background was completely black (void colours). Thus for example, red-purple samples under illuminant C reflect the same light as blue-green samples under illuminant R.

Procedure

The experiments were conducted in a darkened room, with the subject seated 35 cm from the screen. Following 15 min dark adaptation, the subject was pre-adapted for at least 1 min to the uniform N7 background under standard daylight illuminant C. Sequential asymmetric colour matching was adopted. The test sample with the neutral background was presented for 1 second under one of six test illuminants. There followed a 1 s period of re-adaptation to illuminant C and the matching sample then appeared under standard illuminant C. The subject adjusted separately (in small, just discriminable steps) the hue, chroma, and value of the reference sample under illuminant C to match the colour appearance of the test sample as it had appeared under the test illuminant. The presentation sequence is illustrated in Fig. 2. The time to make a match was not restricted, but generally it was between 30 s and 90 s. In this task, only the colour appearance of the central samples (test and match) was important, i.e. the subject matched them in hue, chroma and value. Note that this is not a surface-like match (*cf.* Arend and Reeves, 1986; Arend *et al.*, 1991; Troost and de Weert, 1991; Foster, 2003), which attempts to guess how a given surface would look under a different test illuminant. In a separate session the subject estimated colour appearance of the neutral background (N7) under each of the test illuminants. The hue, chroma and value of the reference sample were adjusted under illuminant C to match the colour appearance of the neutral background as it had appeared under the test illuminant (a colour appearance constancy).

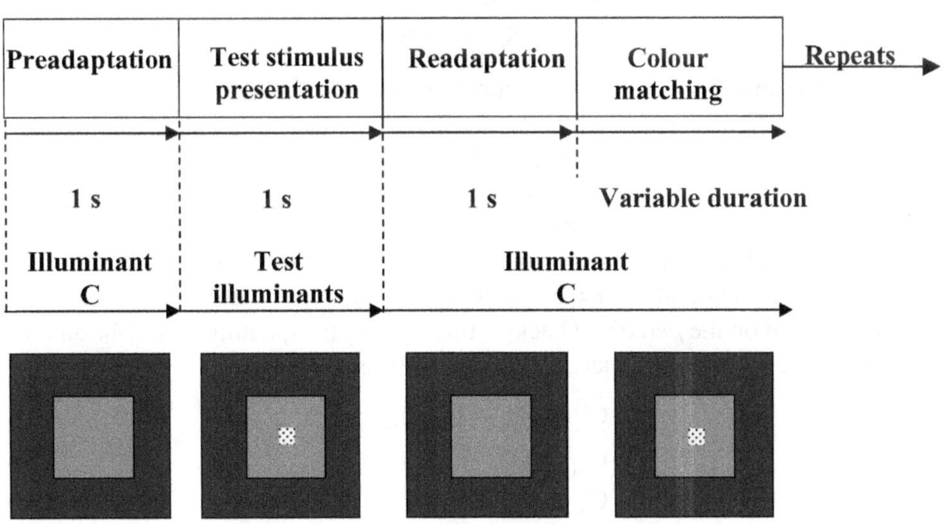

Figure 2. Stimulus configuration and presentation sequence. Note that, apart from the 1 s presentation of the sample under the test illuminant, subjects were continuously adapted to illuminant C.

Subjects

Most subjects were experienced observers with normal acuity and colour vision tested by 100-Hue test and Nagel anomaloscope. The data are reported for main subjects AD (female 32 yrs) and MJ (male 30 yrs). Three additional subjects (one naive) participated in control experiments.

Analysis

Two indices of colour constancy were computed from the results of the experiments.

1. Brunswik Ratio, BR. The simplest index of colour constancy is the Brunswik Ratio, BR, which determines to what extent vision compensates for changes in illumination. For example, illuminant A causes the chromaticity shift of a sample (i), from point Ci to point Ti (Fig. 3). When this sample is matched by an adjustable sample under the reference illuminant C, an intermediate point Mi specifies its appearance.

Brunswik Ratio, BR is computed on the 1976 UCS u′, v′ plane

BR = (|CiTi| − |CiMi|)/|CiTi| = 1 − (|CiMi)/(|CiTi), or

BR = 1 − (perceptual shift)/(physical shift),

BR = 1 is ideal constancy, whereas BR = 0 means no constancy (as with void colours).

2. Cone-contrast definition. This definition specifies how long-, middle- and short-wavelength cones are activated or excited by the contrast between sample and background. Weber's contrast was adopted.

Thus:

$$C_L = (L_s - L_b)/L_b, \tag{1a}$$

$$C_M = (M_s - M_b)/M_b, \tag{1b}$$

$$C_S = (S_s - S_b)/S_b, \tag{1c}$$

where C_L, C_M and C_S are Weber cone contrasts for the −L, −M and −S cone types. L_s and L_b are the L-cone excitations for the sample and background, M_s and M_b are the M-cone excitations for the sample and background and S_s and S_b are the S-cone excitations for the sample and background.

Now, conventionally, the backgrounds for the quantities L_b, M_b and S_b will be illuminant C because our experiments are designed to maintain adaptation to this condition throughout the experiment. However, cone contrasts have been re-computed based on the *perceived* backgrounds giving the quantities $L_{b'}$, $M_{b'}$ and $S_{b'}$ and we then define the inter-match cone contrasts as:

$$C'_L = (L_s - L_{b'})/L_{b'}, \tag{2a}$$

$$C'_M = (M_s - M_{b'})/M_{b'}, \tag{2b}$$

$$C'_S = (S_s - S_{b'})/S_{b'}. \tag{2c}$$

Smith and Pokorny (1975) cone fundamentals (L, M, S) were used in these computations.

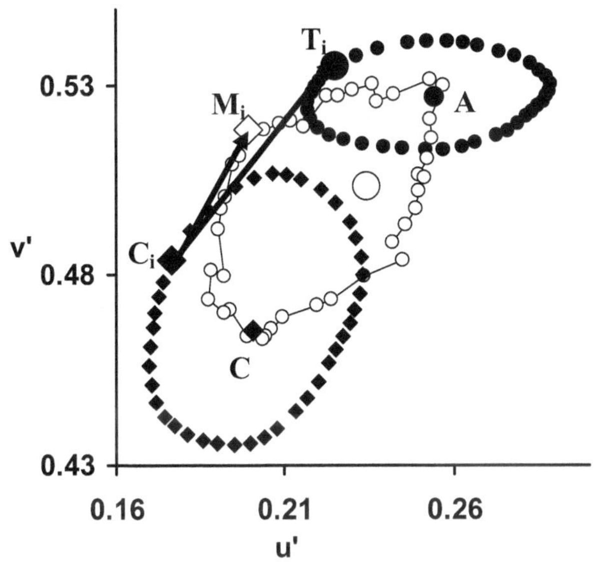

Figure 3. An example of colour matches of 40 test samples on neutral background (N7). Filled diamonds indicate the chromaticity of original reference samples and background. Filled circles indicate chromaticity of these samples and background under test illuminant A (presented for 1 second). Empty circles show the results of matching: a reference sample (under C) is adjusted until it matches a test sample (under A). The large empty circle indicates a match for the neutral background (under A). Large symbols illustrate an example of chromaticity of one sample (2.5G7/4) under C (filled diamond — point Ci), under test illuminant A (filled encircled diamond — point Ti) and when it matched in colour (empty diamond — point Mi).

RESULTS AND COMMENTS

Figure 3 shows the loci of 40 Munsell samples under the reference illuminant C (filled diamonds) and under the test illuminant A (filled circles). Empty circles indicate the results of matching of the test samples on the neutral background, both under illuminant A, by adjusting the reference samples (under C). If colour constancy was ideal, the matches would be coincident with the reference samples. Conversely, an overlap of empty and filled circles would mean that the match is based on the physical parameters of the test and that there is no colour constancy It is clear that the matches (under illuminant C) are roughly mid-way between both loci; they depart markedly from the original colours (see empty circles *versus* filled diamonds), but are not shifted as much as their physical chromaticity coordinates would indicate (filled circles). Note also that the neutrality is also misperceived (large empty circle between C and A).

Large symbols in Fig. 3 illustrate how the colour constancy index, BR, is computed for a green sample 2.5 G. Its locus is C_i (large filled diamond) under illuminant C. It looks greenish-yellow on background N7 when both are illuminated by A and it is matched by a reference sample M_i (about 2.5 GY, of slightly higher chroma than 4; see large empty circle). The perceptual change (C_i – M_i) occurs

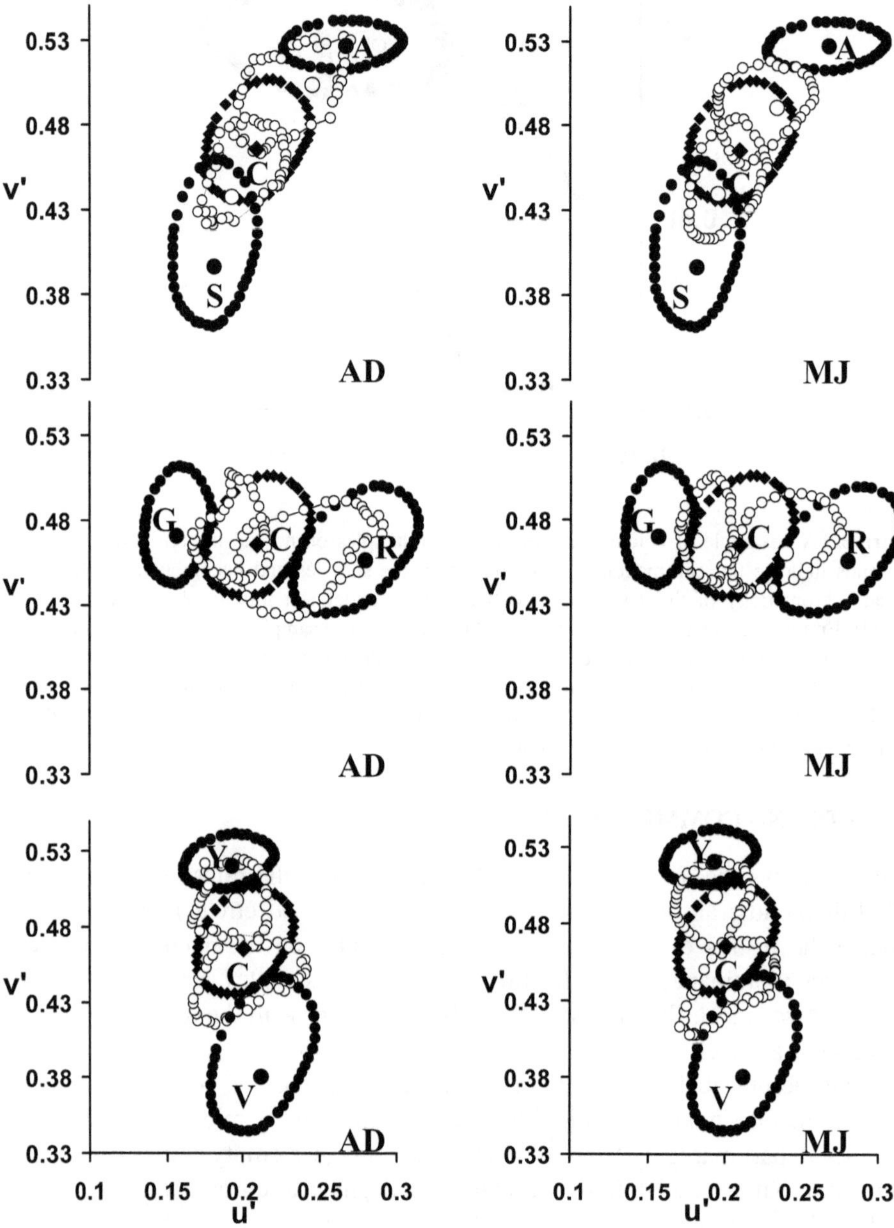

Figure 4. Chromaticity loci of 40 Munsell samples on neural background (N7) under different illuminants (filled symbols: test-circles and reference-diamonds) for two subjects (left- and right-hand columns). The corresponding neutral points are indicated by large filled symbols inside the ellipses. Each graph shows loci under test illuminants arranged in pairs: A and S (Planckian), G and R, Y and V (both cardinal). Empty circles show matches of 40 samples and neutral background (N7).

mainly in hue. However, this change is not as large as indicated by its reflectance T_i (yellow of almost double chroma, depicted by large filled encircled diamond). There is, therefore, some degree of colour (appearance) constancy, as indicated by the Brunswik ratio, BR = 0.4.

Note also that the colour of the neutral background (N7), illuminated briefly by A, has accordingly shifted from its original position (C) to an intermediate point (see the large empty circle in Fig. 3, approximately in the middle of small circles), although not as far as photometric point A. We refer to this point as Lb' above in equations (2a–2c). Thus, neutral grey when presented for 1 second, looks distinctly yellow-reddish (between 10 YR and 7.5 YR) — like a neutral grey sheet of paper briefly illuminated by a tungsten bulb. In this case a change in chroma occurs, but constancy remains low at BR = 0.35.

Figure 4 shows the complete set of chromaticity loci of 40 samples under test illuminants (see Table 1): two Planckian (A, S) and four cardinal (R, G, Y, V) plotted in pairs for clarity (filled circles). The reference ellipse (under C) is shown by filled diamonds. Empty circles in Fig. 4 indicate the results of matching 40 Munsell samples for two subjects (AD-left and MJ-right columns). Likewise the result of matching of a neutral background, when briefly illuminated by each test illuminant, indicates that these points (large empty circles in Fig. 4) are within the corresponding matching loci. A general impression is that both matched samples and background are shifted similarly in their chromaticity.

Figure 5 shows the Brunswik Ratio (colour constancy index) as a function of hue for all illuminants and two subjects (left and right columns of Fig. 5): the illuminant is indicated in the centre of the figure. BR is between about 0.3 and 0.8, which indicates a moderate degree of colour constancy. It will be seen that BRs are similar for both subjects and comparable for both close-to-natural, Planckian illuminants, shown in the upper two panels and highly artificial cardinal R, G illuminants shown in the lower panels.

Cone contrast with respect to illuminant C

In order to test whether von Kries adaptation occurs under these matching conditions cone-contrasts are calculated with respect to illuminant C. Since the matches were made under illuminant C, it would seem plausible to compute cone contrasts with respect to that illuminant. The results for two subjects and two Planckian test illuminants (A and S) are shown in Fig. 6a, b, left columns. The right hand columns illustrate cone contrasts re-calculated using a different reference point and are described below as inter-match cone contrasts.

It is evident that the cone contrast rule does not hold, particularly for the S cones. Cone-contrasts for matches under C are not equal to the corresponding cone contrasts under test illuminants. The data points depart from a straight line and are shifted vertically having a negative y-intercept for illuminant A and a positive y-intercept for illuminant S (see Discussion).

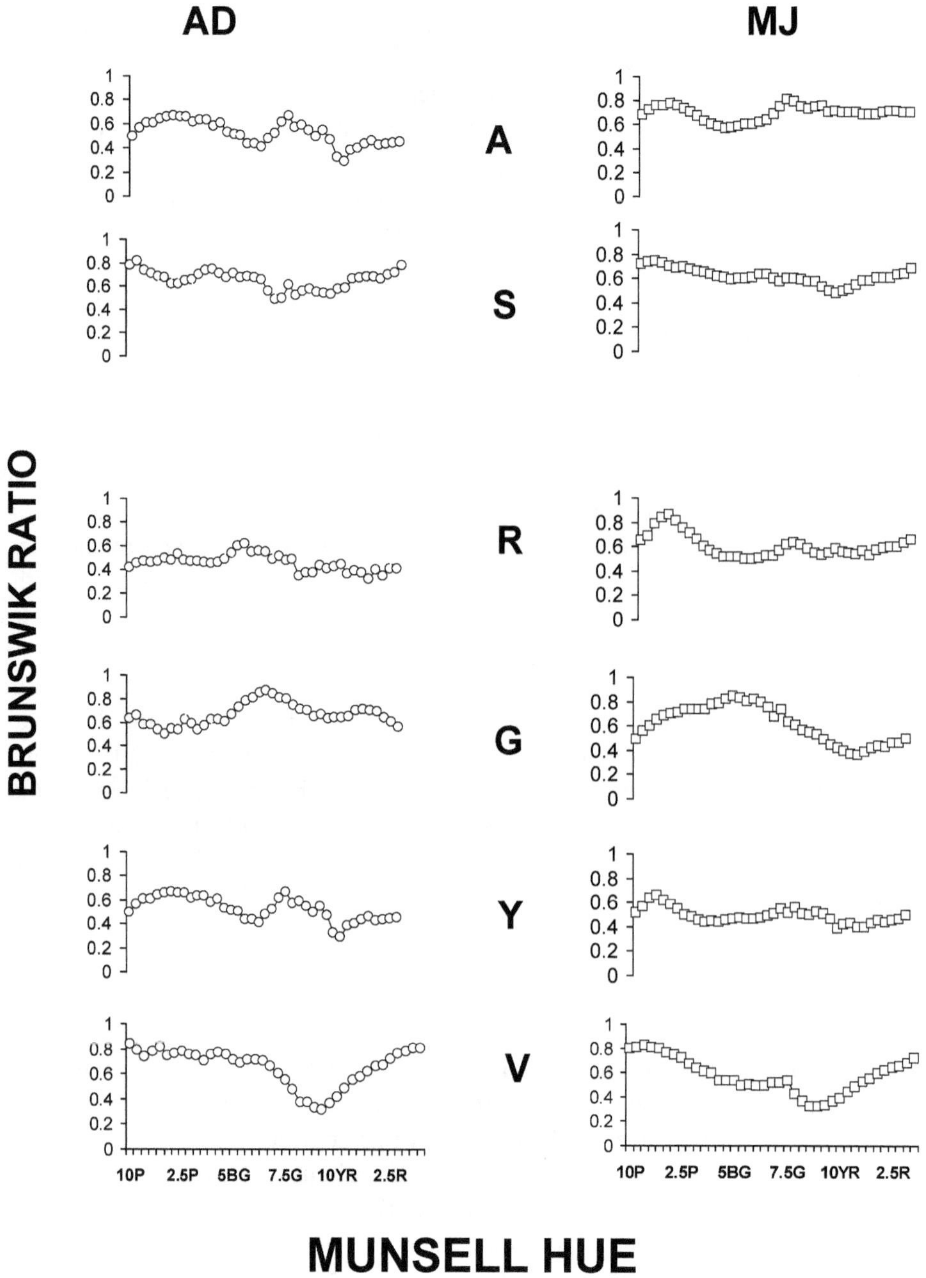

Figure 5. Colour constancy index, Brunswik Ratio (BR) as functions of Munsell hue under different test illuminants for two subjects, as indicated. Conditions are as in Fig. 4.

Inter-match cone contrasts

As neutrality is misperceived under the test illuminant (Fig. 3), it may be that the cone contrast calculation should be based on a new neutral point and in order to test this we have adopted the point $L_{b'}$, which is the perceived neutral point under the different test illuminants. As seen in Fig. 4 it is always inside the ellipses marking chromaticity of the matching samples. This perceived neutral point shifts together with the 40 sample points corresponding to the subjective matches (Daugirdiene *et al.*, 2000). Hence, cone contrasts for any test stimulus (physical samples-background under test illumination) should be compared with contrasts computed for comparable matching stimulus, namely, matching samples and matching background appearance under a particular test illuminant. We call it 'inter-match cone contrast' since cone contrast for any matching sample involves two matches — the apparent colour of this sample and apparent colour of a neutral background.

The results of these computations are shown in right-hand columns of (Fig. 6a, b, right columns) for Planckian illuminants. It is obvious that the cone-contrasts for the matched samples computed with respect to the matched background are a better fit to the cone-contrast model than when computed with respect to illuminant C.

Cone-contrasts and inter-match cone contrasts for cardinal illuminants (R, G, Y, V)

Figure 7 shows cone-contrasts for matches plotted against cone contrasts for test samples under the four cardinal illuminants. Note that cardinal illuminants R and G are a special case (Fig. 7a, b, left columns) since for them S-cone stimulation is constant and S-cone contrast shows minimal shift when compared to the test illumination. It will be seen that this relatively good fit of all cone-contrasts for illuminants R and G does not correspond to a higher colour constancy index, as seen in Fig. 5. The results for cardinal illuminants Y and V resemble those for Planckian (A, S).

On the other hand, all inter-match cone-contrasts (defined above) correspond well to those of the test samples in all cases. Regardless of how good is the fit of direct computation with respect to C, the fits for the inter-match cone contrasts are superior. Thus inter-match cone contrasts closely describe reflecting object surfaces under a variety of conditions under any test illuminant (see Discussion).

DISCUSSION

Cone-contrast versus inter-match cone-contrast and colour constancy

The main finding of these experiments is that colour constancy is generally unrelated to cone contrasts of the matches, whether computed against the physical neutral background, C, or the apparent colour of this background.

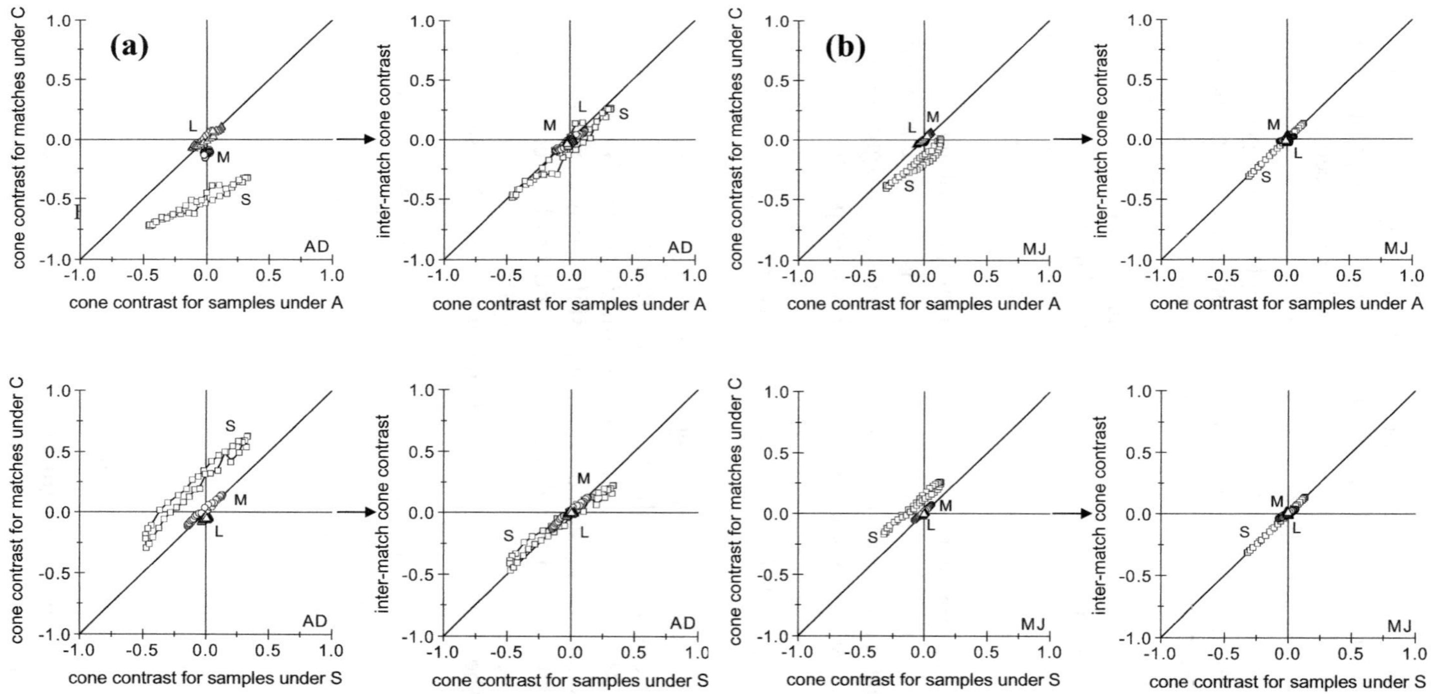

Figure 6. Cone-contrasts for matches under reference illuminant C as functions of cone-contrasts for samples and background N7 under Planckian illuminants (A, S) and two subjects (a, b), as shown on the left; the discrepancy is evident. In the right-hand columns the same matches are re-plotted (see the arrows) in terms of inter-match cone-contrasts, i.e. with respect to matched (perceived) neutrality (N7), not with respect to C. The cone-contrast computed between any matching sample and the matching background closely corresponds to cone-contrast for the test stimulus.

This is clear because we describe conditions in which perceived colour is unstable, that is, colour constancy is poor, despite the fact that the cone-contrast rule is applicable. The poor fit is particularly obvious for S-cones, as noticed by many preceding studies even under conditions that provide a better fit of cone-contrast (e.g. Lucassen and Walraven, 1993; Bauml, 1999). It is possible that the spatial S-cone-contrast computation is poorer and subject to a nonlinear transformation (Lucassen and Walraven, 1993), although in simple detection tasks blue-yellow and red-green contrast sensitivity is approximately equal (McKeefry *et al.*, 2000).

It should be added that there is a vast literature on differences in structure and function of L/M- *versus* S-cones as well as their post-receptoral networks (Dacey and Lee, 1994; Calkins, 2001; Martin *et al.*, 1997; Hendry and Reid, 2000).

Moreover, we show that cone contrasts for the matches of simulated Munsell samples on a perceived neutral background were more likely to be consistent with the cone contrast rule (von Kries, 1905) than cone-contrasts computed against the illuminant C (Fig. 6, left hand columns). The experimental paradigm adopted (test sample/background presented for only 1 s) renders rather a moderate colour constancy index (BR commonly between 0.4 and 0.8), partly due to incomplete adaptation. If presentation is longer and adaptation is more complete, the constancy index (BR) increases and may reach near perfect constancy (Daugirdiene *et al.*, 2000). However, the inter-match cone-contrast (defined above) remains constant under these and all other conditions. This effect is referred to as cone contrast constancy.

The crucial factor is that colour appearance of both the samples and the neutral background need to be evaluated. Under the test illuminant, the neutral background does not appear neutral (or grey) in these experimental conditions, but rather is tinged by the hue of the illuminant. Hence the subjects have two tasks: to match the colour appearance of all the 40 samples and, separately, the appearance of the background. When cone contrasts are computed against this matched background, the inter-match cone contrast constancy emerges and indicates that the sample/background appearance ratio does not vary with illumination (Figs 6 and 7). Thus, these inter-match cone contrasts are a good measure of stability in perception of object's surfaces and of the neutrality of their background. Two separate components are therefore needed to evaluate the stability of surface reflectance — contrast between object and background and background colour (see neural net simulation, Stanikunas *et al.*, 2004). Moreover, it is possible to predict approximately the relative appearance of each sample after estimating the neutral background appearance (see below). The second component — an estimation of the background appearance — is probably a separate process, most likely performed in several later stages. Our experiments probably reveal only the initial or early stages.

Still, these two measures do not tell us about veridical colours *per se*, unless the experiment involves learning object colours prior to colour matching sessions (see Stanikunas *et al.*, 2004). Our related studies (Daugirdiene *et al.*, 2000) showed that changes in experimental conditions affect equally matching loci (ellipses) and

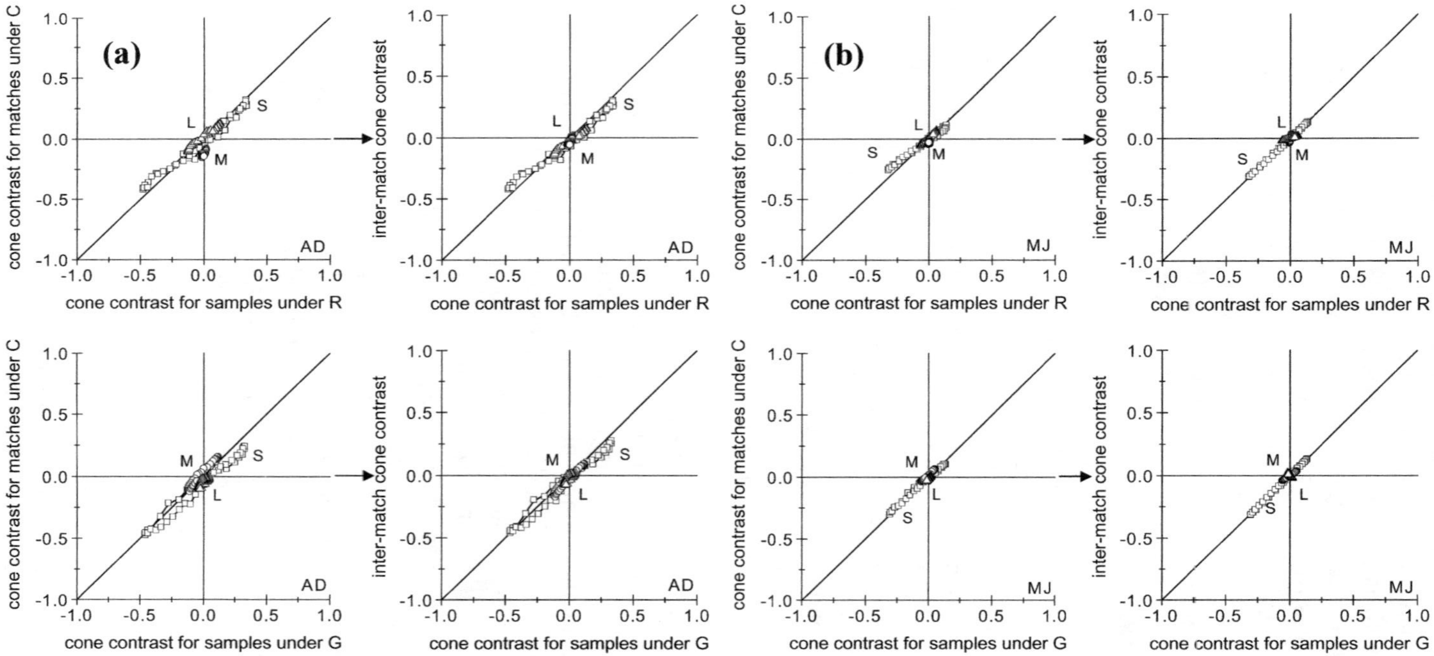

Figure 7. Cone-contrasts for match under C as functions of cone-contrast for samples and background under cardinal illuminants and two subjects. Note that for cardinal G and R cone-contrast for matches under C approximately corresponds to cone-contrasts for the test stimulus (a, b, left-hand columns). Conversely, the cone-contrasts for matched under C (c, d) do not correspond to cone-contrasts for test stimuli under tritanopic cardinal illuminants Y and V (as in Fig. 6a, b). Right-hand columns show that inter-match cone-contrasts (re-computed with respect to matched neutrality for each condition) closely correspond to cone-contrasts for test stimuli under all test illuminants, irrespective whether cone-contrasts for matches under C fit test cone-contrasts or not.

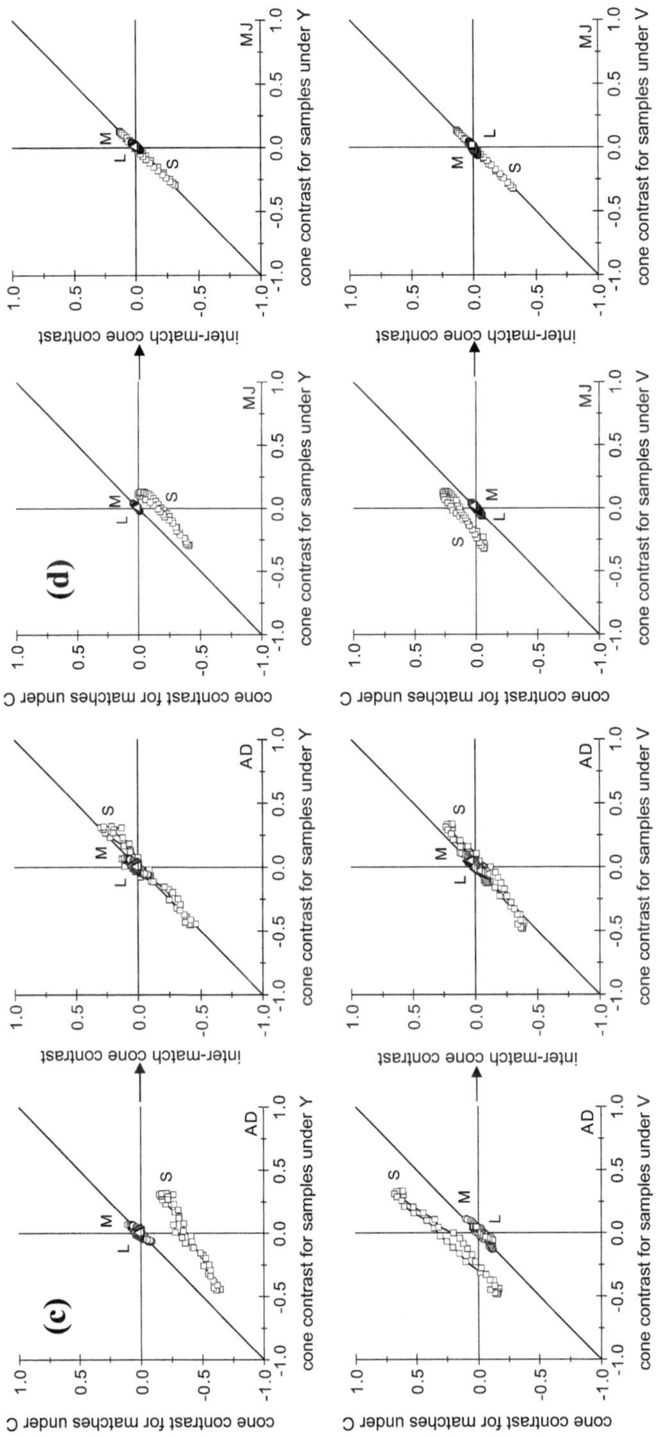

Figure 7. (Continued).

the point corresponding to perceived neutrality. They are always shifted together in colour space. Perceived neutral backgrounds under a test illuminant are always inside the matching ellipses and their colour contrast remains constant (see all loci with 40 points corresponding to subjective matches and large circles in Fig. 4).

Relational colour constancy

The finding of inter-match contrast constancy is consistent with the extensive series of studies on relational colour constancy (Foster and Nascimento, 1994; Nascimento and Foster, 2000; Foster *et al.*, 2001a, b). In these forced choice tasks, subjects decided whether an illuminant change or material change took place in displays presented in quick succession. The subjects ignored colour appearance of individual components, which may change very drastically, and concentrated on reporting whether or not a scene (or illuminant) has changed. A similar situation occurs in real life when a new coloured illumination is introduced for a brief duration and nevertheless constancy of scene is assured, as Foster *et al.* (2001) reported. Yet when the authors introduced conventional colour constancy index they obtained BR about 0.8 (Nascimento *et al.*, 2005), which means that colour perception is not veridical.

A similar situation occurs in our experimental conditions, except that there are only two components, the sample and background, and these have to be matched in separate sessions. This procedure allows an explicit demonstration that apparent colours of both the sample and background are similarly shifted by test illumination, so that their contrast stays constant. Therefore, this is not constancy of colours, but rather 'relative cone-contrast constancy': it determines invariant features of samples and backgrounds, but does not provide information about their real colours.

Obviously, it is possible to predict approximate matches by knowing the reflectance of objects on a neutral background and apparent colour of this background, since the loci of matches are reasonably regular around the 'perceived, or apparent neutral points' (see Fig. 4, empty circles). The reverse procedure of the achromatic adjustment of a stimulus until it appears grey is probably more accurate to predict asymmetric matches (e.g. Speigle and Brainard, 1999).

Task dependence

The effects of colour matching as compared with forced-choices (illuminants changes *versus* material changes) on cone contrasts (or excitation ratios) discussed above may have parallels with other experimental conditions. The early studies (Arend and Reeves, 1986; Arend *et al.*, 1991) reported a much higher colour constancy index when the subjects were instructed to match the samples 'as if they were cut from the same piece of paper'. Likewise Troost and de Weert (1991) reported higher constancy indices in the object matching tasks when the 'subjects had to match what they should see' (rather than 'what they see' in exact colour matching). These authors also found that colour identification tasks yield higher constancy indices, maintaining colours within their categories (Troost and de Weert, 1991). It is interesting that the colour categorization task improves constancy (Kulikowski and Vaitkevicius, 1997) even in simple colour matching tasks, possibly because better memory for categorical, or unique-like, colours, makes it easier to perform object-like matching. Probably, in many of these tasks, which improve constancy, relations between colours guide subjects rather than colour appearance.

Stages of colour processing and constancy

Since colour constancy in practice is below unity, some authors question whether it exists at all (Foster, 2003; Hurlbert, 1999). Certainly, colour perception is a complex multi-stage process, which cannot be embraced by one simple model (Zeki, 1983; De Valois *et al.*, 1999; Moutoussis and Zeki, 2000).

In another study, we have shown that a neural network learns colour recognition tasks faster if it has at least two stages (Stanikunas *et al.*, 2004).

The present study separates the cone-contrast stage, which is equivalent to the cone-opponent stage (see Methods section and Hurlbert *et al.*, 1998), from the higher secondary stages, which deal with colour.

That there is a clear dissociation between functions of these stages was shown in patients with cerebral achromatopsia (i.e. lacking the ability to identify colours). These patients could see differences between colour patches, which means they have what we define as relative cone-contrast constancy, in spite of being colour blind (see Hurlbert *et al.*, 1998). Lesions of visual cortical area V4 produce colour constancy deficit in monkeys (Wild *et al.*, 1985; Walsh *et al.*, 1993), but do not affect simple wavelength discrimination, i.e. cone-contrast discrimination tasks (Wild *et al.*, 1985; Carden *et al.*, 1992; Hurlbert *et al.*, 1998), or indeed basic colour categorization (Walsh *et al.*, 1992).

Cone contrast is not necessarily extracted in the retina. Early post-retinal stages have also cone-opponent characteristics (Derrington *et al.*, 1984) and the input neurons in the primary visual cortex receive these signals (Vidyasagar *et al.*, 2002). In addition, weak binocular interactions suggest possible visual cortical involvement (Nascimento and Foster, 2001). Thus, in this task of colour matching, subjects use the cone-opponent stage of processing (between the retina and input stage of the visual cortex), whereas real colour assessment may require different mechanisms (for discussion see Foster, 2003; Hurlbert *et al.*, 1998, Shepherd, 1997; Walsh 1999).

Effect of adaptation on the sample-background stimulus

Adaptation techniques are powerful tools in separating various processing stages (see Webster, 2003). In particular, Webster and Mollon (1991, 1994) demonstrated two-stage adaptation — receptoral and post-receptoral. Our studies (Daugirdiene *et al.*, 2002, and in preparation) revealed in addition that adaptation to a background, or a surround, is far more complex and cannot be explained in terms of any local adaptation, receptoral or post-receptoral, and various central stages are also involved (Bloj *et al.*, 1999; Shevell and Wei, 2000; Kraft *et al.*, 2002).

In our simple case, the inter-match (relative) cone-contrast (between the sample-background stimulus) is not affected by adaptation. For example, if adaptation to illumination is more effective (Daugirdiene *et al.*, 2002) constancy indices for both samples and a neutral background (see Fig. 3) may be higher, but the inter-match (relative) cone-contrast remains constant. In this sense our concept of the 'relative' cone-contrast constancy resembles relational colour constancy, which also is largely unaffected by illuminant changes (e.g. Nascimento and Foster, 1994). However, our procedure allows us to identify explicitly that the perceived background colour is necessary for estimation of colour contrast between objects, or of a reflecting surface against a background. The problem of effective adaptation to backgrounds will be dealt with in another study.

In summary, using simple sequential asymmetric colour matching, we have shown that subjects can assess sequentially the appearance of samples and backgrounds. This in turn allows a good estimate of the relational reflectance of objects with respect to backgrounds, telling us about their stability in spite of changes in illumination. We propose to call it 'cone-contrast constancy' — a concept closely related to the cone-contrast rule (Whittle, 2003).

REFERENCES

Arend, L. E. and Reeves, A. (1986). Simultaneous color constancy, *J. Opt. Soc. Amer.* **A3**, 1743–1751.

Arend, L. E., Reeves, A., Schirillo, J. and Goldstein, R. (1991). Simultaneous color constancy: papers with diverse Munsell values, *J. Opt. Soc. Amer.* **A8**, 661–672.

Bauml, K. H. (1999). Simultaneous color constancy: how surface color perception varies with illuminant, *Vision Research* **39**, 1531–1550.

Bloj, M. G., Kerten, D. and Hurlbert, A. C. (1999). Perception of thee-dimensional shape influences colour perception through mutual illumination, *Nature* **404**, 877–879.

Buchsbaum, G. and Gottschalk, A. (1983). Trichromacy, opponent colours coding and optimum colour information transmission in the retina. *Proc. Roy. Soc. London* **B220**, 89–113.

Calkins, D. J. (2001). Seeing with S-cones, *Prog. Retinal Res.* **20**, 255–287.

Carden, D, Hilken, H., Butler, S. R. and Kulikowski, J. J. (1992). Lesions of primate visual area V4 produces long-lasting deficits of colour constancy, *Irish J. Psychol.* **13**, 455–472.

Chichilnisky, E. J. and Wandell, B. A. (1995). Photoreceptor sensitivity changes explain color appearance shifts induced by large uniform backgrounds in dichoptic matching, *Vision Research* **35**, 239–254.

Craven, B. J. and Foster, D. H. (1992). An operational approach to colour constancy, *Vision Research* **32**, 1359–1366.

Dacey, D. M. and Lee, B. B. (1994). The 'blue-on' opponent pathway in primate retina originates from a distinct bistratified ganglion cell type, *Nature* **367**, 731–735.

Daugirdiene, A., Jurkutaitis, M., Vaitkevicius, H. and Kulikowski, J. J. (2000). Contrast and shift of perceived background colour in experiments on colour constancy, *Perception* **29**, 43.

Daugirdiene, A., Vaitkevicius, H, Kulikowski, J. J., Stanikunas, R. and Murray, I. J. (2002). The effects of adaptation and surround on colour constancy measurements, *Perception* **31**, S 131.

Derrington, A. M., Krauskopf, J. and Lennie, P. (1984). Chromatic mechanisms in the lateral geniculate nucleus, *J. Physiol.* **357**, 241–265.

De Valois, R. L., Cottaris, N. P., Elfar, S. D., Mahon, L. E. and Wilson, J. A. (2000). Some transformations of color information from lateral geniculate nucleus to striate cortex, *Proc. Natl. Acad. Sci.* **97**, 4997–5002.

Foster, D. H. (2003). Does colour constancy exist? *Trends in Cognitive Science* **7**, 439–442.

Foster, D. H. and Nascimento, S. M. C. (1994). Relational colour constancy from invariant cone-excitation ratios, *Proc. Roy. Soc. London* **B 257**, 115–121.

Foster, D. H., Amano, K. and Nascimento, S. M. C. (2001a). Colour constancy from temporal cues: better matches with less variability under fast illuminant changes, *Vision Research* **41**, 285–293.

Foster, D. H., Nascimento, S. M. C., Amano, K., Arend, L., Linnell, K. J., Nieves, J. L., Plet, S. and Forster, J. S. (2001b). Parallel detection of violantions of color constancy, *Proc. Natl. Acad. Sci. USA* **98**, 8152–8156.

Goltz, J. and McLeod, D. I. A. (2002). Influence of scene statistics on colour constancy, *Nature* **415**, 637–640.

Hendry, S. H. C. and Reid, R. C. (2000). The koniocellular pathway in primate vision, *Annu. Rev. Neurosci.* **23**, 127–153.

Hurlbert, A. C., Bramwell, D. I., Heywood, C. and Cowey, A. (1998). Discrimination of cone contrast changes as evidence for colour constancy in achromatopsia, *Exper. Brain Res.***123**, 136–144.

Hurlbert, A. C. (1999). Colour vision: Is colour constancy real? *Current Biology* **9**, 558–561.

Ives, H. E. (1912). The relationship between the color of the illuminant and the color of the illuminated object, *Trans. of the Illuminating Engineering Soc.* **7**, 62–72.

Judd, D. B. (1940). Hue, saturation and lightness of surface colors with chromatic illumination, *J. Opt. Soc. Amer.* **31**, 2–3.

Kraft, J. M. and Brainard, D. H. (1999). Mechanisms of color constancy under nearly natural viewing, *Proc. Natl. Acad. Sci. USA* **96**, 307–312.

Kraft, J. M., Maloney, S. I. and Brainard, D. H. (2002). Surface-illuminant ambiguity and color constancy: effects of scene complexity and depth cues, *Perception* **31**, 247–263.

Kulikowski, J. J. and Vaitkevicius, H. (1997). Colour constancy as a function of hue, *Acta Psychologica* **97**, 25–35.

Land, E. H. (1986). Recent advances in retinex theory, *Vision Research* **26**, 7–21.

Land, E. H. and McCann, J. J. (1971). Lightness and retinex theory, *J. Opt. Soc. Amer.* **61**, 1–11.

Le Grand, Y. (1948). *Light, Colour and Vision*. Chapman and Hall, London.

Linnell, K. J. and Foster, D. H. (1996). Dependence of relational colour constancy on the extraction of transient signal, *Perception* **25**, 225–228.

Linnell, K. J. and Foster, D. H. (2002). Scene articulation: dependence of illuminants estimates on number of surfaces, *Perception* **31**, 151–159.

Lucassen, M. P. and Walraven, J. (1993). Quantifying color constancy: evidence for nonlinear processing of cone-specific contrast, *Vision Research* **33**, 739–757.

Martin, P. R., White, A. J. R., Goodchild, A. K., Wilder, H. D. and Sefton, A. E. (1997). Evidence that blue-on cells are part of the third geniculocortical pathway in primates, *Eur. J. Neurosci.* **9**,1536–1541.

McKeefry, D. J., Murray, I. J. and Kulikowski, J. J. (2000). Red-green and blue-yellow mechanisms are matched in sensitivity for temporal and spatial modulation, *Vision Research* **40**, 245–255.

McCann, J. J., McKee, S. P. and Taylor, T. (1976). Quantitative studies in Retinex theory: the color Mondrian experiments, *Vision Research* **16**, 445–458.

Moutoussis, K. and Zeki, S. (2000). A psychophysical dissection of the brain sites involved in colour-generating comparisons, *Proc. Nat. Acad. Sci. USA* **97**, 8069–8074.

Nascimento, S. M. C., Ferreira, F. P. and Foster, D. H. (2002). Statistics of spatial cone-excitation ratios in natural scenes. *J. Opt. Soc. Amer.* **A 18**, 1484–1490.

Nascimento, S. M. C. and Foster, D. H. (1994). Illuminant invariants at receptoral and post-receptoral levels as a basis for relational colour constancym *Perception* **23**, 8–9.

Nascimento, S. M. C. and Foster, D. H. (1997). Detecting natural changes of coneexcitation ratios in simple and complex coloured images, *Proc. R. Soc. Lond.* **264**, 1393–1402.

Nascimento, S. M. C. and Foster, D. H. (2000). Relational colour constancy in achromatic and isoluminant images, *J. Opt. Soc. Amer.* **A 17**, 225–231.

Nascimento, S. M. C. and Foster, D. H. (2001). Detecting changes of spatial cone-excitation ratios in dichoptic viewing, *Vision Research* **41**, 2601–2606.

Nascimento, S. M. C., de Almeida, V. M. N., Fiadeiro, P. T. and Foster, D. H. (2006). Effect of scene complexity on colour constancy with real three-dimensional scenes and objects, *Perception* **35**, 947–950.

Parkkinen, J. P. S., Halikainen, J. and Jaaskelainen, T. (1989). Characteristic spectra of Munsell colors, *J. Opt. Soc. Amer.* **A 6**, 318–322.

Shepherd, A. J. (1997). A vector model of colour contrast in a cone-excitation colour space, *Perception* **26**, 455–470.

Shevell, S. K. (1978). The dual role of chromatic backgrounds in colour perception, *Vision Research* **18**, 1649–1661.

Shevell, S. K. and Wei, J. (2000). A central mechanism of chromatic contrast, *Vision Research* **40**, 3173–3180.

Smith, V. C. and Pokorny, J. (1975). Spectral sensitivity of the foveal cone photopigments between 400 and 500 nm, *Vision Research* **15**, 161–171.

Speigle, J. M. and Brainard, D. H. (1999). Predicting color from grey: the relationship between achromatic adjustment and asymmetric matching, *J. Opt. Soc. Amer.* **10**, 2370–2376.

Stanikunas, R., Vaitkevicius, H. and Kulikowski, J. J. (2004). Investigation of color constancy with a neural network, *Neural Networks* **17**, 327–337.

Switkes, E. and Crognale, M. A. (1999). Comparison of colour and luminance contrast, *Vision Research* **39**, 1823–1831.

Troost, J. and de Weert, C. (1991). Naming versus matching in colour constancy, *Perception and Psychophysics* **50**, 591–602.

Vidyasagar, T. R., Kulikowski, J. J., Lipnicki, D. D. and Dreher, B. (2002). Convergence of parvocellular and magnocellular information channels in the primary visual cortex of the macaque, *Eur. J. Neurosci.* **16**, 945–956.

von Kries, J. (1905/1970). In: *Handbook of Human Physiology*, MacAdam, D. L. (English translation edn). Sources of Color Science, MIT Press, Cambridge, Mass.

Walsh, V. (1999). How does the cortex construct color? *Proc. Nat. Acad. Sci. USA* **96**, 13594–13596.

Walsh, V., Kulikowski, J. J., Butler, S. R. and Carden, D. (1992). The effects of V4 lesions on the visual abilities of macaques: colour categorization, *Behavioural Brain Res.* **52**, 81–89.

Walsh, V., Butler, S. R., Carden, D. and Kulikowski, J. J. (1993). The effects of V4 lesions on the visual abilities of macaques: hue discrimination and colour constancy, *Behavioural Brain Res.* **53**, 51–62.

Webster, M. A. (2003). Light adaptation, contrast adaptation, and human colour vision, in: *Colour Perception*, Mausfeld, R. and Heyer, D. (Eds), Chapter 2. Oxford University Press, New York, NY.

Webster, M. A. and Mollon, J. D. (1991). Changes in colour appearance following post-receptoral adaptation, *Nature* **340**, 235–238.

Webster, M. A. and Mollon, J. D. (1994). The influence of contrast adaptation on colour appearance, *Vision Research* **34**, 1993–2020.

Whittle, P. (2003). Contrast colours, in: *Colour Perception*, Mausfeld, R. and Heyer, D. (Eds), Chapter 3. Oxford University Press, New York, NY.

Wild, H. M., Butler, S. R., Carden, D. and Kulikowski, J. J. (1985). Primate cortical area V4 important for colour constancy but not wavelength discrimination, *Nature* **313**, 133–135.

Zeki, S. (1983). Colour coding in the cerebral cortex: the reaction in monkey visual cortex to wavelengths and colours, *Neuroscience* **9**, 741–765.

Color, Line, and Space, pp. 91–116 (2006)

Dynamic texture spreading: probing the mechanisms of surface interpolation

DANIEL WOLLSCHLÄGER * and FRANZ FAUL

Institut für Psychologie Christian–Albrechts–Universität zu Kiel, Olshausenstraße 62 24098 Kiel, Germany

Abstract—Dynamic texture spreading is a filling-in phenomenon where a colored pattern perceptually spreads onto an area confined by virtual contours in a multi-aperture motion display. The spreading effect is qualitatively similar to static texture spreading but widely surpasses it in strength, making it particularly suited for quantitative studies of visual interpolation processes. We first carried out two experiments to establish with objective tasks that texture spreading is a genuine representation of surface qualities and thus goes beyond mere contour interpolation. Two subsequent experiments serve to relate the phenomenon to ongoing discussions about potentially responsible mechanisms for spatiotemporal integration. With a phenomenological method, we examined to what extent simple sensory persistence might be causally involved in the effect under consideration. Most of our findings are consistent with the idea of sensory persistence, and indicate that information fragments are integrated over a time window of about 100 to 180 ms to form a complete surface representation.

Keywords: Texture spreading; surface completion; filling-in; perceptual grouping; multi-aperture viewing.

1. INTRODUCTION

This article introduces dynamic texture spreading (Fig. 3) [see Note 1], a spatiochromatic interaction observable with a new variant of multi-slit view displays (see Section 2.1). Not only does it comprise the formation of strong subjective contours but also the vivid illusory perception of surface qualities like color, texture, and depth stratification. It is closely related to other cases of perceptual completion and bears direct relation to color spreading. It is thus helpful to first consider the well-known effects of static color and texture spreading, and in turn introduce dynamic texture spreading itself.

*To whom correspondence should be addressed. E-mail: dwoll@psychologie.uni-kiel.de

Color spreading demonstrates how certain spatial configurations can give rise to the perception of surface qualities, such as color or texture, attributed to stimulus regions that are themselves devoid of physical information about these features. Figure 1 shows two recently introduced variants of this effect (Wollschläger *et al.*, 2001, 2002). Here, color is seen not only in the physically-colored parts of the display, but also in the regions between them, giving rise to a lightly-colored surface that is bounded by illusory contours. Color spreading can be considered a case of perceptual filling-in (Pessoa *et al.*, 1998), a mechanism that has been associated with, for example, the slow peripheral veiling of figures by their backgrounds during prolonged fixation (Troxler fading, Pirenne, 1962; Troxler, 1804), instantaneous filling-in of blind spots and scotomas (Ramachandran, 1992; Ramachandran and Gregory, 1991), neon spreading (Bressan *et al.*, 1997), long range color spreading (Pinna *et al.*, 2001), texture spreading (Watanabe and Cavanagh, 1991), and color from motion (Cicerone *et al.*, 1995) [see Note 2].

Neon color spreading (Fig. 2a) can typically be observed in line drawings where the color of some segments is abruptly changed to a different one. The color of the embedded segments then seems to disperse onto an area confined by virtual contours, producing a subtle, neon-like veil. Some neon-color-spreading displays can be altered to elicit texture spreading when the uniformly-colored embedded line segments leading to neon spreading are replaced with, for example, segments of a stripe texture (Watanabe and Cavanagh, 1991). In these stimuli, a *non-uniform* color pattern perceptually fills in a homogeneous background region (Fig. 2b), again bounded by subjective contours. "Visual phantoms" (Tynan and Sekuler, 1975; Weisstein *et al.*, 1977) may also be taken as a rudimentary example of texture spreading in static as well as in motion displays, but they occur only with a very restricted class of one-dimensional textures.

While the aforementioned types of filling-in are highly relevant to the study of completion processes and of interactions between spatial stimulus layout and perceived surface qualities, all of them are somewhat vague and perceptually unstable. This elusive character makes it difficult to reliably study them using quantitative methods, such as scaling or matching procedures, and even more so using objective performance tasks.

Figure 1. (See color plate VII) (a) Static flank-transparency demonstration (Wollschläger *et al.*, 2002), including the two-color case. No spreading obtains if the occluding bars are missing. (b) Clipping of one frame from an animated flank-transparency display. The line array is slightly shifted to the right between frames, whereas the virtual ellipse stays put. Small colored flanks are added to the lines in the interior of the virtual ellipse.

Figure 2. (See color plate VII) (a) Neon color spreading in the modified Ehrenstein cross. Color spreading is absent if the embedded segments are not aligned with the inducing lines. (b) Texture spreading in the modified Ehrenstein cross. Texture spreading also vanishes with nonaligned segments.

We believe that our dynamic stimulus changes this situation for texture spreading as it leads to a stable, clearly-defined dispersion of a colored texture. As such, it may provide a useful tool to the experimental study of texture spreading as some recently presented effects did to the study of color-spreading: the watercolor illusion (Devinck *et al.*, 2005; Pinna *et al.*, 2001), color from motion (Cicerone *et al.*, 1995; Ekroll and Faul, 2002), and flank transparency (Wollschläger *et al.*, 2001, 2002) all exhibit much better discernible spreading effects than classical neon-spreading demonstrations.

2. DYNAMIC TEXTURE SPREADING

Dynamic texture spreading can be observed in animated computer displays that simulate a multi-slit view setup (Fahle and Poggio, 1981; Mateeff *et al.*, 1993; Nishida, 2004). These are a generalization of Zöllner's (1862; Rock, 1981) anorthoscopic view apparatus from one to multiple apertures. Suppose you put a disk covered with a stripe pattern on a white background. In front of it, you place a white sheet of paper with black vertical lines painted on it. Seamlessly alongside the lines, you cut narrow rectangular slits, making visible small strips of the background and the textured disk behind (Fig. 3). The animation procedure may be such that the front sheet of paper is translated gradually on a horizontal path, accompanied by a corresponding change in what is visible of the stationary textured shape through the apertures flanking the lines. Alternatively, one may hold the front sheet of paper in a fixed position and horizontally move the striped disk back and forth behind the stationary apertures while tracking the disk with the eyes. At a suitable speed and line density, both animations lead to perceptual integration of non-contiguous image

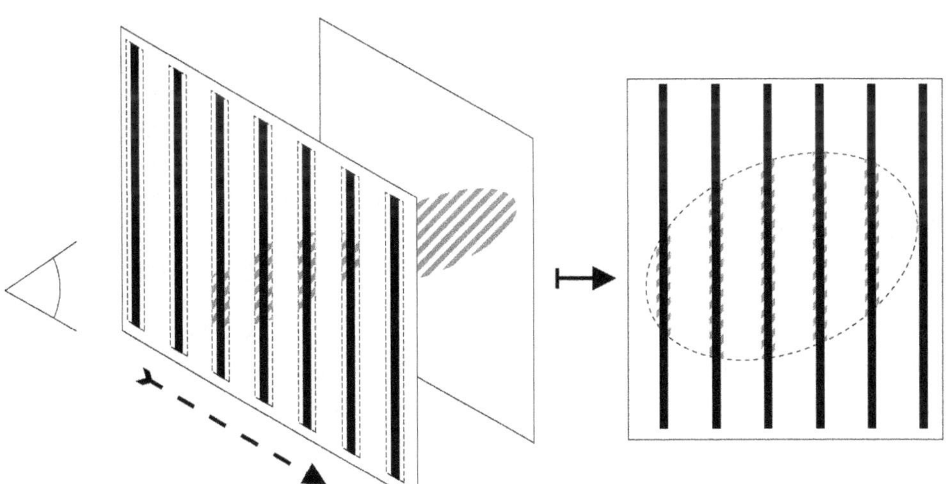

Figure 3. Schematic construction of a dynamic-texture-spreading display (see text). Small striped flanks are added to the lines in the interior of the virtual ellipse.

regions; one has a vivid illusory impression of colored texture in an area defined by crisp subjective contours tracing the outline of the disk.

The described setup differs from standard multi-aperture viewing stimuli in that the latter may have visible or invisible apertures, but they do not incorporate visible lines *between* them. This seemingly subtle distinction has important consequences. Flank transparency generalizes also to static displays (Fig. 1a), allowing for comparisons between static and dynamic perceptual completion with nearly identical stimuli. In addition, it can evoke transparent color spreading even in two-color displays (Fig. 1) — an observation that is currently not handled by models of perceptual transparency (Wollschläger *et al.*, 2001).

2.1. Relation to other phenomena

If the disk is of solid color, the virtual shape also appears uniformly colored, an effect we called flank transparency (Fig. 1b, Wollschläger *et al.*, 2001, 2002) because of the percept's translucent quality, which is similar to the one reported for static neon spreading (Bressan *et al.*, 1997). Flank transparency shares further characteristics with neon spreading, such as a concomitant depth stratification of objects, leading to the perception of the virtual shape as being in front of or behind the background layer. As such, static neon color spreading and texture spreading both have counterparts in the domain of dynamic stimuli, i.e. flank transparency and dynamic texture spreading — each one qualitatively similar but more robust compared to their static analogue.

Filling-in phenomena demonstrate that spatially fragmented input information to the visual system can interact to trigger the perception of complete objects that are defined along the same interdependent perceptual dimensions, such as form, color, motion, and depth, as are "real" objects. In neon color spreading, percepts of glowing and lightly transparent layers lying above the background form in response to only a few embedded segments in a line drawing. This discrepancy between meagre input information available to the visual system in each single frame and a rich final percept is even larger in dynamic texture spreading. In this case, not only the contour but also the object's non-uniform inner structure must be recovered. Yet, in the animated display, it is possible to discern even complex surface layouts like, e.g. a colored Gabor patch (see Section 3.2).

Just as the described stimulus setup is a variant of multi-aperture viewing, it also resembles dynamic occlusion displays using a textured background (Andersen and Cortese, 1989; Bruno and Bertamini, 1990; Hine, 1987; Shipley and Kellman, 1994). These stimuli contain only highly incomplete information about simple objects as they simulate the dynamic accretion and deletion of background texture elements by a moving, and usually invisible mask — the front sheet of paper in the case of dynamic texture spreading.

Even though aperture-viewing and dynamic occlusion displays have been widely used to study perceptual interpolation processes, they appear to form mostly disjoint literature domains (for an exception, see Shipley and Cunningham, 2001). So far, aperture viewing seems to draw most interest from research into motion perception (Anstis and Atkinson, 1967; Morgan, 1980; Morgan *et al.*, 1982; Parks, 1965; Rock *et al.*, 1987), while dynamic occlusion is more associated with studies of object recognition and contour interpolation (Kellman and Shipley, 1991; Kellman *et al.*, 2001).

Investigations in *both* of the aforementioned research contexts have so far been almost exclusively concerned with aspects of boundary integration. Studies in dynamic occlusion that explicitly addressed a surface component (e.g. Cunningham *et al.*, 1998) focused on the perception of a surface quality *per se*, and not on the formation details of particular modal features associated with that surface, like color, texture or degree of transparency. Visual inspection suggests that all these play a vital role in dynamic texture spreading. Our first two experiments thus serve to establish with objective methods that dynamic texture spreading is indeed the result of a genuine perceptual process that goes beyond mere contour completion, and represents a functionally relevant interpolation of surface features.

2.2. Sensory persistence as a possible explanation for visual interpolation

Many of the cited studies on boundary completion tested whether simple sensory persistence or "retinal painting" is a possible explanation for the observed discrepancy between the fragmented stimulus material and the perception of extended contours. Retinal painting asserts that sensory information persists in some form of retinal or post-retinal storage and can thus be integrated by a simple local mechanism over a certain time-frame. If, for example, in the case of aperture viewing, stimulus fragments are presented such that the complete stimulus is subsequently spread over the retina over time, such temporal integration might be able to automatically recover the extended stimulus just by means of sensory persistence. No complex form of shape interpolation would then be required.

Some authors found strong support for a causal role of retinal painting in the perception of outline shapes under aperture-viewing conditions. Anstis and Atkinson (1967) as well as others (for a review, see Morgan *et al.*, 1982; Shipley and Cunningham, 2001) have demonstrated that the shape distortions observed in aperture viewing closely follow the predictions made by retinal painting.

However, other authors claimed that retinal painting is neither necessary nor sufficient to explain the observed phenomena (Bruno and Gerbino, 1991; Fendrich *et al.*, 2005; Rock *et al.*, 1987; Shimojo and Richards, 1986). Already Zöllner (1862), von Vierordt (1868), and Rothschild (1922) tried to show that anorthoscopic perception can occur even when a steady fixation prevents the image of the object from being spread over the retina over time. With modern eye-tracking equipment, Fendrich *et al.* (2005) recently extended that work and found that boundary integration in anorthoscopic perception can still take place with retinally stabilized images. Anstis (2005) reported that anorthoscopic perception in multi-slit viewing also occurs with rotational movement, which is hard to track because the eyes cannot fully rotate around the axis perpendicular to the sagittal plane. Based on motion-direction specific noise masking, Nishida (2004) argued that high-level motion analysis is involved in the perception of patterns viewed through multiple slits, and concludes that simple sensory persistence does not constitute a possible underlying mechanism.

A relevant theoretical issue was pointed out by Dixon and Di Lollo (1994). On the one hand, it is necessary for the visual system to integrate stimuli over time to guarantee perceptual continuity; but on the other hand, it must also detect rapid changes and must therefore segregate those stimulus elements that do not belong to one spatiotemporal object. Retinal painting by itself does not provide a mechanism for distinguishing coextensive fragments in the stream of information from disjoint ones, i.e. for deciding when to integrate and when to segregate.

It appears reasonable to conclude that sensory persistence may be only one among several, and possibly more important, mechanisms involved in boundary integration. Indeed, Shipley and Cunningham (2001) explicitly propose the existence of two different representations with different processing mechanisms. They cite evidence from shape perception tasks supporting on the one hand a retinotopic stimulus representation (here referred to as retinal painting), and on the other hand an additional distal representation with an external frame of reference. Yet, this conclusion does not automatically extend to the integration of surface attributes. Surface filling-in may constitute a different perceptual process that may therefore be mediated by different computational mechanisms. Whether surface completion involves or even completely relies on sensory persistence has not yet been explored. Considering that visual completion phenomena encompass a multitude of complex aspects regarding the reconstruction of boundary and surface information, it seems worthwhile to try to separate them and individually test which mechanisms might be responsible for each one. Retinal painting seems a very simple and natural candidate to explain some characteristics of dynamic texture spreading. Therefore, we take sensory persistence as a first starting point in our investigation on dynamic texture spreading. More specifically, one motivation for our experiments was the question whether retinal painting provides a possible explanation for the interpolation of a surface feature like texture.

2.3. Retinal painting applied to dynamic color and texture spreading

Sensory persistence asserts that a complete surface representation can be sustained by discrete, scattered signal elements provided that each point of the represented surface is frequently enough supported by new input signals consistent with that surface. Applied to our stimulus setup, this means that a complete textured object would be perceived as long as the information signaling the presence of a piece of texture is refreshed often enough at each fixed location, regardless of how this refresh frequency is achieved. As the refresh frequency depends on the density and speed of the moving rectangular apertures through which object information becomes available, these two stimulus parameters should be important determinants of the texture-spreading effect.

This consideration is backed by empirical results about the relevance of signal element density for object recognition in animated sparse displays. Shipley and Kellman (1993; Cunningham *et al.*, 1998) found that shape identification improved with more signal elements in their displays used in research on spatiotemporal boundary formation. This result was also reported various times for experiments on dynamic occlusion (Andersen and Cortese, 1989; Bruno and Bertamini, 1990; Bruno and Gerbino, 1991; Shipley and Kellman, 1994) as well as by Mateeff *et al.* (1993) and Bull *et al.* (2003) for multi-aperture stimuli. Fidopiastis *et al.* (2000) showed that the strength of color from motion is modulated by the particular layout of the dot array which provides signal elements for form and color information. Specifically, it was established that the color-spreading effect increases with a denser and more regular dot array. For color from motion (Cicerone *et al.*, 1995) and for flank transparency (Wollschläger *et al.*, 2002), a close dependence on the amount of motion has been observed.

It therefore seems natural to analyze dynamic texture spreading by asking about the specific interplay between the density of the line array and its speed in determining the filling-in effect. This will be the starting point for Experiments 3 and 4.

3. EXPERIMENTS

With Experiments 1 and 2, we intend to show that texture spreading is a functionally relevant effect that involves more complex processing and provides more information than contour interpolation. To this end, we used highly fragmented stimuli in performance tasks whose successful completion required not only boundary information but, ideally, a complete surface representation. Experiments 3 and 4 then derive quantitative information about the texture interpolation process from a phenomenological task and relate the findings to the concept of sensory persistence.

3.1. Experiment 1

In Experiment 1, we used a 2AFC discrimination task to objectively measure the degree to which texture spreading can help to successfully recover the texture composition of a fragmented surface. The stimuli were designed not to contain local features that would be diagnostic for their discrimination. Instead, we wanted to be able to infer from a successful task completion the full perceptual presence of the surface layout. We therefore tried to construct stimuli whose identification rests on features that become apparent only in full view. This, however, proved to be no trivial goal. In particular, two commonly used stimuli, meaningless line drawings and letters, are inappropriate. The former lacks any surface qualities at all, and the latter contains — by the very virtue of its construction — many local cues which aid in its discrimination.

The aforementioned issue reminds us of a general problem when trying to draw inferences from results in performance tasks as they are now commonly used in the study of perceptual completion: in contrast to purely introspective reports, these tasks often have the advantage of separating genuine perceptual aspects from cognitive reasoning influences (Kellman *et al.*, 2001). But while they can provide an objective measure of the level of some functional relevance of a subjective percept, inferences concerning the qualities of that percept are indirect and based on certain assumptions: Shape discrimination tasks used in studies on boundary interpolation often implicitly presume that a fully completed contour was at the observer's disposal for making a successful shape categorization. To what extent shapes are also discriminable based on local features is seldom discussed. The development of adequate stimulus material thus remains a crucial prerequisite for the study of surface interpolation.

3.1.1. Subjects. Five students from the University of Kiel (two female, three male) participated in Experiment 1, all were naïve as to the purpose of the experiment. We carried out a separate control experiment with five subjects, among them one (JJ) who also participated in the experimental condition, and one of the authors (DW). All subjects had normal color vision and normal or corrected-to-normal visual acuity.

3.1.2. Apparatus and stimuli. The stimuli were displayed on a Sony Multiscan 500 PS 21 inch color monitor with a vertical refresh rate of 100 Hz using the OpenGL graphics library. The timing of the stimulus frames was synchronized with the vertical refresh of the monitor. Color resolution for each pixel was 8 bits for each channel R, G, B. Following a standard procedure (Brainard, 1995), the monitor was calibrated and gamma corrected using a colorimeter (LMT C1210). Observers placed their heads on a chin rest and binocularly viewed the display in a darkened room at a distance of 80 cm.

The background of the display was grey (CIE $x = 0.30$, $y = 0.31$, $L = 38\ \mathrm{cd\,m^{-2}}$) and subtended a visual angle of 17.4 by 13.3 deg. An array of vertically-oriented black lines (CIE $x = 0.30$, $y = 0.31$, $L < 0.2\ \mathrm{cd\,m^{-2}}$), 6.75 min of arc wide and 10 deg long, was placed on the background.

The target stimulus was placed in the interior of a virtual disk with a radius of 2.9 deg. Throughout the experiment, only thin stripes (1.35 min of arc wide) of the target stimulus became visible as flanks alongside the black vertical lines.

The target stimuli were non-meaningful images (Fig. 4) which were constructed in the following way: First, the disk was completely covered with randomly placed circles of varying radii. Next, we defined three invisible, non-intersecting Bézier curves with a thickness of 0.25 deg which started and ended at the disk's boundary. As a final step, we colored those circles whose center lay outside the curves green

Figure 4. (See color plate VIII) Examples of the two stimulus classes used in Experiment 1. Circles are colored red if their center falls within one of three imaginary Bézier curves, else green. (a) Target stimulus with a continuous line. (b) Target stimulus without a continuous line.

(CIE $u = 0.16$, $v = 0.47$, $L = 12$) and then those whose center fell within one of the curves red (CIE $u = 0.24$, $v = 0.45$, $L = 12$). Gaussian red-green chromaticity noise as well as Gaussian luminance noise was added to each circle's color to make the curves' boundaries less detectable by simple local edge operators.

Subjects had to discriminate between two classes of the described stimulus. In one class, all three Bézier curves were interrupted once within the disk such that a gap would separate two parts of the curve. In the other class, only two curves were interrupted such that there remained one continuous curve without a gap. The rationale behind the stimulus construction was that in order to verify the presence of a continuous line, the observer must have a complete (i.e. gapless) perceptual representation of that line. The two classes should not be distinguishable based on local features. Since both classes contained continuous line segments as well as gaps between segments, the detection of a continuous segment or of a gap did not suffice to make a reliable discrimination. In addition, the number of circles within the curves did not vary systematically between classes, such that the total amount of redness in a stimulus was not diagnostic for a correct classification.

We designed five different stimuli for each stimulus class. In order to reduce the chance of recognition of a particular exemplar, the target stimulus was randomly rotated on each trial.

We used two "motion conditions": in the first one (subjects JJ, DA), the line array rigidly moved to the right at a speed of 1.13 deg s^{-1}, whereas the virtual target stimulus remained stationary at the center of the screen. In the second one (subjects TG, MH, KR), the line array was stationary, and the virtual target stimulus moved at a speed of 1.13 deg s^{-1}. Subjects were instructed to track the target which made the second condition equivalent to the first one with respect to the resulting retinal image (see Section 2). In the control experiment, neither lines nor the virtual disk moved, providing a static reference case which indicated the achievable performance without temporal integration.

3.1.3. Procedure. We varied the spacing of the vertical lines from 0.25 to 1.1 deg, corresponding to line densities of 80 to 16 lines visible. The dependent variable was the correctness of the subject's answer to the question whether the target stimulus contained a continuous line or not. Each target stimulus class had an equal probability of being chosen on a given trial, as was the case for all five exemplars within one such class.

The spacing of the vertical lines was varied according to a custom adaptive method. We used this in order to minimize the number of trials necessary to reliably reconstruct the function relating independent and dependent variable for a given subject. Unlike methods such as QUEST (Watson and Pelli, 1983), we did not assume any characteristics of the function other than monotonicity, and wanted to reconstruct the function over the whole range of the independent variable, as opposed to only estimating the parameters of a fixed function. Details of the employed algorithm are given in Appendix A.

Each stimulus presentation lasted for 8.4 seconds, after which subjects pressed one of two cursor keys to indicate their classification of the target stimulus. This keystroke automatically started the next trial. Subjects were encouraged to make pauses at their own disposal. Prior to the experiments, all subjects familiarized themselves with the task in 20 training trials which were not included in the analysis. It took subjects from 260 (TG) to 433 (MH) trials to complete the described procedure.

3.1.4. Results. Figure 5a displays the individual results for subject JJ in the motion condition as well as in the static control. The two motion conditions are equivalent with respect to the retinal image and, as expected, did not produce systematically different results. The same was true for the data of individual subjects, which warranted their pooling in the two motion conditions and the static control, respectively. Figure 5b shows the averaged data. For plotting the data, we divided the full parameter range of the independent variable into intervals. The data points represent the mean percent correct of the subject's classification in these intervals. Both curves were visually fitted with psychometric functions [see Note 3].

The discrimination performance is inversely linked to line distance in the motion as well as in the control condition. It continuously varies from near perfect discrimination at low distances to about chance performance at high distances. In the static control condition, the drop in performance with wider line spacing happens faster than in the motion condition, the difference becoming more pronounced with line spacings wider than 0.56 deg. At a line distance of, for example, 0.67 deg, the performance of JJ in the motion condition is still at nearly 90% correct, whereas in the static control condition, it has already dropped to about 65%. For the averaged data, mean performance in the static condition at a line distance of, for example, 0.72 deg is 61% correct, while in the motion condition it is still 79%. The relative performance benefit of the motion conditions over the static control is plotted in Fig. 5c for the averaged data as well as for the fitted psychometric functions. Mean discrimination was up to 30% better in the motion condition than in the static control.

3.1.5. Discussion. The obtained data show that discrimination performance for a fragmented texture pattern objectively benefits from spatiotemporal integration. The performance difference to a static control condition may be seen as a measure for the functionally relevant amount of surface feature information that is gained through integration. The results also imply that the integration of a complex texture does not simply lead to its spatiotemporal averaging or blending into a uniform color, thus setting dynamic texture spreading apart from homogeneous color spreading.

The stimuli were constructed such that the necessary information for a successful discrimination could not come from simple contour integration alone, as was the case with most discrimination tasks reported in the literature, which used line drawings or letters. We were thus able to show that spatiotemporal integration of a

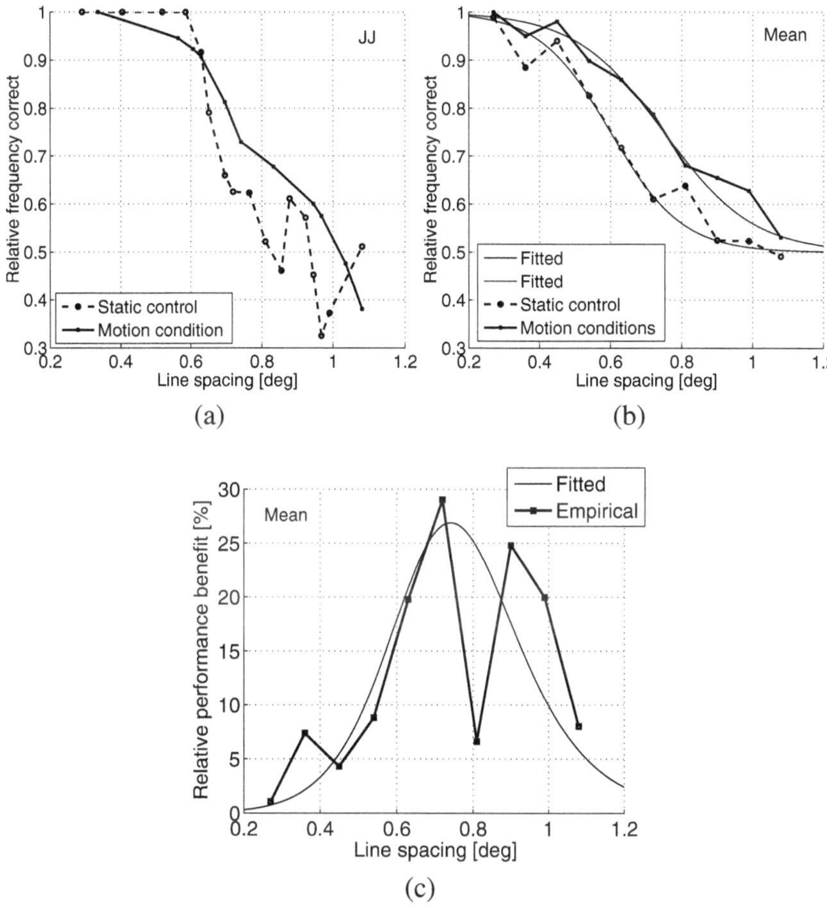

Figure 5. Results from Experiment 1. (a) Individual data from JJ in the motion as well as in the static control condition. (b) Combined results averaged over subjects in the motion as well as in the control conditions. The data was visually fitted with psychometric functions (see text). (c) Mean relative performance benefit of the motion conditions over the static control for the empirical data as well as for the fitted psychometric functions.

surface feature with nonlocal properties, such as a complex texture, can indeed take place.

Our goal was to design a task that, to be successfully accomplished, requires a *complete* surface representation available to the visual system. However, as is evident from the static control condition, information relevant to the discrimination task was, up to a certain degree, also available with incomplete stimuli. One explanation for this finding is that with a line distance smaller than the gaps within a Bézier curve, detection of gaps becomes trivial. Gap information is then available in spatial frequencies low enough to be fully recovered with the available spatial sampling frequency without the need for temporal integration. Another explanation may be required to account for the relatively smooth performance drop with increasing line spacing in the static control. It is possible that diagnostic local features were still present in the target stimuli to a certain extent. This illustrates the difficulty in designing appropriate stimuli for the objective assessment of the degree to which a complete object representation has formed. Nevertheless, the fact that spatiotemporally integrating the stimulus leads to a measurable performance gain, once line distance becomes larger than curve gaps, means that non-local surface feature information were relevant and could be recovered in the integration process.

3.2. Experiment 2

In Experiment 2, we wanted to use a matching procedure to obtain another objective measure of the subjects' ability to perceptually integrate a fragmented stimulus. To this end, we had subjects match the frequency of a vertically oriented Gabor patch to that of a test Gabor which was shown in a multi-slit view setup with visible slits. We chose a Gabor patch as it constitutes a textured surface without sharp boundaries, thereby reducing the diagnostic value coming from simple boundary integration mechanisms.

3.2.1. Subjects. Five male subjects participated in Experiment 2, among them three students from the University of Kiel who were naïve to the purpose of the experiment, and the authors (FF, DW). Subjects had normal color vision and normal or corrected-to-normal visual acuity.

3.2.2. Apparatus and stimuli. The same technical setup was used as in Experiment 1. We constructed the stimuli such that in the top half of the display, subjects were presented a multi-slit view setup which made partially visible a Gabor patch within the region covered by the slits. The background color was black, the slits were 1.35 min of arc wide, 5.7 deg long and grey (CIE $x = 0.30$, $y = 0.31$, L = 12 cd m^{-2}). The vertically oriented Gabor patch (Fig. 6) of radius 2.9 deg was horizontally centered on the screen. It was equiluminant and varied in its chromaticity according to the Gaussian-windowed cosine-wave on the axis defined by the colors green (CIE $u = 0.16$, $v = 0.47$, $L = 12$ cd m^{-2}) and red (CIE $u = 0.24$, $v = 0.45$, $L = 12$ cd m^{-2}).

The slit array varied in its density. It rigidly moved to the right at a speed of 4.5 deg s^{-1}, whereas the virtual Gabor patch remained stationary. In the bottom half of the display, subjects were presented a fully visible Gabor patch of the same size as the test Gabor. Its chromaticity pattern was identical to that of the test Gabor, but its luminance was reduced to 8 cd m^{-2}. Its luminance was also windowed with a two-dimensional Gaussian to make it blend in with the black background. The match Gabor was adjustable with respect to the frequency of the underlying cosine-wave.

Figure 6. (See color plate VIII) Example Gabor patch used as a target stimulus in Experiment 2. Please note that the colors may be incorrectly reproduced in this print.

The vertical orientation of the Gabor together with the narrow slit width ensured that, in a given frame, no frequency information would be available within each slit. As such, the highest frequency that could be reliably recovered from a static version of the display is given by the Nyquist limit, equal to one half of the slit frequency.

3.2.3. Procedure. The frequency of the test Gabor was set to 0.69, 1.05, 1.39, 1.74, 2.09, 2.79, 3.48 or 4.18 cycles deg^{-1}, corresponding to 4, 6, 8, 10, 12, 16, 20 and 24 cycles in the area of the Gabor. The slit frequency was one of 2.44, 3.14, 3.83 or 4.53 slits deg^{-1}, corresponding to 14, 18, 22 and 26 slits in the area of the Gabor. The two independent variables were combined in a fully crossed 8×4 design.

Subjects adjusted the frequency of the match Gabor with the keyboard's cursor keys. Ten repetitions for each setting were carried out, resulting in a total of $8 \times 4 \times 10 = 320$ trials. If a setting could not be made because the Gabor was not seen and its frequency therefore indeterminable, subjects could flag this trial as invalid using the keyboard's space bar.

A separate control condition with a fully visible test Gabor patch was used to yield individual assessments of the adjustment method's inherent inaccuracy. All of the aforementioned frequencies had to be matched by each subject in the control condition, resulting in $8 \times 10 = 80$ additional trials. The results provided a reference case for the adjustments in the experimental conditions.

3.2.4. Results. Trials that were flagged as invalid, a total of 193, equalling 12% of the trials, were excluded from the following analysis. Since the subjects did not exhibit systematic differences in their frequency settings, we pooled their data. The resulting data are shown in Fig. 7a. The graph depicts the mean relative deviations of subjects' settings from the respective frequency of the test Gabor. The relative deviation is measured in percentage of the test frequency. The graph also contains the respective Nyquist limit for each slit frequency. In addition to the experimental conditions, the graph provides the range of relative deviations in the control condition together with their standard errors.

The settings were in general rather precise: mean relative deviation of the subjects' settings typically ranged from only 10 to 15% of the test frequency. Most importantly however, mean deviations in the experimental conditions were about as large as, or only slightly larger than, in the control condition — with a clear exception of the lowest line density. A general trend for all four slit frequencies was that mean relative deviation was higher for the lowest and highest Gabor frequencies. This also means that relative deviations did not simply increase with Gabor frequency in a monotonic way. With regard to slit frequency, it is apparent that lower slit frequencies are associated with higher mean relative deviations of Gabor frequencies.

Figure 7b depicts the data in a way that makes explicit the relationship between mean relative deviation in cases where the Gabor frequency was below the Nyquist limit imposed by the slit frequency, compared to cases where the slit frequency

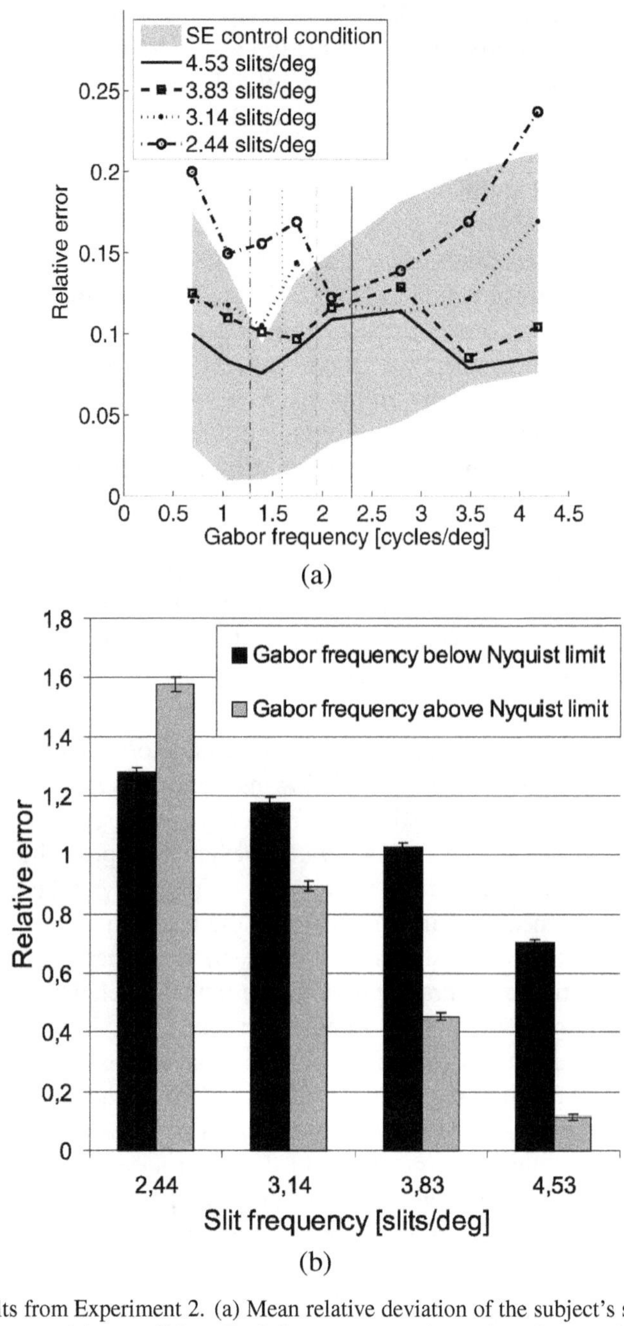

Figure 7. Results from Experiment 2. (a) Mean relative deviation of the subject's settings is plotted against the frequency of the test Gabor. Each line represents the settings for one slit frequency. The grey background area shows the range of the subject's settings in the control condition, i.e. the mean relative deviation ±1 SE. Vertical lines indicate the respective Nyquist limits induced by the corresponding slit frequencies. (b) Mean relative deviation of the frequency settings for all tested slit frequencies below and above their respective Nyquist limit. Error bars indicate ±1 SEM.

was above that limit. Contrary to intuition, mean relative deviation was not higher beyond the Nyquist limit; in fact it was lower, with a decreasing difference between them towards the lowest slit frequency. Only here subjects showed worse performance beyond the Nyquist limit. The figure also shows more directly that adjustment inaccuracies generally decreased with higher slit frequencies.

3.2.5. Discussion. Results from Experiment 2 again show that spatiotemporal integration of a texture does not lead to its blending into a uniform color. Instead, the surface structure is preserved as indicated by the fact that the fragmented Gabor pattern's frequency could be reliably recovered well beyond the Nyquist limit, theoretically imposed by the frequency of the slits. In fact, for the two highest slit frequencies, a subject's performance was just as good over nearly all tested frequencies as in the reference case with an unobstructed test Gabor. This means that due to the possibility of temporal integration, surface feature perception was almost perfect in these two cases.

Based on these results, one may ask whether the employed task is at all sensitive to the amount of surface information that is available in the display in each frame. This however follows from the pattern of results which indicates that lower slit frequencies indeed led to higher mean inaccuracies. The analysis of the performance differences with respect to test frequencies below and above the Nyquist limit (Fig. 7b) also suggests a similar conclusion: For the lowest slit frequency, the Nyquist limit had the expected effect of separating better from worse performance. For all higher slit frequency, the Nyquist limit lost this theoretical property, hinting at the fact that there is no resolution limit when spatiotemporal integration is possible (Nishida, 2004). Still, it remains elusive as to why the performance above the Nyquist limit was actually better for the three highest slit frequencies.

3.3. Experiments 3 and 4

Experiments 3 and 4 have complementary designs to address the question to what extent quantitative properties of texture spreading are compatible with the mechanism of sensory integration as delineated in Section 2.2. This is done using a dynamic texture spreading stimulus to study the interplay between the density of the line array and its speed in determining the filling-in effect. Although Experiments 3 and 4 do not directly test how much proper surface information is used to complete the task [see Note 4], results from Experiments 1 and 2 attest that the dynamic texture spreading stimulus indeed provides relevant information about the surface layout. Since Experiments 3 and 4 employ the same stimulus class, there is no reason to assume otherwise for their case.

3.3.1. Subjects. Six subjects (three female, three male) participated in Experiments 3 and 4, all of them students from the University of Kiel who were naïve as to the purpose of the experiment. Subjects had normal color vision and normal or corrected-to-normal visual acuity. In a previous run of these two experiments with four subjects, we used slightly different parameter ranges for the independent variables. The obtained results did not differ systematically from the ones reported here.

3.3.2. Apparatus and stimuli. The same technical setup was used as in Experiments 1 and 2, with the exception of the monitor's vertical refresh rate, which was set to 75 Hz.

The background of the display was white (CIE $x = 0.30$, $y = 0.31$, $L = 55\ \mathrm{cd\,m^{-2}}$) and subtended a visual angle of 26.7 by 21.9 deg. An array of vertically-oriented black lines (CIE $x = 0.30$, $y = 0.31$, $L < 0.2\ \mathrm{cd\,m^{-2}}$), 6.75 min of arc wide and 3.37 deg long, was placed on the background.

In the interior of a virtual ellipse, red (CIE $x = 0.63$, $y = 0.34$, $L = 12.6\ \mathrm{cd\,m^{-2}}$) striped flanks of 1.35 min of arc width and 2.78 cpd stripes were added parallel to the lines on either side. The virtual ellipse was centered on the screen and had main radii of 1.69 and 1.46 deg (Fig. 3). From trial to trial, its orientation was varied randomly in steps of 45 deg.

A fixation cross was placed on the horizontal center line at an eccentricity of 0.23 min of arc to the disk boundary, as informal observations had indicated that texture spreading was strongest under parafoveal viewing conditions, similar to what has been reported for anorthoscopic perception (Hecht, 1924), and static texture spreading (Watanabe and Cavanagh, 1991). The line array rigidly moved to the right, whereas the virtual disk remained stationary at the center of the screen. The velocity of the moving lines was varied by changing the length of the discrete "jumps" of the lines from one frame to the next. The trailing side of the display was continuously replenished with lines.

3.3.3. Procedure. In both experiments, we used a method very similar to that employed by Haber and Nathanson (1968), and asked subjects to find the threshold for perceiving a complete textured ellipse using an interactive adjustment procedure. Subjects were given two criteria for the perceptual presence of a *complete* ellipse. First, they should be able to discern the striped texture, and second, they should be able to identify the ellipse's orientation, asserting that they saw the complete boundary. The suggested strategy was to first explore the perceptual outcome of the setting over a wide range of the values of the dependent variable, and then to close in on threshold level.

In Experiment 3, the speed of the moving line array served as the independent variable. It was varied in five steps, 3.37, 5.06, 6.75, 8.44, and 10.12 $\mathrm{deg\,s^{-1}}$. The initial value of the line spacing was set to 0.67 deg. Subjects interactively controlled the spacing of the vertical lines in steps of 1 pixel using the arrow keys of the computer keyboard until the illusory ellipse was just noticeable.

In Experiment 4, the spacing of the vertical lines was varied in five steps (0.27, 0.54, 0.81, 1.08, 1.35 deg), corresponding to line densities of 106, 53, 35, 26, and

21 lines visible. The initial speed was set to 6.75 deg s^{-1}. Subjects interactively controlled the speed of the line array in steps of one pixel displacement per frame until they just perceived the illusory ellipse.

The chosen ranges for the two independent variables were derived from results in preliminary studies with the goal of obtaining easily comparable functions in the two experiments relating line density and speed of the line array. That is, the values were selected such that the values for the independent variable in Experiment 4 would be similar to the dependent-variable settings in Experiment 3, and *vice versa*.

Experiments 3 and 4 were carried out separately with the order of the independent variables counterbalanced across subjects. In each experiment, subjects performed 10 repetitions for each of the five conditions in pseudo random order, resulting in a total of 50 trials.

Prior to the experiments, all subjects familiarized themselves with the task of adjusting the speed and spacing of the line array, respectively. There was no time limit for either task — subjects pressed a key to confirm their setting, and continued with the next trial.

3.3.4. Results. Figure 8 displays the combined results for the two experiments for an individual subject as well as averaged across subjects. Line spacing is plotted against the speed of the moving line array. In both experiments, these two variables seem to be linked by a positive, approximately linear relationship. In the following account of the results, we will refer to the data in terms of line density, inversely related to line spacing. It is evident that for both types of threshold determination, the density of the moving line array and its speed are in a compensatory relationship. This means that in Experiment 3, line density could be reduced with higher speeds while maintaining the illusory percept, or, conversely, line density had to be increased with slower speeds to obtain the illusory percept. For Experiment 4, results indicate that with lower line density, speed had to be increased to reach threshold, and with denser line spacing, speed could be reduced.

As intended, the functions relating speed and line spacing in the two experiments have mostly overlapping domains and ranges, and they can thus be easily compared. The functions are separated by a small horizontal offset, but are quite similar in shape. This suggests that the results of the two experiments indeed describe the same relationship, and can be combined into one data set. Yet, the two functions do not coincide exactly. The observed difference between them may be attributed to small shifts of the subjective criterion between experimental sessions.

As is apparent in Fig. 8, there is some inter-subject variability in the absolute settings of the dependent variables. The narrow error bars indicating the SEM however show that subjects' settings were in general quite consistent within conditions.

Based on these data, it is possible to calculate the delay between two consecutive appearances of a flank at a given fixed location within the virtual ellipse at threshold level. This delay between two refreshes of texture information at a stationary point

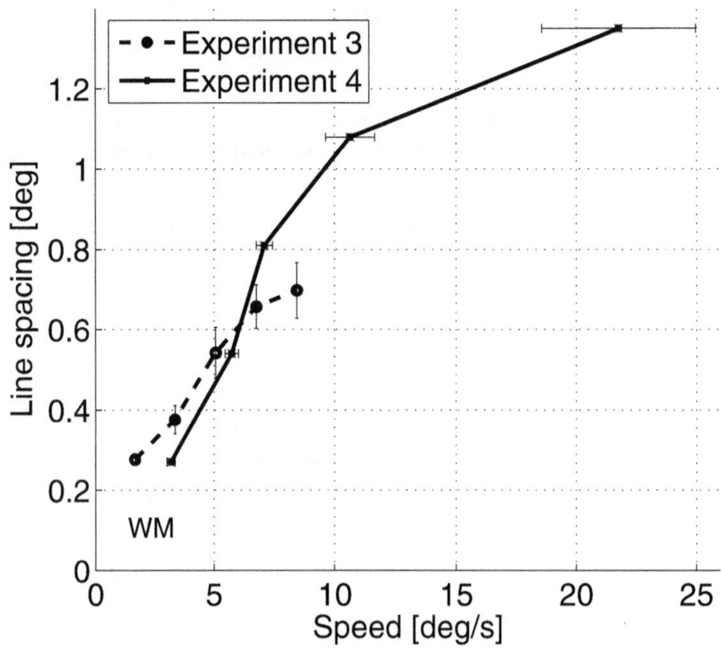

(a)

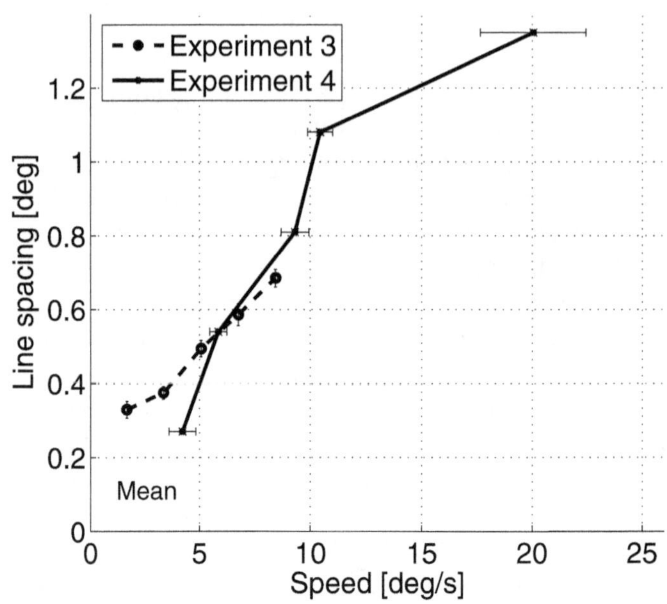

(b)

Figure 8. Results from Experiments 3 and 4. Distance of the moving lines is plotted against the speed of the moving lines. (a) Individual data from WM. Error bars indicate ±1 SE. (b) Combined results from all subjects. Error bars indicate ±1 SEM.

is plotted in Fig. 9 against the speed of the moving line array for both experiments. The pattern of results slightly differs between subjects, but in each case the data points varied within a close range from 100 to 180 ms. This consistence of the maximally-tolerated delay is noteworthy because its calculation involves a division operation and is therefore especially sensitive to changes in its input values. The tolerated delay increases non-linearly with very low speeds in Experiment 1, indicating that the line spacing setting never fell below a fixed value. This is to be expected from the fact that a small distance between lines can be bridged by color or texture spreading even in static displays.

3.3.5. Discussion. The results from both experiments show a fair degree of consistency within as well as across subjects. This seems worth noting, considering the peculiar perceptual nature of most illusory percepts on the one hand, and the employed adjustment method of threshold determination on the other hand. Any vagueness in the percept would have necessarily resulted in highly variable threshold settings. Since the individual settings actually stayed in a reasonably close range, one may infer that the percept itself is stable — a prerequisite for any further data analysis with regard to possible perceptual mechanisms.

In line with results presented by Morgan (1980) and Morgan *et al.* (1982), the idea of simple temporal integration through sensory persistence (see Section 2) was able to capture the data from both experiments surprisingly well: the density of the line array and its speed seem to act together in a compensatory manner in determining the strength of the illusory percept. Within the studied parameter range, this means that a change in the strength of the illusory percept due to, for example, a reduction in line density could be compensated by a corresponding increase in the speed of the line array.

That sensory persistence accounts well for our data warranted a further analysis in terms of one of its important parameters: the maximum delay between two input signal refreshes that is tolerated by the visual system while still yielding a complete percept. This delay was not exactly constant over experimental conditions as would be expected from retinal painting, but it stayed in a narrow range from around 100 to 180 ms (Fig. 9), indicating the maximum time window for integration processes. This estimate is remarkably similar to results obtained in a number of other experiments on object recognition: Burr (1980) estimated that information in the visual system is preserved for about 120 ms, similar to the integration timeframe of 150 ms that Morgan *et al.* (1982) found necessary for perceiving a complete circle that is successively defined through single small dots. Studying the integration of local element changes in the perception of subjective contours in spatiotemporal boundary formation, Shipley and Kellman (1994) came to the conclusion that signal integration takes place within 165 ms. In EEG studies on the time course of neural signals in object recognition, Johnson and Olshausen (2003) found recognition-correlated ERP components between 150 and 300 ms. Thorpe *et al.* (1996) reported an EEG component at 150 ms linked to the detection of objects in complex natural scenes.

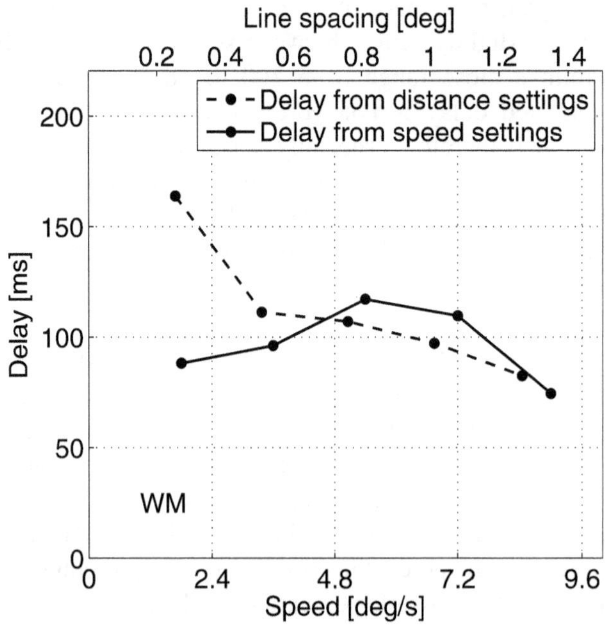

(a)

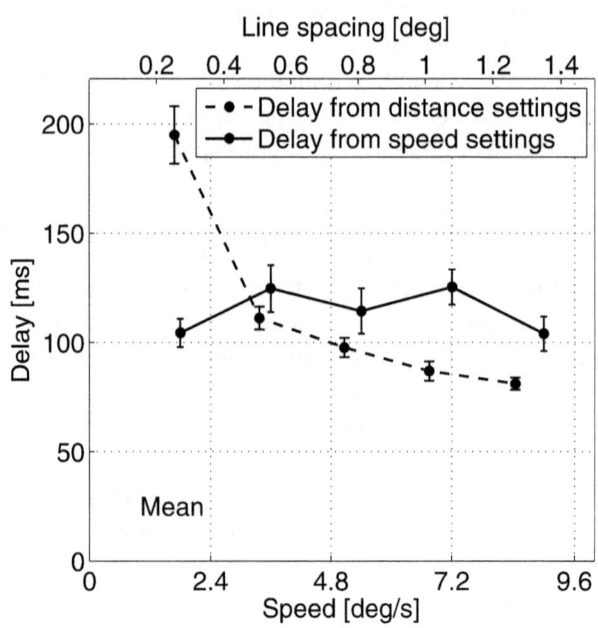

(b)

Figure 9. Results from Experiments 3 and 4, re-analyzed in terms of the maximally tolerated delay between two consecutive appearances of a textured flank at threshold level. (a) Individual data from WM. Error bars indicate ±1 SE. (b) Combined results from all subjects. Error bars indicate ±1 SEM.

4. GENERAL DISCUSSION

The results obtained in all four experiments were systematic, reliable, and of sufficient inter-individual consistency. This warrants the use of dynamic texture spreading as a tool to investigate visual completion processes with phenomenological (Experiments 3 and 4) as well as with more rigorous quantitative methods (Experiments 1 and 2). We believe that this possibility constitutes a significant advantage over static color or texture spreading, which are often perceptually unstable and therefore prove elusive to grasp with quantitative methods.

Experiments 1 and 2 served to illustrate that texture spreading allows observers to recover significantly more information about the surface layout of complex and highly fragmented stimuli than just their boundaries. This was done using performance tasks whose successful completion required that a rich perceptual surface pattern representation was at the observer's disposal. Texture spreading brought with it a measurable performance benefit and thus proved to be a functionally relevant effect.

Experiments 3 and 4 then tried to relate the phenomenon under consideration to sensory persistence, a simple perceptual mechanism whose involvement in spatiotemporal integration is frequently discussed (see Section 2.2). The data from these two experiments were in good agreement with a retinal painting account of dynamic texture spreading. In addition, the results of Experiments 1 and 2 may also be interpreted as being compatible with retinal painting. The performance drop associated with a progressive reduction of diagnostic stimulus fragments was gradual rather than abrupt.

Still, some qualitative observations indicate that there are several aspects of the phenomenon that are not accounted for by sensory persistence. First, since sensory persistence and the corresponding integration mechanism are strictly local, the effect should not be sensitive to global changes in the stimulus layout, as long as these changes do not affect the refresh frequency of texture information. However, it can be observed that texture spreading does not occur if the black lines, which do not directly contribute to the sensory persistence of texture cues, are removed from the display (Fig. 1a). While this spatial interaction is not readily explained by retinal painting, it can be naturally integrated into a high-level perspective which emphasizes the global stimulus layout and scene organization. The Bregman–Kanizsa display (Bregman, 1981; Kanizsa, 1979; Nakayama *et al.*, 1989) for example illustrates that fragmented letters are immediately recognized only if the occluding material is also visible. Applying this reasoning to dynamic texture spreading, we may interpret the lines as occluders, signaled by the T-junctions they form with the flanks. Their presence seems to constitute a cue that determines which fragments in the stream of information are classified as coextensive, and are therefore integrated, and which ones are classified as disjoint and duly segregated (Dixon and Di Lollo, 1994).

As a second issue, it is unobvious how to apply the temporal integration scheme to static neon-spreading displays. In this case, it is not plausible to assume that the gaps between corresponding stimulus elements are temporally joined with eye movements which recurringly spread the complete stimulus onto the retina. However, it should be noted that, in general, the color and texture spreading is a lot weaker in static displays compared to their dynamic counterparts.

Finally, one of the main objections against a major involvement of sensory persistence was that anorthoscopic contour perception can still occur when experimental conditions preclude the possibility that a complete image is temporally spread over the retina. This argues against a critical role of retinal painting in boundary integration. However, apart from the possibility that a retinotopic and a distal representation may co-exist (Shipley and Cunningham, 2001), it does not touch the question whether other perceptual attributes, such as surface filling-in, trace back to sensory persistence. Taken together, this suggests that, unlike certain spatial and figural stimulus aspects, sensory persistence by itself may not be a *necessary* prerequisite for all spreading effects. Nevertheless, when stimulus conditions favor its occurrence, it seems to be able to provide performance-relevant information about surface attributes and to determine crucial characteristics of the percept. In particular, it may be responsible for the particular vividness of the spreading effect in dynamic stimuli.

5. CONCLUSION

Dynamic texture spreading is a filling-in effect arising from motion displays that constitutes another example of the interaction of color, motion and spatial stimulus configuration (Hoffman, 2003). It exhibits a large discrepancy between the small amount of input information available to the visual system in each separate frame and the rich perceptual interpretation of the animated stimulus in terms of a complete object featuring strong illusory contours as well as a genuine surface qualities like a colored texture and depth stratification.

Apart from constituting an interesting phenomenon in its own right, dynamic texture spreading is a particularly robust effect that proved to be amenable to standard psychophysical methods. It may therefore be used as a reliable tool to study models of object-completion processes.

We established that texture spreading is a genuine representation of surface qualities that goes beyond mere contour interpolation. We further related the phenomenon to ongoing discussions about putative mechanisms for spatiotemporal integration: We examined the adequacy of retinal painting as a possible cause for the filling-in effect, a process that has often been proposed as an explanation for anorthoscopic perception. Our data are consistent with this hypothesis and showed that motion and the density of the moving apertures can act together in a compensatory manner in sustaining the texture spreading. This suggests that the interpolation in dynamic texture spreading partly takes place at a rather basic processing stage that temporally integrates sensorily persistent input signals. We discussed several findings which indicate that there are some aspects of dynamic texture spreading that are not captured by this simple integration mechanism. We thus conclude that an elementary integration process may well play an important role, but is not sufficient to explain dynamic texture spreading in full detail. Further work is required to better isolate the respective contributions of low and high-level mechanisms in dynamic spreading phenomena.

Acknowledgements

The authors are indebted to two anonymous reviewers for their thoughtful comments that contributed significantly to the development of the manuscript. The authors are grateful to Vebjørn Ekroll for helpful comments on a previous draft of this work. Franz Faul was supported by the Deutsche Forschungsgemeinschaft (FA 425/1-1).

NOTES

1. For an online demonstration of the display, please see http://www.psychologie.uni-kiel.de/psychophysik/demos/ts/
2. As was rightly pointed out by one referee, the terminology of filling-in effects is currently quite heterogeneous and does not induce a convincing classification. Until a conceptual clarification of the terms has taken place, we try to adopt the terminology used by the respective original authors.
3. Motion conditions: $0.5/(1 + 8x) + 0.5$, static condition: $0.5/(1 + \exp(-6 + 10x)) + 0.5$.
4. We thank one anonymous reviewer for directing our attention to this issue.

REFERENCES

Andersen, G. J. and Cortese, J. M. (1989). 2-D contour perception resulting from kinetic occlusion, *Perception and Psychophysics* **46**, 49–55.

Anstis, S. (2005). Local and global segmentation of rotating shapes viewed through multiple slits, *J. Vision* **13**, 194–201.

Anstis, S. M. and Atkinson, J. (1967). Distortions in moving figures viewed through stationary slit, *Amer. J. Psychol.* **80**, 572–585.

Brainard, D. H. (1995). Colorimetry, in: *OSA Handbook of Optics*, Bass, M. (Ed.), pp. 26.1–26.54. McGraw-Hill, New York, NY.

Bregman, A. L. (1981). Asking the 'what for' question in auditory perception, in: *Perceptual Organization*, Kubovy, M. and Pomerantz, J. R. (Eds), pp. 99–118. Lawrence Erlbaum, Hillsdale, NJ.

Bressan, P., Mingolla, E., Spillmann, L. and Watanabe, T. (1997). Neon color spreading: a review, *Perception* **26**, 1353–1366.

Bruno, N. and Bertamini, M. (1990). Identifying contours from occlusion events, *Perception and Psychophysics* **48**, 331–342.

Bruno, N. and Gerbino, W. (1991). Illusory figures based on local kinematics, *Perception* **20**, 259–274.

Bull, N. J., Hunter, M. and Finlay, D. C. (2003). Cue gradient and cue density interact in the detection and recognition of objects defined by motion, contrast, or texture, *Perception* **32**, 29–39.

Burr, D. C. (1980). Motion smear, *Nature* **284**, 164–165.

Cicerone, C. M., Hoffman, D. D., Gowdy, P. D. and Kim, J. S. (1995). The perception of color from motion, *Perception and Psychophysics* **57**, 761–777.

Cunningham, D. W., Shipley, T. F. and Kellman, P. J. (1998). The dynamic specification of surfaces and boundaries, *Perception* **27**, 403–415.

Devinck, F., Delahunt, P. B., Hardy, J. L., Spillmann, L. and Werner, J. S. (2005). The watercolor effect: quantitative evidence for luminance-dependent mechanisms of long-range color assimilation, *Vision Research* **45**, 1413–1424.

Dixon, P. and Di Lollo, V. (1994). Beyond visible persistence: An alternative account of temporal integration and segregation in visual processing, *Cognit. Psychol.* **26**, 33–63.

Ekroll, V. and Faul, F. (2002). Perceptual transparency in neon color spreading displays, *Perception and Psychophysics* **64**, 945–955.

Fahle, M. and Poggio, T. (1981). Visual hyperacuity: Spatiotemporal interpolation in human vision, *Proc. Roy. Soc. London B* **213**, 451–477.

Fendrich, R., Rieger, J. W. and Heinze, H.-J. (2005). The effect of retinal stabilization on anorthoscopic percepts under free-viewing conditions, *Vision Research* **45**, 567–582.

Fidopiastis, C., Hoffman, D. D. and Prophet, W. D. (2000). Constructing surfaces and contours in displays of color from motion: the role of nearest neighbors and maximal disks, *Perception* **29**, 567–580.

Haber, R. N. and Nathanson, L. S. (1968). Post-retinal storage? Some further observations on Parks' camel as seen through the eye of a needle, *Perception and Psychophysics* **3**, 349–355.

Hecht, H. (1924). Neue untersuchungen über die Zöllnerschen anorthoskopischen Zerrbilder, *Zeitschrift für Psychologie* **94**, 153–194. (Herausgegeben von F. Schumann: I. Die simultane Erfassung der Figuren)

Hine, T. (1987). Subjective contours produced purely by dynamic occlusion of sparse-points array, *Bull. Psychonomic Soc.* **25**, 182–184.

Hoffman, D. D. (2003). The interaction of colour and motion, in: *Colour — Mind and the Physical World*, Mausfeld, R. J. and Heyer, D. (Eds), pp. 361–378. Oxford University Press, Oxford, UK.

Johnson, J. S. and Olshausen, B. A. (2003). Time course of neural signatures of object recognition, *J. Vision* **3**, 499–512.

Kanizsa, G. (1979). *Organization in Vision*. Praeger, New York, NY.

Kellman, P. J. and Shipley, T. F. (1991). A theory of visual interpolation in object perception, *Cognitive Psychol.* **23**, 141–221.

Kellman, P. J., Guttman, S. E. and Wickens, T. D. (2001). Geometric and neural models of object perception, in: *From Fragments to Objects: Segmentation and grouping in vision*, Shipley, T. F. and Kellman, P. J. (Eds). Elsevier Science Publishers, Oxford, UK.

Mateeff, S., Popov, D. and Hohnsbein, J. (1993). Multi-aperture viewing: perception of figures through very small apertures, *Vision Research* **33**, 2563–2567.

Morgan, M. J. (1980). Analogue models of motion perception, *Phil. Trans. Roy. Soc. London B* **290**, 117–135.

Morgan, M. J., Findlay, J. M. and Watt, R. J. (1982). Aperture viewing: a review and a synthesis, *Quart. J. Exper. Psychol.* **34A**, 211–233.

Nakayama, K., Shimojo, S. and Silverman, G. H. (1989). Stereoscopic depth: Its relation to image segmentation, grouping and the recognition of occluded objects, *Perception* **18**, 55–68.

Nishida, S. (2004). Motion-based analysis of spatial patterns by the human visual system, *Current Biology* **14**, 830–839.

Parks, T. E. (1965). Post-retinal visual storage, *Amer. J. Psychol.* **78**, 145–147.

Pessoa, L., Thompson, E. and Noë, A. (1998). Finding out about filling-in: a guide to perceptual completion for visual science and the philosophy of perception, *Behav. Brain Sci.* **21**, 723–802.

Pinna, B., Brelstaff, G. and Spillmann, L. (2001). Surface color from boundaries: a new 'watercolor' illusion, *Vision Research* **41**, 2669–2676.

Pirenne, M. H. (1962). Light adaptation. I. The Troxler phenomenon, in: *The Eye*, Davson, H. (Ed.), pp. 197–199. Academic Press, London, UK.

Ramachandran, V. S. (1992). Blind spots, *Scientific American* **266**, 85–91.

Ramachandran, V. S. and Gregory, R. L. (1991). Perceptual filling-in of artificially induced scotomas in human vision, *Nature* **350**, 699–702.

Rock, I. (1981). Anorthoscopic perception, *Scientific American* **244**, 145–153.

Rock, I., Halper, F., DiVita, J. and Wheeler, D. (1987). Eye movements as a cue to figure motion in anorthoscopic perception, *J. Exper. Psychol.: Human Perception and Performance* **13**, 344–352.

Rothschild, H. (1922). Untersuchungen über die sog. Zöllnerschen anorthoskopischen Zerrbilder, *Zeitschrift für Psychologie* **90**, 137–166.

Shimojo, S. and Richards, W. (1986). 'Seeing' shapes that are almost totally occluded: a new look at Park's camel, *Perception and Psychophysics* **39**, 418–426.

Shipley, T. F. and Cunningham, D. W. (2001). Perception of occluding and occluded objects over time: spatiotemporal segmentation and unit formation, in: *From Fragments to Objects: Grouping and Segmentation in Vision*, Shipley, T. F. and Kellman, P. J. (Eds), pp. 557–585. Elsevier Science, New York, NY.

Shipley, T. F. and Kellman, P. J. (1993). Optical tearing in spatiotemporal boundary formation: When do local element motions produce boundaries, form, and global motion? *Spatial Vision* **7**, 323–339.

Shipley, T. F. and Kellman, P. J. (1994). Spatiotemporal boundary formation: boundary, form, and motion perception from transformations of surface elements, *J. Exper. Psychol.: General* **123**, 3–20.

Thorpe, S. J., Fize, D. and Marlot, C. (1996). Speed of processing in the human visual system, *Nature* **381**, 520–522.

Troxler, D. (1804). Ueber das Verschwinden gegebener Gegenstände innerhalb unseres Gesichtskreises, in: *Ophthalmologische Bibliothek II*, Himly, K. and Schmidt, J. A. (Eds), pp. 1–119. F. Frommann, Jena, Germany.

Tynan, P. and Sekuler, R. (1975). Moving visual phantoms: a new contour completion effect, *Science* **188**, 951–952.

von Vierordt, K. (1868). *Der Zeitsinn nach Versuchen: §27. Scheinbare Verzerrung bewegter Gegenstände*. Laupp, Tübingen, Germany.

Watanabe, T. and Cavanagh, P. (1991). Texture and motion spreading, transparency and the aperture problem, *Perception and Psychophysics* **50**, 459–464.

Watson, A. B. and Pelli, D. G. (1983). QUEST: A Bayesian adaptive psychometric method, *Perception and Psychophysics* **33**, 113–120.

Weisstein, N., Maguire, W. and Berbaum, K. (1977). A phantom-motion aftereffect, *Science* **198**, 955–958.

Wollschläger, D., Rodriguez, A. M. and Hoffman, D. D. (2001). Flank transparency: transparent filters seen in dynamic two-color displays, *Perception* **30**, 1423–1426.

Wollschläger, D., Rodriguez, A. M. and Hoffman, D. D. (2002). Flank transparency: the effects of gaps, line spacing, and apparent motion, *Perception* **31**, 1073–1092.

Zöllner, F. (1862). Ueber eine neue Art anorthoskopischer Zerrbilder, *Annalen der Physik und Chemie* **117**, 477–484.

APPENDIX A. ADAPTIVE ALGORITHM USED IN EXPERIMENT 1

The algorithm we used in Experiment 1 to determine the tested line distances first divided the complete parameter range into four intervals. It then randomly drew and tested values from each interval until all intervals would contain at least 40 data points. It then calculated the mean percentages correct for each interval and compared the results of adjacent intervals. If two adjacent intervals would differ by more than 10%, their spanned range would be sub-divided into four intervals. Values from each new interval were then randomly drawn and tested until they all contained at least 40 data points. Mean percentages correct for all existing intervals were (re-)calculated and compared between adjacent intervals. This process of recursive subdivision continued until all differences between the mean percentages correct of adjacent intervals were less than 10%, or until the minimum interval width of 1 px was reached. The method ensured that the sampling rate was increased in areas where the rate of change of the underlying function was high, thus leading to an approximately constant smoothness of the reconstructed curve. Conversely, the sampling rate was lowered and trials saved in areas where the underlying function did not change very rapidly.

The described method may be seen as a breadth-first-search algorithm. The nodes of the search tree correspond to pairs of tested intervals, and the depth of the search tree represents the resolution of tested intervals. The decision criterion used at each node to decide whether to descend into a deeper level of the tree corresponds to the question whether the difference between adjacent intervals was greater than the 10%.

Color, Line, and Space, pp. 117–159 (2006)

Lightness, illumination, and gradients

DEJAN TODOROVIĆ *

Laboratory of Experimental Psychology, Department of Psychology, University of Belgrade, Čika Ljubina 18-20, 11000 Belgrade, Serbia

Abstract—The illumination interpretation approach claims that lightness illusions can be explained as misapplications of lightness constancy mechanisms, processes which usually enable veridical extraction of surface reflectance from luminance distributions by discounting illumination. In particular, luminance gradients are thought to provide cues about the interactions of light and surfaces. Several examples of strong lightness illusions are discussed for which explanations based on illumination interpretation can be proposed. In criticisms of this approach, a variety of demonstrations of similarly structured control displays are presented, which involve equivalent lightness effects that cannot readily be accounted for by illumination interpretation mechanisms. Furthermore, a number of known and novel displays are presented that demonstrate effects of gradients on the qualitative appearance of uniform regions. Finally, some simple simulations of neural effects of luminance distributions are discussed.

Keywords: Lightness; illumination; luminance gradients.

INTRODUCTION

It has been known since the times of Aristotle, Alhazen, and Leonardo (see Wade, 1998), that the visual appearance of a surface patch does not depend solely on the material composition of the patch itself and the light it reflects towards the observer, but also on the properties of its surround. For example, early in the 19th century, Johann Wolfgang von Goethe noted that 'a gray image on black ground appears much lighter than the same image on white ground. When the two cases are placed next to each other, one can hardly persuade oneself that both images were painted from the same pot' (see Trunz, 1981, p. 337). This particular effect was later named 'simultaneous contrast', and is depicted in many perception textbooks. A large variety of other context-induced achromatic color illusions were described and studied subsequently.

*E-mail: dtodorov@dekart.f.bg.ac.yu

There are several types of theoretical accounts of illusory contextual effects in perception. A popular approach is to regard them as results of inappropriate applications of visual mechanisms generally subserving veridical perception. Such mechanisms are needed to recover distal attributes of objects, because the corresponding proximal attributes are affected by various extraneous factors. For example, the projected, proximal size of an object depends not only on its distal size but also on its distance from the observer; proximal shape depends not only on distal shape but also on orientation; proximal motion on the retina depends not only on distal motion of the observed object but also on the motion of the eyeball; etc. In particular, the luminance of a surface patch, that is, the amount of visually effective light it projects onto the retina (which may be labeled as its 'proximal achromatic color'; see Todorović, 2002), depends not only on its reflectance (the percentage of light it reflects, or its 'distal achromatic color'), but also on its illumination (the extraneous factor involving the amount of light currently incident on the patch). However, in spite of such effects of extraneous factors on proximal attributes, usually perceptual constancies obtain, that is, distal properties are generally perceived more or less veridically. This suggests the existence of special visual mechanisms that compensate for the effects of extraneous factors. In color perception, these are the mechanisms that are supposed to recover constant surface reflectances from the incoming luminance distributions by discounting the effects of illumination.

One example of a theoretical account in which constancy mechanisms are hypothesized to generate non-veridical percepts involves size illusions. In this account it is assumed that depth cues in pictorial displays may trigger processes in the visual system that in everyday perception are activated by real 3D depth, so that the same mechanisms that standardly generate veridical percepts may, in this case through misapplication, cause visual illusions (see Gillam, 1998). For example, in the well-known Ponzo illusion, two target lines of equal size, flanked by two converging slanted lines, are perceived as of different length. The explanation is that perspective depth cues induced by the slanted lines indicate that the target lines are located at different distances, causing the further appearing target to look longer, as indeed it would be, had it been actually located further from the observer (see Palmer, 1999). In this way, constancies and illusions are elegantly and parsimoniously accounted for within the same theoretical framework. In the following I will first sketch such an account of achromatic color illusions, and will then proceed to criticize it.

In color perception this type of approach essentially goes back to Helmholtz. Although his original idea was formulated mainly with chromatic colors in mind, it can be readily applied in the achromatic domain as well. The achromatic color illusions under discussion here involve flat images exhibiting various reflectance distributions, observed under uniform illumination. However, they may be *interpreted* by the perceiver as depicting non-uniformly illuminated scenes. The basic assumption of this approach is that cues of illumination may trigger reflectance recovery mechanisms appropriate for corresponding real-world scenes, and in this way generate non-veridical percepts of image reflectances. For example, if two image patches of equal luminance are interpreted as receiving different amounts of illumination, then they may be perceived as having different reflectances, as indeed they would have in an actual non-uniformly illuminated scene. Such circumstances may provide the explanation of the phenomenon of simultaneous contrast, described above. If the observer of the display involving two equiluminous gray patches interprets one gray patch, together with its white background, as receiving higher illumination than the other gray patch and its dark background, then it follows that the first patch must have lower reflectance, in order to reflect the same amount of light as the second patch (see Hurvich and Jameson, 1966; Kingdom, 1999).

This type of explanation is formally closely analogous to the above account of the Ponzo size illusion: in both cases objects that have equal proximal attributes (such as size or color) are perceived as being different, supposedly due to effects of cues that convey the presence of different extraneous factors for the two objects (such as different distance or illumination). One potentially complicating factor for this type of account, that I will not address here in detail (but see Todorović, 2002, 2003c), is the need to distinguish between perceived distal attributes (such as perceived distal size or lightness) from perceived proximal attributes (such as perceived proximal size or brightness). Note that this type of account does not necessarily presuppose that scene interpretation is based on explicit, conscious deliberations by observers. Helmholtz already talked about 'unconscious inferences', implying that although such processes may be formally described as involving premises and conclusions, this does not mean that they need to be accessible to conscious awareness, or that they necessarily take the same form as logical reasoning or tap into higher cognitive processes. I will refer to putative processes of this type as *illumination interpretation mechanisms* (Kingdom, 2003a, b; Todorović, 2003a, b).

In recent years a number of authors have in various ways discussed the possibility that lightness phenomena in relatively complex flat displays may, at least in part, be due to mechanisms of this type (see Adelson, 1993, 2000; Agostini and Galmonte, 2002; Kingdom, 1997, 1999, 2003a, b; Knill and Kersten, 1991; Logvinenko, 1999, 2003; Logvinenko *et al.*, 2005; Purves and Lotto, 2003; Purves *et al.*, 2004; Schirillo and Shevell, 1997, 2002; Williams *et al.*, 1998a, b). I will deal with some of these approaches in a little more detail below. This class of accounts is only one among several different types of lightness theories. For example, an account of simultaneous contrast popular in textbooks bases this effect on actions of lateral inhibition processes or, equivalently, properties of neurons with concentric-antagonistic receptive fields. Although such a proposal in its simplest form has serious shortcomings (Gilchrist *et al.*, 1999), several much more complex computational models, based on and inspired by physiological data, have accounted for a number of lightness effects (Blakeslee and McCourt, 1997, 2003, 2004; Hong and Grossberg, 2004; Grossberg and Todorović, 1988; Kingdom and Moulden, 1992; Pessoa *et al.*, 1995). A variety of other theoretical approaches to lightness phenomena are also available (see e.g. Adelson, 2000; Anderson, 1997; Anderson and Winawer, 2005; Bressan, 2001; Gilchrist, 1988; Gilchrist and Economou, 2003; Gilchrist *et al.*, 1999; Whittle, 1994a, b). Although these different explanatory frameworks are generally theoretically opposed, they are not necessarily completely mutually exclusive in all respects, and any given effect might, in principle, be due to a combination of causes.

In this paper my aim is to critically examine the illumination interpretation account by starting from original stimulus images containing equiluminant target patches that look different, and whose difference may be plausibly accounted for through illumination interpretation mechanisms. The next step is to modify the structure of these stimuli in order to modulate particular illumination cues. If such a stimulus modification should be accompanied by corresponding changes in lightness, the illumination interpretation approach is corroborated; however, if strong manipulations of essential illumination cues fail to affect the lightness of critical targets, then the illumination interpretation account for these displays becomes problematic. It is this latter possibility that I will pursue in the following.

I have previously (Todorović, 1997, 2003a) used this strategy to challenge the illumination interpretation account of particular stimulus displays designed by Adelson (1993) and Kingdom (2003a); see also Kingdom's (2003b) reply and my reply to his reply Todorović (2003b), contained on the same CD as my first comment. In the present paper I will continue with such criticisms by addressing three among the strongest and most attractive recent lightness phenomena for which explanations based on the illumination interpretation can be and have been offered. These explanations turn out to be formally similar to the above account of classical simultaneous lightness contrast, except that the new displays are much more effective in conveying the represented scenes. As counter-examples, I will present a variety of similarly structured control displays that involve similar lightness effects but which, I will argue, cannot readily be accounted for by illumination interpretation mechanisms. I will not offer any comprehensive theory of lightness, but will claim that these particular effects can be better understood by concentrating on relatively local luminance relations and their effects on the visual system.

Natural scenes often contain *luminance gradients*, that is, regions in which luminance varies gradually over space. For example, gradients may appear at the edges of shadows (penumbras), as shading patterns on curved surfaces (due to smooth variations of surface orientation with respect to the light source), as consequences of the gradual decrease of illumination intensity with the increase of distance from the light source, or due to variations of light source intensity. Most of the displays that will be presented here involve luminance gradients, constructed using graphics software. According to the illumination interpretation account, these gradients may serve as image cues of illumination, which help trigger the corresponding interpretative mechanisms that are presumed to underlie the lightness effects. An alternative approach, which I will advocate, is that the lightness phenomena in these images are due to other structural aspects of the stimuli. This is not to deny that luminance gradients help establish the impression of illumination in these displays, but rather to question the causal role of illumination interpretation mechanisms in the generation of lightness effects exhibited in the images. Such arguments will be illustrated with displays presented in Figs 1 to 5, which involve detailed analyses of the illumination structure of a number of displays. Following that, I will discuss some additional phenomena involving gradients, which demonstrate that they may not only affect lightness, but also induce qualitative differences in surface appearance. Such analyses will be illustrated with displays presented in Figs 6 to 10. Figure 11 will present some effects involving chromatic gradients. Finally, some simple simulations of neural effects of luminance gradients will be illustrated with diagrams in Fig. 12. The presented lightness effects are informal demonstrations that have not yet been confirmed in published experiments. However, I hope that readers will find most of them robust enough to merit consideration.

THE CHECKERED SHADOW AND ITS DECONSTRUCTION

Figure 1a is a close replica of a now well-known image designed by Adelson (1995). It depicts a scene containing a horizontal checkered plane and a vertical cylinder, illuminated from the right. The two target patches A and B are uniform and have the same luminance, but their gray shades appear rather different. Within the illumination interpretation approach the explanation of this effect is straightforward, and involves the same logic as the above mentioned account of simultaneous contrast, but the argument for Adelson's image is more plausible because the illumination structure of the represented scene is more perceptually salient: in the conveyed scene patch B is readily interpreted as located within the shadow cast by the cylinder, whereas patch A appears normally illuminated. Based on information about the interaction of light and surfaces (accumulated individually or inherited through evolution), it can be concluded that since the patches are differently illuminated but have the same luminance, they must have different reflectances. The visual system automatically implements this conclusion, and therefore patch B comes to appear lighter gray but poorly lit, and patch A as well illuminated darker gray.

Figure 1b was designed as a control display to point out some problems with the above account. It conveys a similar scene as Fig. 1a, but with the shadow region, as it were, forcibly detached and artificially displaced from its natural position. This region in Fig. 1b contains all Fig. 1a patches that exhibit luminance gradients, preserving their exact luminance profiles, shapes, and mutual local spatial relations. I will call these two identical multi-patch regions in Figs 1a and 1b 'gradient regions'. Note that the two equiluminant target patches A and B in Fig. 1b appear differently light. However, in contrast to the gradient region in Fig. 1a, the gradient region in Fig. 1b lacks a number of features that are standardly associated with shadows.

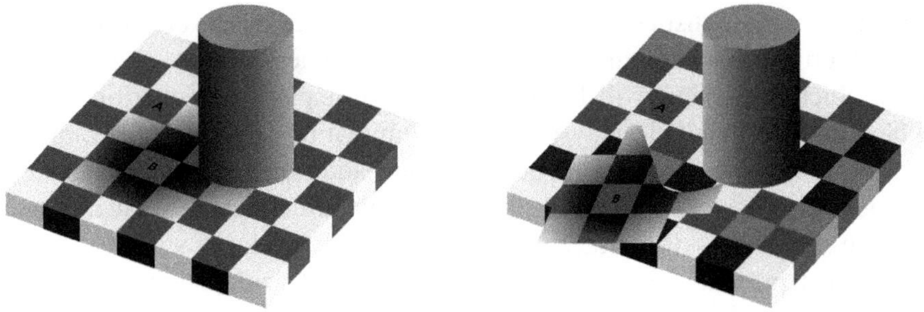

(a) (b)

Figure 1. Lightness and depicted cast shadow. The target patches A and B have the same luminance, but A looks darker than B. In (a) this difference may be accounted for by the presence of the shadow, but in (b) this explanation is compromised. See text for details.

To begin, in Fig. 1a the gradient region surrounding patch B has the natural appearance of a shadow cast upon the ground, reminiscent of an immaterial transparent veil spread across a surface, whereas in Fig. 1b the corresponding region looks more like a thin, graspable material textured sheet lying upon or hovering above the plane. In consequence, whereas in Fig. 1a patches A and B are seen to belong to the same surface but to be differently illuminated, in Fig. 1b they are seen as belonging to different surfaces. Furthermore, in Fig. 1a the luminance gradients along the margins of the gradient region appear as the characteristic penumbra of a cast shadow, whereas in Fig. 1b the appearance of the same gradients as the shadow penumbra is disturbed, due to the presence of sharp luminance discontinuities all along the gradient region borders. A salient difference between the two gradient regions is that in Fig. 1a the checkered pattern outside of the gradient region continues smoothly into the checkered pattern within the region, which is a signature feature of shadows cast upon textured surfaces, whereas in Fig. 1b the patterns within and outside the gradient region clash with each other at the border of the region, both geometrically and photometrically. Furthermore, in Fig. 1a the average luminance within the gradient region is less than outside of it, which is a regular consequence of diminished illumination within shadows, whereas in Fig. 1b, due to the darkening of a few tiles on the plane outside of the gradient region, this is not necessarily the case. Whereas in Fig. 1a the two patches appear to belong to the same chess-board pattern, patch A being one of the many black checks and patch B one of the many white checks, but accidentally exposed to different illumination, in Fig. 1b the two patches belong to two different patterns. Also, in Fig. 1a the shadow casting cylinder appears as illuminated from the right, so that its darker, left side, interpreted as the attached shadow, is positioned near the gradient region, interpreted as the cast shadow, whereas in Fig. 1b the cylinder shading is inverted, its darker portion lying to the right, suggesting that illumination is coming from left, and that the gradient region is at the wrong side to be the shadow of the cylinder. Furthermore, in Fig. 1a the shape of the gradient region is in rough projective agreement with the shape of the shadow casting cylinder and the direction of illumination, whereas in Fig. 1b these projective relations are violated.

The main point of interest here is that although many characteristics of shadows which are respected in Fig. 1a are violated in Fig. 1b, the contrast between the lightnesses of target patches in Fig. 1b does not appear much weaker than in Fig. 1a. This poses a problem for the illumination interpretation account. For example, Purves *et al.* (2004), who support a variant of this approach, claim that '... any change relevant to the probable contribution of illumination, reflectance, and transmittance — no matter how subtle — should influence the apparent brightness and/or lightness of the elements in the scene' (p. 145). If the lightness illusion in Fig. 1a were indeed caused by processes based on information about the various types of interactions of light with objects, then one would expect the strength of the effect to vary proportionally with the degree of similarity of the image structure to ecologically valid features of shadows. The fact that in Fig. 1b many of these features are absent, indicating that the gradient region could not be a generic shadow, but that the lightness effect does not appear to be much diminished, casts doubt upon the relevance of the illumination interpretation account for these displays. One might imagine, though, that the gradient region in Fig. 1b is cut out from a similar pattern as the plane on which it rests, placed upon it, and then somehow subjected to shadowing precisely aimed to coincide with its boundaries, so that patch A would indeed be dark but well illuminated and patch B light but poorly illuminated; however, such a causal scenario would hardly be generalizible from everyday situations, that is, based on normal experience with lights and objects.

How can the lightness effects in these figures be accounted for, if the illumination interpretation is questioned? As pointed out by Adelson, in Fig. 1a the darker appearing patch A is surrounded by lighter patches, whereas the lighter appearing patch B is surrounded by darker patches; note that the same is true for Fig. 1b. I will call this type of relation between target patches and their backgrounds 'surround contrast constellations'. With respect to this feature, the structure of these displays is similar to classical simultaneous contrast, as described by Goethe, and may have a similar cause. I will not at this point enter into a discussion of various explanations of simultaneous contrast and its variants, but there are plenty alternatives to the illumination interpretation account, since most every theory of lightness is expected to address this classical effect. It should be noted that there are three differences of these displays from Goethe's variant. First, in the checkered displays the target patches are not completely surrounded by contrasting patches. Second, the background is not uniform but articulated, a feature known to enhance the strength of the contrast effect (Adelson, 2000). Finally, the displays involve luminance gradients in the vicinity of target patches. These features are shared with most subsequent figures discussed below.

A difficulty for the proper evaluation of the lightness effects in the preceding examples is the relative complexity of the transformation of Fig. 1a into 1b, which may affect not only the potential illumination cues but also many local luminance relations. Thus any eventual difference of appearance of targets in the two figures may not only be due to effects of stimulus variations on putative processes of illumination interpretation but also due to changes of the structure of local contrasts. Such potential confounds of illumination cues and structural features present serious threats to validity for many studies aiming to disentangle the sources of lightness phenomena (see Kingdom, 2003a, b; Todorović, 2003a, b). In the following sections I will use simpler and more controlled manipulations of both aspects of stimuli, thus enabling more focused analyses of their effects on the appearance of target patches.

THE GRADIENT WALL OF BLOCKS AND ITS SHADOW-INCOMPATIBLE VERSIONS

Figure 2a is a close replica of an ingenious display designed by Logvinenko (1999). It is composed of uniform patches and gradient patches (patches that contain vertically oriented luminance gradients). The figure was produced with graphics software by first constructing a layer of adjacent rectangular gradient patches, and then superimposing five rows of equiluminant uniform patches on top of the grid. There are two types of gradient patches: the 'dark' patches in which the shading varies from black to light gray (directed top to bottom and bottom to top in alternate rows), and the 'light' patches in which shading varies from white to dark gray (also directed top to bottom and bottom to top in alternate rows), arranged in a chessboard-like pattern. The uniform patches are diamond-shaped and have equal luminance. The point of the display is that the diamonds in odd rows look much darker than those in even rows.

How can this lightness difference be explained? Logvinenko (1999, 2003) has pointed out that this 2D display can be interpreted as depicting a 3D scene containing a wall of blocks exposed to a wave of illumination which changes intensity sinusoidally along the vertical axis, with the diamonds in odd rows receiving the strongest illumination and the diamonds in even rows receiving the weakest illumination. Given such an illumination structure, the explanation of the lightness effect within an illumination interpretation account is straightforward: the diamonds in odd rows are perceived as highly illuminated dark surfaces, whereas those in even rows are perceived as poorly illuminated light surfaces. This is essentially the same type of reasoning described above in the Helmholtzian account of simultaneous contrast and Adelson's checkered shadow. The difference involves only the particular geometric and photometric structure of Fig. 2a, in which the luminance gradients on the sides of the blocks presumably serve as cues of variable illumination. As noted before, such an explanation, and the term 'interpretation' that I use in describing this account, does not include the claim that observers actually consciously perform geometric and photometric calculations involving luminances, reflectances, spatial orientations, and illumination direction and intensity; instead,

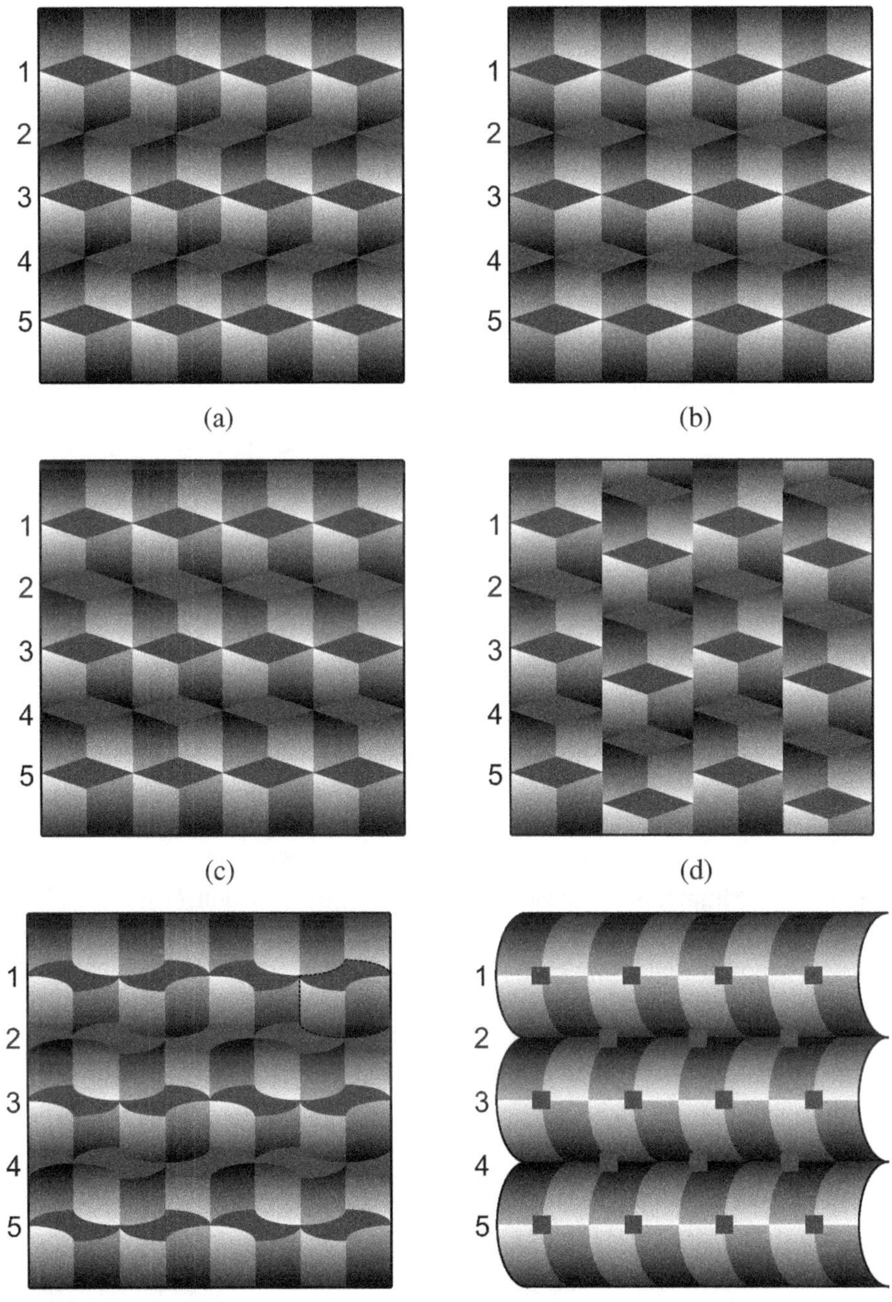

Figure 2. Lightness and depicted wave of illumination. The patches with the shapes of horizontally oriented diamonds in (a), (b), (c), and (d), as well as blades in (e) and small squares in (f) all have the same luminance, but those in odd rows look darker than those in even rows. The lightness difference in (a) may be accounted for by the presence of a wave of illumination falling upon a wall of blocks, but in other figures this type of explanation is less likely. See text for details.

their visual systems may be assumed to engage in analogous automatic, complex but reflex-like processing, only the results of which are accessible to consciousness.

The gradient wall of blocks and some related displays are labeled by Logvinenko (1999, 2003) as being *shadow-compatible*, meaning that there exists a pattern of illumination that can produce such a luminance pattern. However, this is not a very restrictive definition since 'every luminance pattern can, generally speaking, be produced by adjusting an illumination', so that this notion is to be applied 'from the human visual system point of view', such that shadow-incompatible displays are those for which 'the human visual system 'does not believe' that the pattern has been produced by the illumination' (Logvinenko, 2003, p. 723). An example of a shadow-incompatible display involves an array of blocks for which the illumination direction would have to be different for each adjacent block (Logvinenko, 1999, Fig. 11b). Although this feature could perhaps be defined more precisely, I will take shadow-compatibility to refer to representations of physically reasonably plausible 3D scenes involving a generally coherent illumination structure, the likes of which one might generically encounter in everyday environments. In contrast, shadow-incompatibility is a feature of 2D displays whose 3D interpretation would involve impossible or non-generic constellations of spatial shapes and illumination sources, highly unlikely to be produced in naturally occurring conditions. Logvinenko (1999, p. 812) has proposed that lightness illusions, such as in Fig. 2a, are generated only when luminance gradients are shadow-compatible. He has presented examples of displays related to Fig. 2a, which also contain luminance gradients, but which are not shadow-compatible, and in which the lightness illusion is reduced or lost, and has proposed that the decline of the strength of the illusion in such images is caused by the decline of shadow-compatibility (Logvinenko 1999, 2003).

Logvinenko's figure, itself based on a display by Adelson (1993), and used by him to challenge Adelson's theories of lightness phenomena, is a fruitful display paradigm which was, in turn, used by others to criticize Logvinenko's own explanation of his effect (Bressan, 2001; Todorović, 1999). For example, Fig. 2b, designed by Bressan (2001), is a variation of Logvinenko's figure in which the sides of blocks in every other row are interchanged. She pointed out that this figure portrays a shadow-incompatible wall of blocks, but that the illusory lightness difference between the diamonds in alternate rows is retained, challenging the explanation based on illumination interpretation. The wall of blocks in this figure is indeed shadow-incompatible, in the sense that its setup would necessitate a complex and unlikely constellation of illumination sources that change direction between different rows. Logvinenko (2003) has countered that Bressan's display can also be interpreted as conveying an alternative, shadow-compatible scene, which does not involve a wall of blocks. In this version the diamonds are patches freely floating in front of a corrugated background, and the scene is illuminated by a sinusoidal wave of light falling slightly from the side.

However, there is still a difficulty for the illumination interpretation account. The display in Fig. 2b is bi-stable, and with a little effort the observer may switch from one geometrical interpretation to the other, similar to observing a Necker cube. But note that the perceptual switch from Logvinenko's shadow-compatible scene to Bressan's shadow-incompatible one does not seem to have any effect on the perceived shades of the uniform diamonds. They appear equally light and dark in alternate rows, regardless of which of the two perceived 3D scenes they are part of. This is a problem, because one would expect that the presence/absence of shadow-compatibility should be accompanied by a corresponding presence/absence of the illusory lightness effect. Note that Logvinenko (2003) has insisted that this display is shadow-compatible by definition, simply because there exists at least one pattern of illumination that can produce the luminance pattern in question. Nevertheless, it seems that in order to be perceptually effective, it is not enough for a shadow-compatible interpretation to be just a valid theoretical possibility, but that it should also be actually realized in some way in the visual system of the perceiver. We know that it is realized when the observer perceives Fig. 2b accordingly, and in this case the lightness illusion might be accounted for as due to the illumination interpretation. But what happens when the percept switches to the shadow-incompatible variant and how can the unchanging lightness illusion be explained in that case? One might speculate that while consciously perceiving the shadow-incompatible interpretation, the visual system could also at the same time entertain the shadow-compatible one at some unconscious processing level, and use it for conscious lightness assignment in accord with the illumination interpretation approach, but this does not strike me as plausible.

I will now discuss some additional variations of Logvinenko's display (some variants of which were presented in Todorović, 1999) and argue that they also challenge the illumination interpretation account, because the lightness illusion persists unchanged in spite of shadow-incompatibility. Figure 2c was generated from Fig. 2a by simply vertically aligning the diamonds in even rows with those in odd rows. In this way portions of the underlying gradient patches were occluded and others were disoccluded, thus changing their visible shapes from parallelograms to trapezoids. Note that the lightness difference between the diamonds in alternate rows in Fig. 2c remains much the same as in the original Fig. 2a, and that, as in the original, there may exist an impression of a vertically oriented wave of illumination. However, the global geometrical structure of the conveyed scene has changed, and it is not easy to make sense of. Attending to regions around the diamonds conveys partial views of protruding or hollow blocks viewed either from above or below, or some bulging solids with trapezoidal lateral sides, but these impressions are fleeting and unstable because the stimulus geometry does not support them consistently. Therefore the display does not seem to be shadow-compatible, the reason being that, unlike in the previous two displays, there does not appear to exist a coherent 3D spatial structure for the shadow to be compatible with. Nevertheless, the lightness illusion remains steady, apparently oblivious to the inconsistent and fluctuating geometry. Figure 2d is obtained from Fig. 2c by vertically shifting alternate columns, a manipulation which, while maintaining the inconsistent geometry of

Fig. 2c, in addition destroys the sense of coherent illumination as well, and induces a more flat, mosaic-like impression. Nevertheless, the same lightness difference between the target diamonds is retained as in previous displays, again suggesting a dissociation of lightness and conveyed spatial structure in this class of displays.

Figure 2e is a modification of Fig. 2a in which the original diamonds were transformed by caving in and bulging out their sides, making them look somewhat like blades. Note that the lightnesses of the blades are much the same as the lightnesses of the corresponding diamonds in Fig. 2a and in the other preceding figures. However, the 3D spatial interpretation of the represented scene is again different, as the original cubical components of the wall in Fig. 2a have morphed into various curved 3D blocks. Closer inspection reveals that, unlike the blocks in Fig. 2a, many of the curved blocks are geometrically and photometrically ambiguous. For example, consider the block near the top right corner of the display, indicated by the dotted outline, which is replicated at several other places in the figure. Its visible portions include the blade-shaped top and a left side of higher average luminance than the right side. However, its top and bottom portions are incongruent: note that in the bottom part its shape appears to be cylindrically convex, so that its left and right halves seem to have different reflectance, whereas in the top part the shape appears partly convex and partly concave, so that the two halves seem to be illuminated differently. Note also that although much of the time the blocks are perceived as observed from above, thus exposing blade-shaped tops to view, occasionally portions of the display exhibit perceptual inversions, appearing as blocks with blade-shaped bottoms observed from below, generated by 'borrowing' their sides from three neighboring normally oriented blocks; similar effects may also be noted in previous figures. In sum, neither the geometric structure or spatial orientation, nor the nature of illumination, nor the reflectance pattern of the components of the scene appear to be well defined or to remain steady, and yet the lightness illusion involving the blades seems unchanged and fairly constant in prolonged viewing, although one might have expected some fluctuations, were it to depend on the unstable structure of the perceived scene.

Figure 2f is still another variation of Fig. 2a, in which the diamonds have shrunk into small squares centered at the same locations as the original diamonds, and the underlying gradient patches were transformed by curving their sides. Note that the lightness difference between the alternate rows of squares appears much the same as the difference between the diamonds in other figures. However, the spatial structure of the represented scene is not only rather different from previous displays, but unlike them it is also clear and coherent: the figure dominantly conveys the appearance of three semi-cylindrical convex checkered surfaces, with a set of vertically oriented small squares pasted on or located directly in front of them; there is also an alternative, also coherent spatial interpretation, in which the three surfaces are concave, but it is fleeting and hard to maintain. Note that the lightnesses of the squares are the same in both spatial interpretations. Closer inspection reveals that the conveyed illumination of the dominant interpretation of the scene is rather straightforward: the lightest portions of the semi-cylinders are the ones most in front, and they get darker as they curve back, all of which can be accounted for by assuming a spatially uniform light field illuminating the surface from the front and falling perpendicularly upon the horizontal midlines of the three surfaces; a similar account applies for the alternate spatial interpretation. Given such a pattern of illumination, the increased average luminance corresponding to highlights at the three horizontal midlines (where the darker appearing squares are located), and the reduced average luminance corresponding to the joints of the surfaces (where the lighter appearing squares are located) are not due to the variation of the intensity of the light source, as in Fig. 2a, but due to the curvature of the surfaces. The important point to note is that given that the illumination is uniform and that all the small squares appear to have the same spatial pose, it follows that all of the squares in all rows receive the *same* amount of illumination from the light source. This is in contrast to the diamonds in Fig. 2a, in which alternate rows were interpreted as receiving *different* amounts of illumination. The upshot of this analysis is that the lightness differences between the squares in this figure, in either of the two spatial interpretations, cannot be accounted for in the same way as the lightness differences between diamonds in Fig. 2a, that is, that the dark ones appear well illuminated, and the light ones poorly illuminated.

In sum, the displays in Fig. 2 suggest a dissociation of shadow-compatibility and lightness. Although the lightness illusion is present in shadow-compatible displays in accord with the illumination interpretation account (Fig. 2a and one interpretation of Fig. 2b), it is retained in a shadow-incompatible display (Fig. 2b in the other interpretation), in spatially ambiguous displays which are also shadow-incompatible (Figs 2c, d, e), as well as in a display which *is* shadow-compatible and non-ambiguous (Fig. 2f), but in which the illusion cannot be explained through illumination interpretation mechanisms. The implication is that shadow-compatibility as such does not appear to be a relevant determining factor of lightness perception in this class of displays.

How then can the lightness effects in these figures be accounted for? The structure of the luminance relations of the target patches and their immediate surrounds in Fig. 2a was pointed out by Logvinenko (1999) and Bressan (2001). Each diamond patch is bordered by two dark and two light gradient patches, such that the lighter appearing diamonds in even rows are flanked at all four sides by the *darker* ends of the gradient patches (black ends of the dark gradient patches and dark gray ends of the light gradient patches), whereas the equiluminant darker appearing diamonds in odd rows are flanked at all four sides by the *lighter* ends of the gradient patches (white and light gray ends). Note that this is another variant of a surround contrast constellation, which characterizes both the classical simultaneous contrast display and the target patches in Figs 1a and 1b. This local feature is shared by all target patches in all six displays in Fig. 2, despite their differences in other respects.

If the presence of the lightness illusion in Fig. 2 displays depends on the existence of the surround contrast constellation, then manipulations of this feature should affect the strength of the illusion. This is indeed the case. For example, Bressan (2003b) has noted that a wall of block display constructed by Logvinenko (2003), which involves a much smaller lightness difference between two sets of diamonds than in Fig. 2a (an effect which Logvinenko attributed to shadow-incompatibility), may instead be related to the fact that in this figure one set of diamonds is flanked at only three sides by lighter gradient ends and by the darker gradient end at the remaining side, and *vice versa* for the other diamond set. In other words, weakening of the surround contrast is associated with a weakening of the lightness difference.

This type of manipulation of the surround contrast is illustrated here in three displays in Fig. 3, which develop further the observations by Bressan (2003a), presented in her Fig. B. All three displays consist of two columns of diamond-shaped uniform patches, superimposed upon three columns of gradient patches, with horizontally oriented luminance gradients. The three diamonds in the left column of Fig. 3a are all flanked at all four sides by darker (two black and two

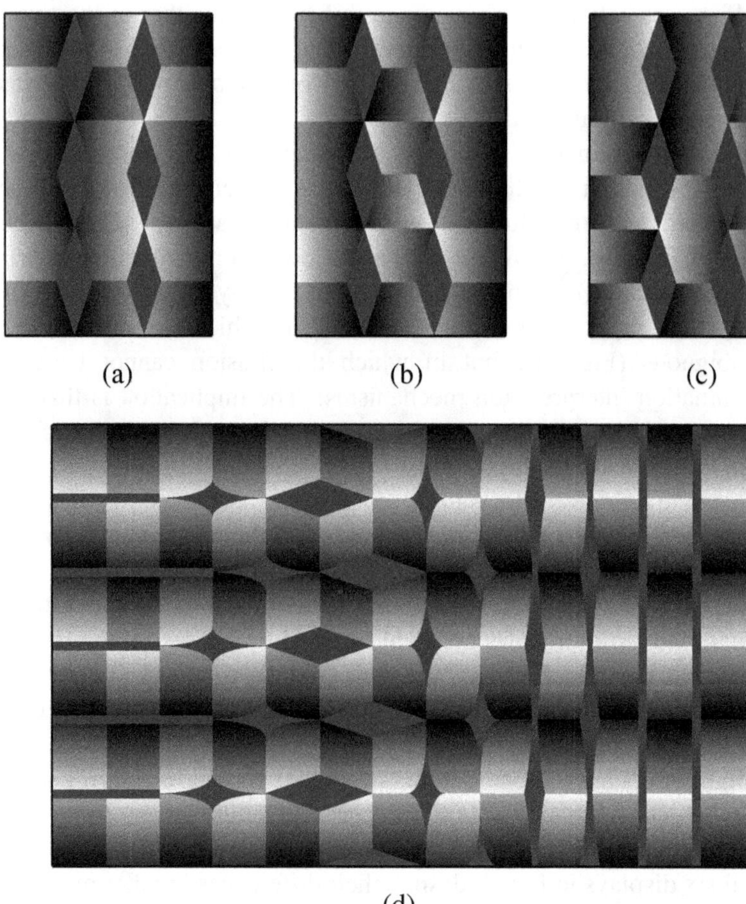

Figure 3. Further variations of displays from Fig. 2. See text for details.

dark gray) gradient ends, in three different spatial arrangements, whereas the three diamonds in the right column are all flanked by four lighter (two white and two light gray) ends, also in three different spatial arrangements. Note that in this figure there is a strong lightness difference between the two diamond columns, similar to the corresponding difference in displays in Fig. 2. Figure 3b was obtained from Fig. 3a by flipping the direction of three out of six gradients in the middle column of gradient patches. Therefore, each of the three diamonds in the left diamond column is now flanked at only three sides by darker (two black and one dark gray) gradient ends and at one side by a lighter (white) end, in various spatial arrangements, whereas the three diamonds in the right column are flanked at three sides by lighter (two light gray and one white) ends and at one side by a darker (dark gray) end, also in different arrangements. Thus here the surround contrast constellations for the diamonds are mixed, whereas in Fig. 3a the contrast has consistent direction at all four sides. In consequence, the difference in the appearance of the diamonds is appreciably smaller than in Fig. 3a, although the diamonds in the left column still look somewhat lighter than those in the right column. Finally, in Fig. 3c the diamonds in the left column are flanked by two white and two black gradient ends, in different arrangements, and the diamonds in the right column are flanked by two dark gray and two light gray ends. The differences between their lightnesses are negligible, in accordance with the fact that all diamonds involve opposed surround contrast constellations.

Figure 3d is constructed in the same general manner as the displays in Fig. 2, and consists of a bottom layer involving a chessboard-like arrangement of gradient patches, over which a top layer involving several types of uniform patches is superimposed. The figure contains a portion of the original Logvinenko wall-of-blocks display with uniform diamond tops, located slightly to the left from the central vertical axis; the rest of the figure demonstrates the consequences of certain geometrical transformations of the diamonds, directed towards the two ends of the figure. Thus in the direction towards the right end of the figure the diamonds are transformed into physically uniform vertical stripes, whereas towards the left end they are transformed into physically uniform horizontal stripes. Note how the 3D relief in the left-center of the figure gives way to different flat mosaic-like impressions towards the left and right end. In the left half of the figure the patches in different rows are kept segregated from each other, whereas proceeding towards the right end the patches get increasingly closer, then touch, and finally fuse into vertical stripes. This geometrical difference is accompanied by different lightness effects. In the left half, the patches exhibit discretely different uniform lightness levels, whereas in the right half there appears a non-uniformity in the perceived color of the patches, as they keep crossing increasingly more surround luminance levels, culminating in the wavelike variation of lightness in the vertical stripes.

The described manipulations serve to relate the wall of blocks illusion to some other lightness effects reported in the literature. The strong lightness illusion in the equiluminant horizontal stripes in the left end is reminiscent of another display by Logvinenko (1999, his Fig. 8a), as well as a figure studied by Agostini and Galmonte (2002), although they all used a single type of inducing gradients, whereas the present effect uses two types (dark and light), and a somewhat different geometric constellation. On the other hand, the appearance of induced illusory gratings in the uniform vertical stripes at the right end is closely related to the grating induction effect of McCourt (1982), except, again, for differences in grating types and geometric configurations.

How can the lightness effects in Fig. 3d be accounted for? Three possibilities may be considered. One approach is to regard them as versions of the grating induction effect and classical simultaneous contrast, basically involving variants of surround contrast constellations. The apparently somewhat weaker lightness illusion in the vertical than in the horizontal stripes may be due to the smaller area of contact of the vertical stripes with extreme luminance levels of the gradients. Another approach is to explain all effects on the basis of shadow-compatibility. One problem for this approach is how to plausibly account for the appearance of illusory gratings in the physically uniform vertical stripes. Finally, both principles of explanation may together be applied to the figure, with shadow compatibility, say, dominating in the left portion and grating induction in the right portion. However, it seems to me that a uniform principle of explanation of all lightness effects in this figure is preferable, especially as it was deliberately constructed with the aim of showing how effects otherwise considered as diverse may in fact be regarded as variations on basically the same theme, involving interactions of gradients with uniform patches.

THE CURVED STAIRCASE AND ITS VARIATIONS

Figure 4a is based on a remarkable display published by Purves and Lotto (2003, their Fig. 3.9), but uses a slightly different geometrical layout. The depicted object looks like a smoothly curved striped staircase. The dominant appearance is as if it is viewed from above, but the surface may also perceptually invert and appear as if viewed from below; in the first spatial interpretation the illumination also appears as coming from above, and in the alternative interpretation as coming from below. Figure 4b is my variation that has the same structure, except that it is checkered rather than striped. Both staircases have two vertical planar portions and two horizontal ones ('steps') and three rounded portions ('knees'); in the dominant spatial interpretation, the two outer knees appear convex and the middle, inner knee appears concave, but the directions of curvature are reversed in the alternative percept. The effect of interest here is that the four target regions indicated by arrows have the same uniform luminance, but the two located on the horizontal steps look darker than the two located on the vertical steps.

The luminance distributions in the two displays are clarified in Figs 4g to 4i. As depicted in Fig. 4g, the surfaces can be described through a curved grid, with six zigzagging columns, numbered 1 to 6, and seven rows, denoted by A to G. The luminance distributions along odd-numbered columns are mutually identical, as are the luminance distributions along even-numbered columns. Thick lines are depicted along two columns — a dashed one along column 2 as a representative of even numbered columns, and a full line along column 5 as a representative of odd numbered columns. The luminance distributions along these lines are schematically represented in Fig. 4h (for Fig. 4a), and in Fig. 4i (for Fig. 4b). The seven dashed line segments in the graph in Fig. 4h depict the luminance distribution along column 2 as the dashed line passes across seven rows (A to G), whereas the full line segments correspond in the same way to column 5. It can be seen that luminance along the steps (rows A, C, E, G) is uniform, involving three levels denoted as low, mid, and high, and that it is monotonic along the knees (rows: B — bottom outer knee, D — inner knee, F — top outer knee), which exhibit luminance gradients between the uniform levels. The two equiluminant, medium luminance patches at positions E2 and C5 in the grid are indicated by arrows in the graph in Fig. 4h. Turning to the graph in Fig. 4i, it can be seen that it is the same as in Fig. 4h, except that the full and dashed monotonic segments are switched. In other words, Fig. 4b was produced from Fig. 4a by interchanging the 'dark knees' (involving the gradient from low to medium luminance) and the 'light knees' (gradient from medium to high luminance) throughout the display, with the uniform patches forming the steps retaining their luminance levels.

Note that both the striped and the checkered staircase are shadow-compatible, as they evoke clear and consistent impressions of objects having a particular 3D geometry, painted with a particular achromatic texture, and exposed to uniform illumination (coming from above or below, depending on the spatial interpretation). Attending to the structure of the lighting, it is readily apparent that the horizontal steps receive a high level of illumination, that the vertical steps receive a lesser amount (consistent with subtending a sharper angle with respect to the direction of illumination), and that the luminance gradients across the knees correspond reasonably to the gradual change of surface orientation between the horizontal and the vertical steps. Of course, photographs of similar real surfaces under real illumination would likely have luminance structures of higher complexity, involving more complicated gradients, inter-reflections between surfaces, highlights, irregularities, etc. Note also that Fig. 4b is related to Fig. 1a, in that both involve checkered patterns which contain equiluminant regions that appear to receive different amounts of illumination; the difference is that in Fig. 1a this impression is due to conveyed variation of incoming illumination (shadow), whereas in Fig. 4b the impression is due to conveyed 3D surface curvature.

Purves and associates have developed a theoretical framework in which lightness percepts are based on the relative frequency of occurrence of all possible sources of the stimuli witnessed in the past (for detailed accounts of theory and data see Purves and Lotto, 2003; Purves *et al.*, 2004; Williams *et al.*, 1998a, b). Within this framework, the explanation of the lightness effect in the Purves and Lotto (2003) figure, referred to above, of which of Fig. 4a is a close variant, can essentially be based on the same idea as the illumination interpretation accounts of effects in

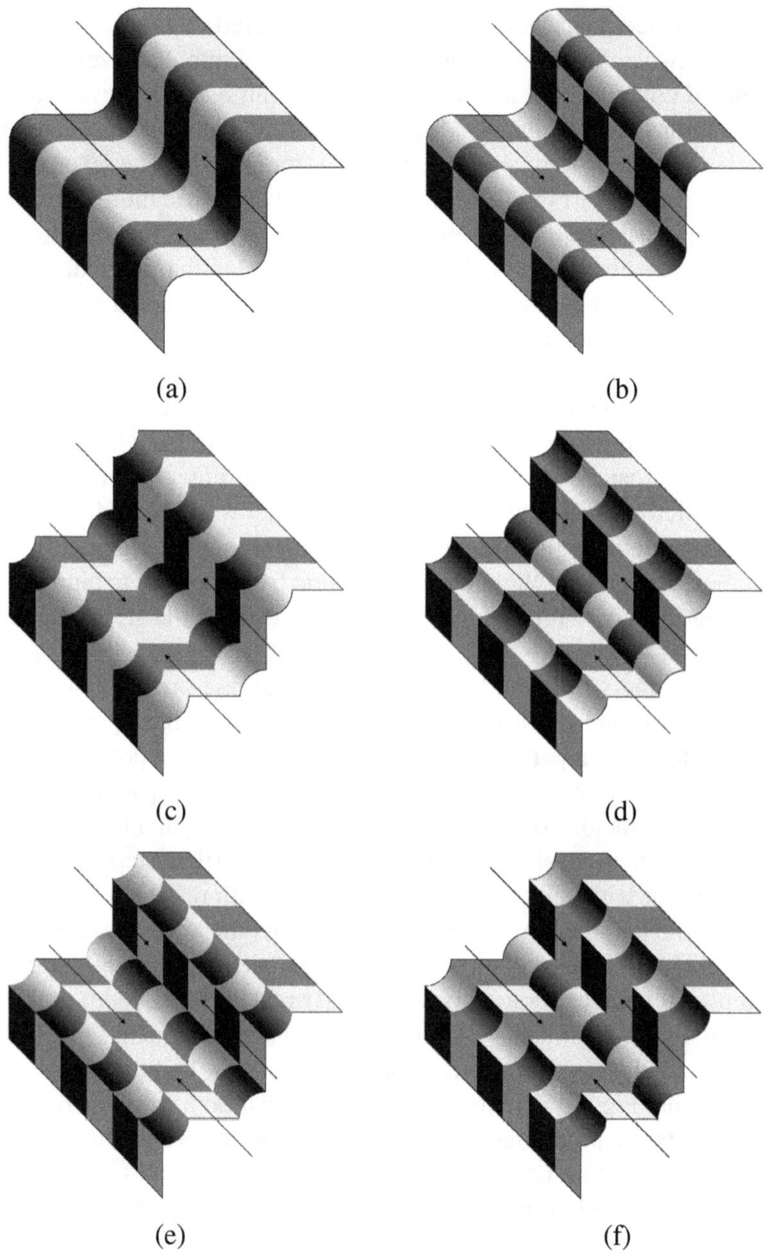

Figure 4. Lightness and depicted 3D shape. The target patches, pointed at by arrows, all have the same luminance, but patches in the horizontal plane look darker than the patches in the vertical plane. In (a) and (b) this difference may be accounted for by the presence of shading on a curved body illuminated by a homogeneous light source, but in (c), (d), (e), and (f) this explanation is less likely. Display (g) is a spatial scheme of figures (a) and (b), used for subsequent graphs that represent the luminance distributions of the previous displays, along two indicated lines. Display (h) is the graph of the luminance distribution of (a) and (c), display (i) is the graph for (b) and (e), display (j) is the graph for (d), and display (k) is the graph for (f). See text for details.

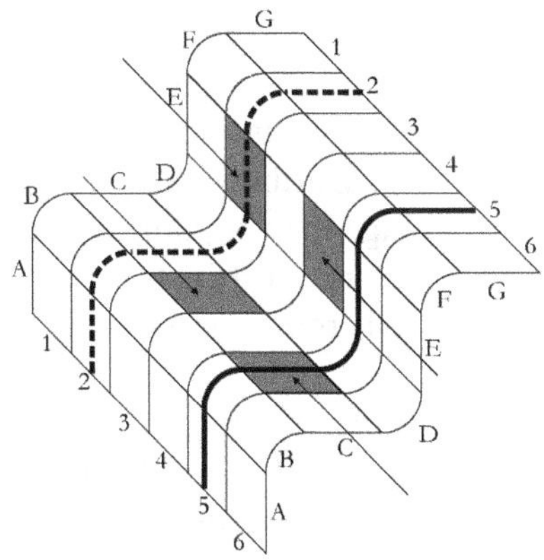

(g)

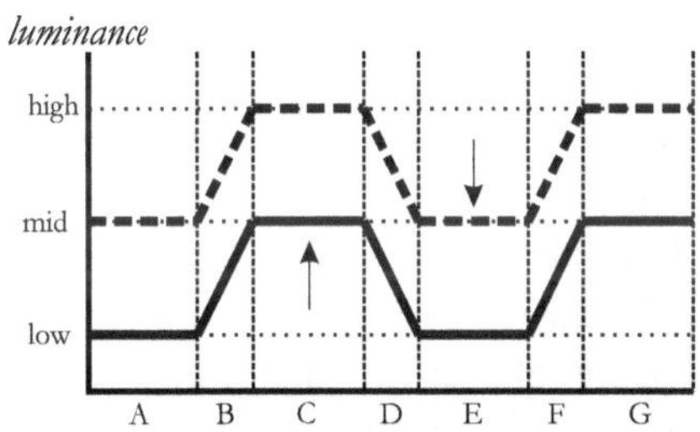

(h)

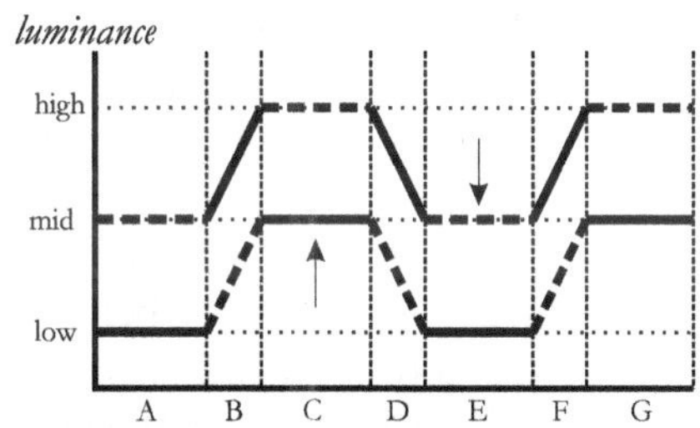

(i)

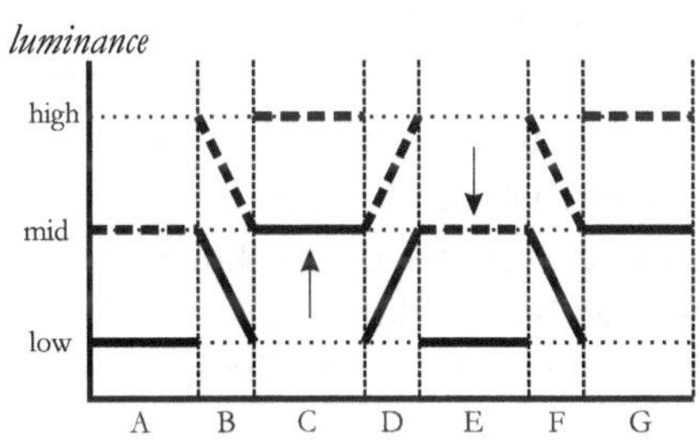

(j)

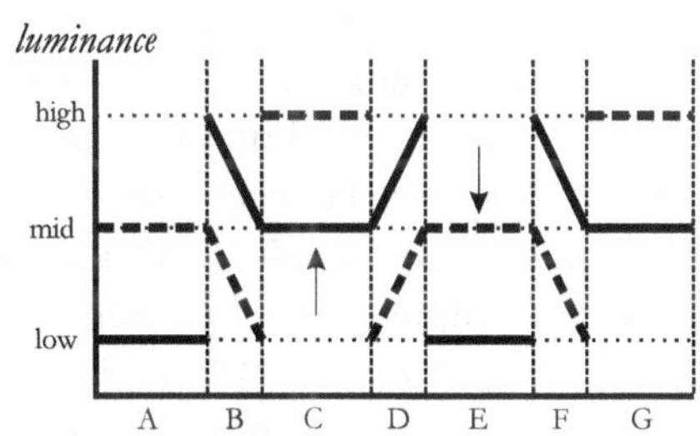

(k)

Figure 4. (Continued).

Figs 1a and 2a: the lightness difference between the targets can be explained by supposing that observers interpret the horizontal targets as well illuminated dark regions, and the vertical targets as poorly illuminated light regions. The same explanation can be applied to my version, Fig. 4b.

I will now present some simple geometric and photometric transformations of these two displays that convey objects with less straightforward shape and illumination, and challenge the illumination interpretation account of the lightness effects. These transformations apply only to the knees of the figures, and leave the steps unchanged. Figure 4c was generated from Fig. 4a by a geometric change, involving flipping the curvature of the contours of all knees. In Fig. 4d a photometric change was added, involving flipping the directions of the luminance gradient across the knees as well. Figures 4e and 4f were generated in analogous manner from Fig. 4b. The luminance distributions of Figs 4c and 4e are the same as for Figs 4a and 4b, respectively. The luminance distribution for Fig. 4d is presented in the graph in Fig. 4j, and was obtained from Fig. 4h by flipping the orientation of all monotonic segments of the graphs. The luminance distribution for Fig. 4f is presented in the graph in Fig. 4k, and was obtained from Fig. 4i in the same way, by flipping all monotonic segments.

In the following I will analyze the structure of displays 4c to 4f and relate it to the appearance of the target patches. Consider Fig. 4d first. It represents a different geometric object than Fig. 4a, because the formerly convex knees appear concave, and *vice versa* (although persistent observers may fleetingly re-invert them). The transitions between the steps and knees, which are geometrically and photometrically smooth in Fig. 4a, are abrupt in Fig. 4d. Nevertheless, because both the curvatures of the knees and the directions of their luminance gradients were inverted concomitantly, the conveyed object is just as shadow-compatible as the one in Fig. 4a, and can in all details be consistently interpreted as a striped stair-like (or perhaps butterfly-like) structure, with horizontal planar portions being well lit and vertical planar portions being poorly lit. In particular, note how the variation of luminance across all knees fits with this interpretation, in that the conveyed illumination is consistent with the way light falls upon the planar portions: the knees are most illuminated along those portions in which their orientation is most nearly horizontal, and as they curve away from this orientation, they gradually become least illuminated along those portions which are most nearly vertical. However, in spite of the same geo-photometrical logic, it appears that, compared to Fig. 4a, the lightness difference between target regions is somewhat diminished.

Turning now to Fig. 4c, note that flipping the curvatures of the knee contours without also flipping the directions of gradients within the knees (as was done in Fig. 4d) induces a deterioration of cohesion of both the spatial and the illumination structure of the conveyed scene. Due to the change in geometry, the inner knee which in the dominant interpretation appears concave in Fig. 4a, tends to appear convex in Fig. 4c. In consequence, the luminance variation across it, being the same as in Fig. 4a, conveys an illumination change inconsistent with the illumination of the adjoining horizontal and vertical steps (in contrast to Figs 1a and 1d, in which the illumination is consistent across steps and knees). The same conflict, but in reverse, involves the two outer knees. The global shape of the depicted body appears less well defined than in Fig. 4d, because the abrupt change of spatial orientation between planar and curved portions of the surface is not accompanied by correspondingly abrupt changes in luminance. Furthermore, the percept of the depicted object is unstable, in that the outer knees may relatively easily invert and appear convex. Nevertheless, in spite of these geo-photometric inconsistencies, the lightness illusion seems stable and appears much the same as in Fig. 4a. In sum, in comparison to Fig. 4a, Fig. 4c appears less shadow-compatible but exhibits a similar lightness illusion, whereas Fig. 4d appears equally shadow-compatible, but exhibits a diminished illusion.

Consider now Fig. 4e, and note that the lightness difference between the targets appears relatively strong, and much the same as in Fig. 4b. However, the display poses similar interpretational problems as Fig. 4c, presenting them in a perhaps more salient form. In the dominant spatial interpretation, the inner knee appears convex and the outer knees concave, just as in Fig. 4d, but inverted compared to Fig. 4b. However, the luminance variations across the knees are inconsistent with the luminance across the steps: although for the steps the illumination seems to come from above, so that the horizontal steps appear well illuminated and the vertical steps appear more weakly illuminated, for knees the illumination seems to come from left rather than from above, because they appear least illuminated when they are most nearly horizontal and best illuminated when they are most nearly vertical. Furthermore, similar to Fig. 4c, the percept of the depicted object is unstable, in that the outer knees may relatively easily invert and temporarily appear convex, just as the inner knee. When all knees thus appear convex, the spatial structure of the object looks contorted and its parts seem oddly geometrically incongruent with each other, since the outer knees and the inner knee, although parallel in 2D, do not appear parallel in 3D. However, note that these geometrical instabilities do not affect the lightness illusion in the least. In sum, in contrast to Fig. 4b which is shadow-compatible, in that its geometry and photometry are mutually consistent, Fig. 4e is shadow-incompatible, but the lightness illusion in both figures has the same strength.

Finally, consider Fig. 4f. It differs from Fig. 4e only in that the gradient directions across all knees are inverted. This seemingly modest change has caused a surprisingly drastic difference in the global structure of the figure, as well as in the local appearance of the target regions, whose lightness difference has greatly diminished, compared to Fig. 4e. Because the spatial interpretation of this display is somewhat unclear, it does not present a challenge to the illumination interpretation account.

To summarize the preceding discussions, although the illumination interpretation account applies rather naturally to Figs 4a and 4b, which are both shadow-compatible and both show the lightness illusion, it is challenged in the related Figs 4c, 4d, and 4e, because the presence of shadow-compatibility is dissociated from the strength of the illusion. Contrary to the predictions based on shadow-compatibility, according to which the strength of the illusion should be retained in the shadow-compatible Fig. 4d and reduced in the shadow-incompatible Figs 4c and 4e, the illusion in fact persists largely unchanged in the shadow-incompatible displays, but appears somewhat diminished in the shadow-compatible display.

How may the lightness effects in these figures be accounted for? Consider first Figs 4b and 4e. Taken together, they present perhaps the most direct challenge to the illumination interpretation account in this figure set, because they exhibit clear and strong identical lightness effects, although one figure is shadow-compatible and the other is not. Note that in both these figures the lighter appearing targets are surrounded at all four sides by darker patches, whereas the darker appearing targets are surrounded at all four sides by lighter patches. This is another case of a surround contrast constellation. Figure 4e differs from Fig. 4b through the flip of knee curvatures, a manipulation which affects the interpretation of the global spatial structure and illumination of the represented object, but does not change the surround relations of the target patches.

Consider now Figs 4a and 4c. In both of them the vertical, lighter appearing targets are flanked at two longer sides by darker patches, whereas the horizontal, darker appearing targets are flanked at two longer sides by lighter patches. These configurations are *partial* surround contrast constellations, as they involve only two sides of the targets. At the other two sides the targets blend continuously into gradients, the lighter appearing targets into light gradients and the darker appearing targets into dark gradients. One way in which such constellations might contribute to the lightness illusion could be as follows. Suppose that there is a tendency towards increased uniformity of lightness within continuous regions, such as the stripes in Figs 4a and 4c that may contain gradients but no abrupt luminance changes. The neural basis for such a tendency might consist in lateral facillitory interactions, similar to filling-in processes, confined within stimulus regions delineated by luminance steps (see Grossberg and Todorović, 1988; Pessoa and DeWeerd, 2003). Note that the horizontal targets have the highest luminance within dark stripes, and that the vertical targets have the lowest luminance within the light stripes. If such a homogenizing tendency were at work, its effect would be to shift the lightness of all portions within a stripe towards the mean of the stripe, thus decreasing the lightness of the horizontal target patches and increasing the lightness of the vertical target patches. I will label such conditions 'surround assimilation constellations'. Thus the target patches in these figures exhibit both partial surround contrast and assimilation constellations. These constellations are congruent, in that the vertical targets are expected to lighten and the horizontal targets to darken both through contrast and assimilation of this type. Note that neither of the two postulated contributing factors depends on the curvature of the knees, which means that no difference is predicted between the effects in Figs 4a and 4c, unlike in the illumination interpretation account, which should predict a smaller effect in Fig. 4c, because of shadow-incompatibility.

Consider now Fig. 4d. Note that the vertical targets are bordered at their two longer sides by darker patches and at their two shorter sides by lighter patches, and that opposite relations hold for the horizontal targets. Therefore all targets are adjoined on their short and long sides by regions of oppositely directed contrasts, thus exhibiting incongruent partial surround contrast constellations. In consequence, compared to Figs 4a and 4c, a decrease of the lightness difference between the targets is expected, its exact amount depending on the relative strength of the two opposing tendencies.

Finally, consider Fig. 4f. Note that, as in Figs 4a and 4c, the surround constellation of the targets involves both contrast and assimilation. However, here these two factors are incongruent, in that the horizontal targets should be darkened by contrast but lightened by assimilation, and *vice versa* for the vertical targets. Therefore the prediction is that the difference in appearance of the vertical and horizontal targets should be small, as is indeed the case.

Figures 5a to 5d are geometrical variants of Figs 4a and 4b, using rectangular (5a, 5b) and circular (5c, 5d) arrangements of similar basic elements, both in striped and in checkered versions. As in Figs 4a and 4b, the equiluminant target regions, indicated by arrows, look different, though the effect appears weaker for Fig. 5a. Compared to Figs 4a and 4b, the shapes and local relations between individual patches are changed only modestly, but the global constellations are rather different, and Figs 5a to 5d look much flatter. However, there may still exist impressions of surface undulation, coupled with impressions of some linear or circular variations of surface illumination. Therefore one might still argue for an illumination interpretation account of the lightness effects, involving light but poorly lit and dark but well lit targets.

I will challenge the illumination interpretation account for these figures in a different way than in Fig. 4, by using stereoscopic presentations. Figures 5e and 5f provide stereoscopic versions of Figs 5b and 5c. In Fig. 5e the two rows containing the target patches are shifted laterally and occupy a different stereoscopic depth plane than the rest of the figure. As the reader may confirm using free fusion, the lightness effect remains much the same as in non-stereoscopic observation, although the impression of illumination upon a common plane is disrupted, a fact which should disrupt the illumination interpretation. In Fig. 5f, some of the rings are shifted laterally, generating a radically changed geometrical structure of the represented scene, as observed stereoscopically. As the reader may confirm, stereoscopic views provide impressions of upright coaxial cylinders or truncated cones viewed from above, resting on a plane disk, with the cylinders/cones looking light and the disk dark, or *vice versa*. However, the lightness difference between the target regions appears much the same as in the non-stereoscopic view, although it is hard to think of a plausible way to account for it, in the stereoscopically conveyed scene, in terms of some interaction of illumination and geometry.

INTERIM DISCUSSION

To summarize briefly the preceding displays and arguments: I have presented a number of close replicas or slight variations of three influential displays — Adelson's checkered shadow (Fig. 1a), Logvinenko's gradient wall of blocks (Fig. 2a), and Purves and Lotto's smooth staircase (Fig. 4a) — and have also contributed a few designs of my own (Figs 4b, and 5b to 5d), which all involve strong lightness illusions for which plausible illumination interpretation explanations can be proposed.

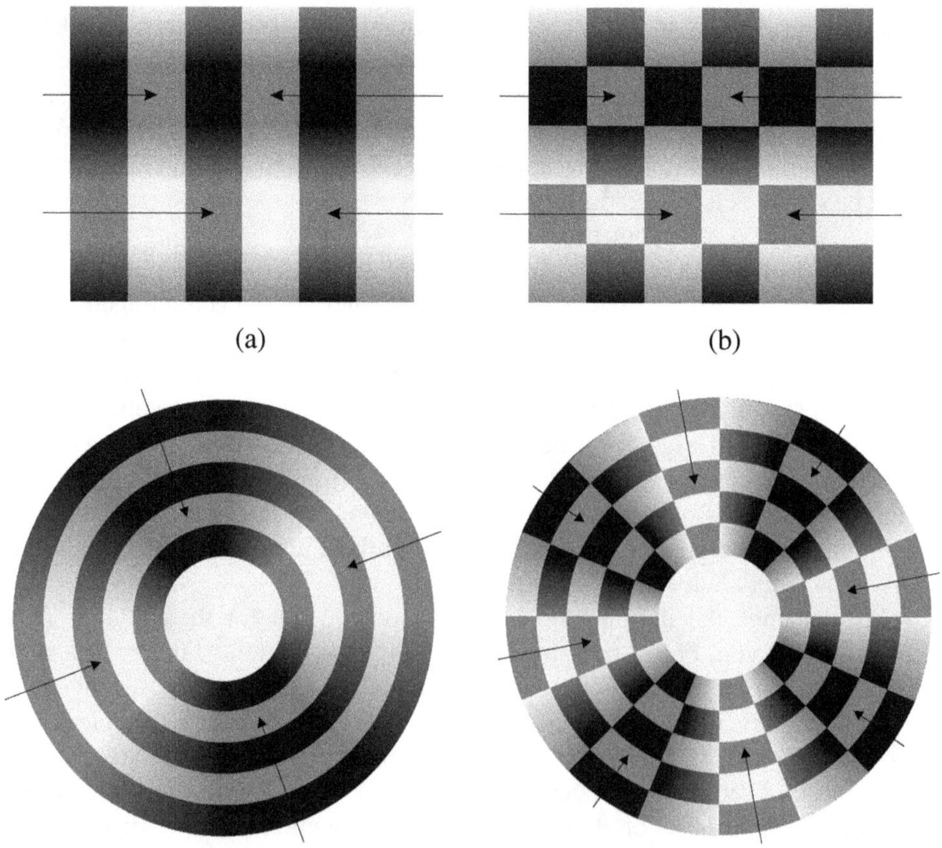

(a) (b) (c) (d)

Figure 5. Displays (a) and (c) are geometrical variants of display 4(a), and displays (b) and (d) are geometrical variants of display 4(b). The target patches, pointed at by arrows, all have the same luminance, but look different. Displays (e) and (f) are stereoscopic variants of displays (b) and (c), in which the perceived geometry is different than in non-stereoscopic displays but the lightness effect is retained. See text for details.

I have also mustered a host of carefully constructed 'counter-illusions', that is, control variants of these images whose purpose was to challenge such explanations in various ways. I believe that the presented evidence, although it certainly would need to be backed by experiments, argues against the possibility that this class of illusions could be accounted for solely based on the notion of illumination interpretation, at least in the way that I have interpreted this notion. However, this conclusion does not mean that the general relevance of illumination in the perception of lightness is brought into doubt. It is well established that appreciation of illumination is a decisive factor for lightness determination in 3D scenes involving real depth and real light sources (Kardos, 1934; Gilchrist, 1977, 1988; Gilchrist *et al.*, 1999). The topic under consideration here is not the overall role of illumination but rather whether processes activated in real scenes contribute to lightness determination in pictorial

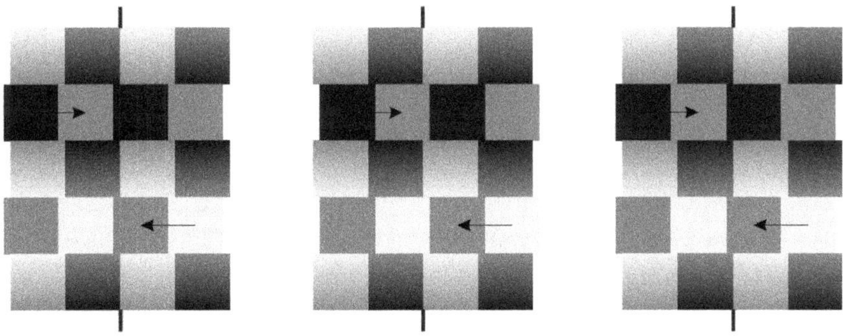

(e)

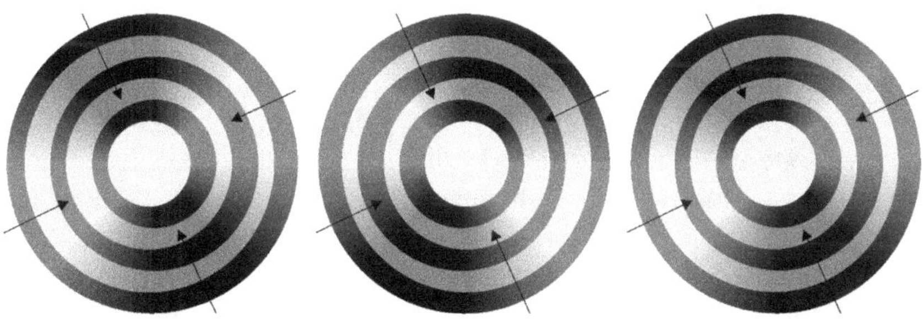

(f)

Figure 5. (Continued).

displays such as those presented in this paper, which are flat and uniformly illuminated, but convey depth and variable illumination through static pictorial cues. Note that a robust cue for illumination variation in real scenes, in contrast to pictorial displays, is the fact that the value of the ratio of maximum to minimum luminance may far exceed the value found in images, which may rise only up to about 30 : 1.

One might argue that whereas illumination interpretation may not be the sole cause of these lightness illusions, it still may be one of its determinants, among other contributing factors. The idea is that while one may accept that factors like the surround contrast constellation may affect lightness, still illumination interpretation would add something over and above other causes of appearance of image patches. While this remains a general possibility, according to the arguments I have presented even this more modest proposal is brought into doubt for the displays analyzed here. Note that I have discussed a number of cases in which the illusion in the original display appears to be plausibly accounted for through illumination interpretation, but where in the modified, control display this interpretation is challenged, although the illusion itself remains unaffected. I would argue that in such a case, given that both the structural aspects of the display as well as the illumination cues are well controlled, it is not only that the illusion in the modified display cannot be accounted for as being caused by illumination interpretation; in addition, its presence in the original display cannot have been, even in part, due to this cause either. This is because, had illumination interpretation played any role there, the reduction of illumination cues in the modified display should have caused a reduction in the illusion strength, but it did not. For an example of analogous reasoning, suppose that it is first established that a mixture of substances X and Y causes drowsiness, but that a subsequent study shows that X alone causes the same amount the drowsiness; then one would not argue that in the first experiment Y might have affected the outcome to some extent, over and above X, but would more parsimoniously conclude that Y had no effect on drowsiness whatsoever.

A different line of argument against my criticisms would be to abandon the general notion that illumination interpretation is based on detailed appreciation of ecologically valid, mutually consistent cues for illumination, acquired through prolonged observations of all environmental regularities concerning the interactions of light and surface shapes and their reflectance patterns over the whole visual field. Alternatively, the putative illumination interpretation mechanisms might use only some and not all potentially valid cues, which could be rough rather than optically accurate in detail, and / or provide spatially local rather than globally consistent heuristic indications of illumination. In that case, the manipulation of illumination cues which may be physically appropriate but are not actually used by the visual system would have little perceptual relevance, detailed agreement with optical regularities would not be necessary for such mechanisms to do their job, and they might not be bothered by displays containing shadow-incompatibilities of the sort that I have analyzed here. Thus, as I have argued in more detail in discussing another set of images in Todorović (2003a, b), rather than refuting the illumination interpretation account, the counter-examples that I have marshalled against it might only serve to amend it, pointing out the limitations of the global, all-encompassing variant of illumination interpretation, by suggesting which image features such mechanisms are less likely to respond to.

It is indeed possible that illumination interpretation mechanisms might not be based on elegant, comprehensive, and interconnected statistics but on rough, primitive, and isolated bag-of-tricks type heuristics. However, I see several problems with this explanatory strategy. For example, the visual system is known to otherwise exhibit exquisite sensitivity to some aspects of interactions of luminance and geometry, such as in shape-from-shading interpretation processes, and it seems surprising that it should forsake this sensitivity in the process of lightness assignment. Furthermore, the original simplicity, elegance, and parsimony of the illumination interpretation notion, such as in explaining the effects in Figs 1a, 2a, 4a and 4b, would be compromised by such a move, which would make the theory more flexible but less testable. A premier task for its proponents would be to propose and check which among the many potentially available optical features of illumination might actually be causally effective in lightness perception in flat images, and which might not. It would also be interesting to know why the visual system should use some cues rather than others, instead, say, the other way around, given that all are potentially available. The cues actually used would presumably be those which are most robust and useful, but the question is which ones would those be, and whether they are indeed picked up by the visual system. In the preceding discussions I have argued that a number of plausible illumination cues do not actually affect lightness. It remains to demonstrate which ones do. Furthermore, an additional task would be to show that a putative heuristic illumination cue is in fact based on regularities concerning illumination, and is not just spuriously correlated with such regularities in particular displays, so that it could be disassociated from them in suitable control displays, as I have shown here. If in assessing lightness the visual system is not thought to respond to illumination in a comprehensive manner, but only particular illumination heuristics are deemed effective, whereas other, potentially similarly informative, are not, then one would need additional arguments to back up the claim that the effective ones are actually based on interpreting illumination.

QUALITATIVE EFFECTS OF LUMINANCE GRADIENTS

The preceding displays have, in one way or the other, involved the role of luminance gradients in the determination of lightness of uniform regions. In the following I will present replicas, variants, as well as some new designs which demonstrate that gradients may affect not only the lightness but also the mode or quality of appearance of uniform regions (Beck, 1972). Basic effects of this type were reported by Evans (1959), Fry (1931), Kanizsa (1980), MacLeod (1947), and Zavagno (1999). Here the phenomenon will be demonstrated by comparing similar displays with and without gradients, as in Fig. 6. Figure 6a presents a white disk on black ground with a sharp, step-shaped luminance border. Figure 6b presents the same disk, but surrounded by a ramp-shaped luminance ring, in the form of a luminance gradient between the white figure and the dark background; faint Mach rings bordering the ramp region may be noted. Approximate luminance profiles of the two displays are presented in Fig. 12 as stimuli labeled Inc1 and Inc2. Analogous images involving luminance inverses of these figures are presented in Figs 6c and 6d.

The phenomenon of interest here is that disks of identical luminance appear qualitatively somewhat different, depending on whether they are delimited by a sharp border or a gradient. However, it is not straightforward to express this difference verbally. To contrast their appearances, describing the interior surfaces of the step-bordered disks (Figs 6a and 6c) one might use terms like *clear* or *opaque* (and perhaps also *hard*, *cold*, *compact*, *smooth*, and *planar*), whereas the interior surfaces of the ramp-bordered disks (Figs 6b and 6d) may be said to look more *hazy* or *fuzzy* (and perhaps also, in contrast to Figs 6a and 6c, as *softer, warmer, diffuse, wooly*, and *ruffled*). In the following I will mainly refer to this effect as the 'clear *vs.* hazy' distinction. Furthermore, in the white ramp-bordered disk there is also some sense of *translucency* and *luminosity* or *glow*, whereas in the dark ramp-bordered disk the corresponding quality is hard to label. There also appears to be a lightness difference between the equiluminant step-bordered and ramp-bordered disks, but because of the qualitative differences in their appearance, its direction is somewhat

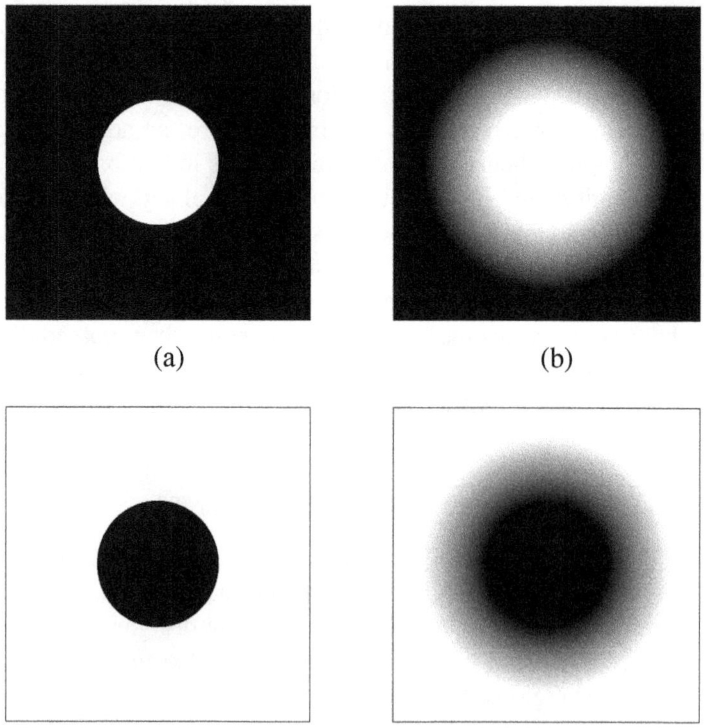

(c) (d)

Figure 6. Effects of luminance gradients on qualitative appearance, part I. Sharp-bordered patches with full (a, c) or partial (e, g) borders look qualitatively different than corresponding ramp-bordered patches with full (b, d) or partial (f, h) borders. See text for details.

hard to pin down unambiguously. However, MacLeod (1947), who used gray disks, found that ramp-bordered disks on white background looked lighter and on black background looked darker than corresponding gray step-bordered disks. This is in accord with the homogenizing tendency, postulated in analyses in previous sections. I will now draw connections between these and several other effects, some related to ones reported before and some new.

Figures 6e and 6g present modifications of Figs 6a and 6c, involving the introduction of trapezoidal stripes oriented perpendicular to the disks, which have the same luminance as the disks, that is, they are white in Fig. 6e and black in Fig. 6g. The disks are now physically defined only along 50% of their perimeters, but perceptually one still has a relatively clear impression of their presence, although half of their contour is virtual. This is a variant of the well-known Ehrenstein illusion (Ehrenstein, 1941; Spillmann *et al.*, 1976). Identical uniform trapezoidal stripes are also introduced in Figs 6b and 6d to form Figs 6f and 6h, respectively, making the corresponding disks semi-virtual as well.

The effect to be noted here is that the difference in appearance between step-bordered and ramp-bordered semi-virtual disks is rather similar to the difference

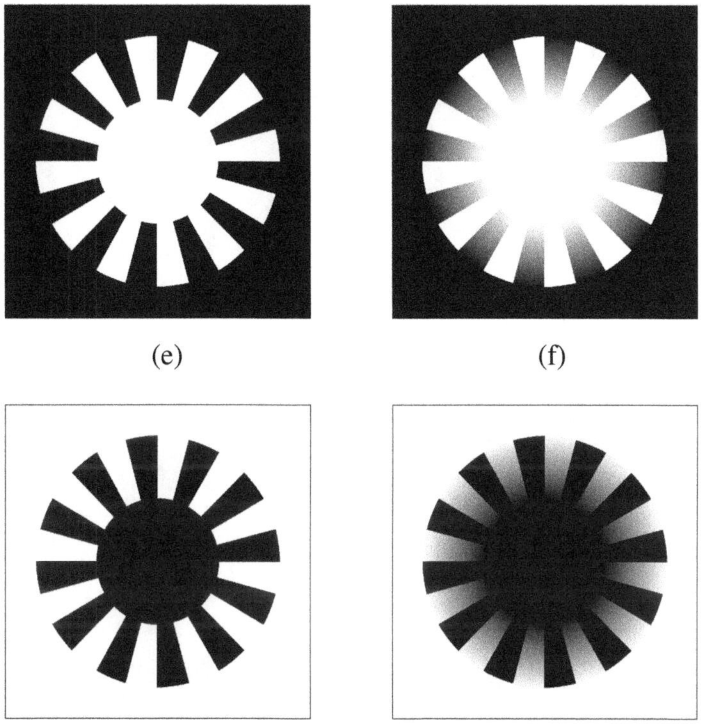

(e) (f) (g) (h)

Figure 6. (Continued).

between corresponding real disks, in that the clear *vs.* hazy distinction again applies, and the white disk again looks luminous. The effects present in Figs 6f and 6h are versions of phenomena originally reported in several variations by Zavagno (1999), and are related to a phenomenon described earlier by Kennedy (1976). The juxtaposition of Figs 6f and 6h with Figs 6b and 6d serves to point out the close similarity between these effects, suggesting similar mechanisms behind them. Note also how the virtual contours of the disks in Figs 6e and 6g are replaced by some dust-like impressions within the physically uniform trapezoidal stripes in Figs 6f and 6h. One may speculate that both the sharp virtual contours and the spread-out virtual dust are based on the same neural mechanisms, triggered in the first case by the real sharp luminance edges and in the second by the real luminance gradients.

Another format for demonstrating the effects of gradients on qualitative appearance, as well as for linking them to effects on lightness, is used in Fig. 7. It contains three sets of figures, each composed of three displays. Each display consists of five columns of three non-uniform oval disks; columns 1, 3, and 5 are identical, and columns 2 and 4 are their mirror images. The effects of interest in all displays do not concern the disks themselves but the uniform backgrounds surrounding them, in the four interspaces between the five columns, numbered 1 to 4 and pointed at

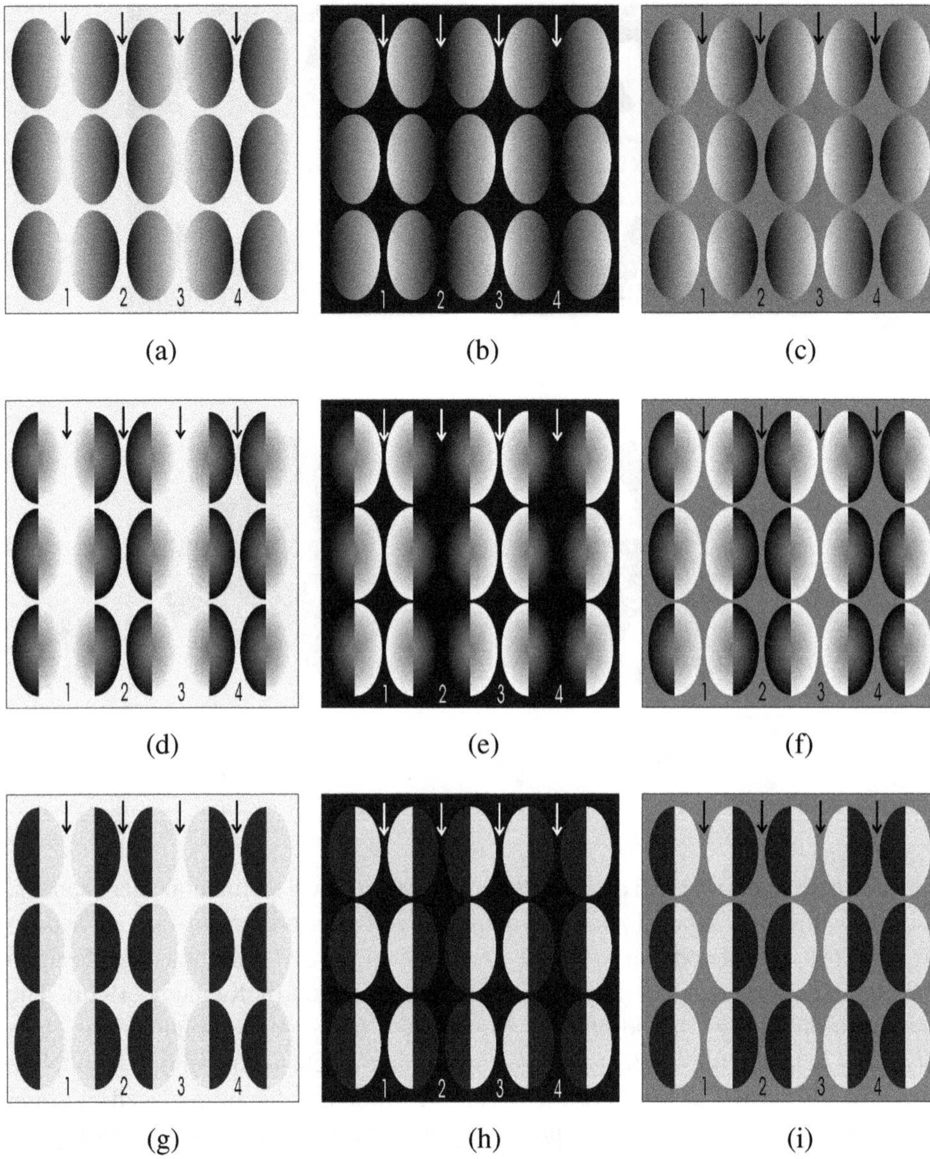

Figure 7. Effects of luminance gradients on qualitative appearance, part II. Three types of inducing elements are placed on white, black, and gray backgrounds. The clear versus hazy contrast is strong on white backgrounds and somewhat weaker on black backgrounds, but it disappears on gray backgrounds, which only involve lightness effects. See text for details.

by arrows. Note that some of the effects may appear more salient if viewed on the computer screen.

The disks in the first set of displays, Figs 7a, 7b, and 7c, contain horizontal luminance gradients spanning the range from black to white; the directions of the gradients alternate between neighboring columns of ovals, being (from left to right) dark to light in the first column, light to dark in the second, etc. The disks and their arrangement in all three figures are identical, and only the shade of the background varies, being white in Fig. 7a, black in Fig. 7b, and gray in Fig. 7c. The difference in the appearance of disks, which is not of immediate concern in this section, is likely due to surround contrast constellations. What is of interest is that although all backgrounds have uniform luminance they do not appear uniform, and this non-uniformity is different in different figures. In particular, note that in Fig. 7a, where the background is white, impressions of whitish haze as well as glow appear within the odd-numbered interspaces, at places where the white ends of the disk gradients blend into the uniform white background, whereas within the even-numbered interspaces the background looks clear. In Fig. 7b, which contains identical ovals, but on black background, the locations of appearance of hazy and clear portions are switched, compared to Fig. 7a: blackish haze appears within the even-numbered interspaces, at places where the black gradient ends blend into the black background, whereas the background looks clear within the odd-numbered interspaces; however, the appearance of haze may be less salient than in Fig. 7a. In Fig. 7c the background has intermediate luminance and the gradient ends do not blend into it. In consequence, the clear *vs.* hazy difference has vanished, giving way to a lightness difference, such that the background within the odd-numbered interspaces looks darker than the background within the even-numbered interspaces. Note that this is still another variant of a surround contrast constellation, in that the background regions bordering the lighter gradient ends appear darker than those bordering the darker gradient ends. Thus the lightness effects in Fig. 7c are related to the corresponding effects in several previous figures involving surround contrast constellations.

The second set of displays, Figs 7d, 7e, and 7f, is similarly structured, but involves ovals with luminance gradients directed radially instead of horizontally, starting out as middle gray in the center and proceeding towards black in one half of the oval and towards white in the other half. This feature enables the disks to blend into the background along half of their perimeter, rather than only at one point, as in Figs 7a and 7b. As in the first set of figures, all three displays contain identical ovals and only the background is different, being white in Fig. 7d, black in Fig. 7e, and gray in Fig. 7f (in which the structure of the ovals may be appreciated most easily, since they do not blend into the background at either end). The induced haze effects in the backgrounds in Figs 7d and 7e appear at the same locations as in Figs 7a and 7b, but are uniform throughout and rather strong, although haze may appear more saliently on the white background. Furthermore, in both Figs 7d and 7e the impression of haze is accompanied by impressions of transparency, in the form of translucent vertical stripes. In contrast, on the intermediate luminance background (Fig. 7f) both the haze and the transparency have disappeared, and are replaced by a lightness illusion involving the interspaces, which has a similar structure as the effect in Fig. 7c, and is associated with a similar type of surround contrast constellation.

The luminance gradients in all preceding figures were relatively smooth, involving dozens of gray levels. The question arises whether similar effects can be produced by coarser gradients. I will only present one case of the coarsest possible 'gradient', which is a simple luminance step involving only two gray levels. Figures 7g, 7h and 7i use bi-colored ovals involving luminance steps whose direction corresponds to the global direction of luminance change of the gradients in preceding sets of figures, being dark to light in the first column of ovals, light to dark in the second, etc. The ovals are identical in all three figures, but the background is white in Fig. 7g, black in Fig. 7h, and gray in Fig. 7i. The effects of interest here are as follows. In Fig. 7g uniform whitish haze as well as impressions of transparency appear, located in the same interspaces as the corresponding effects in Figs 7a and 7d; however, the impression of glow is absent. Similar, though less salient effects, may be noticed by inspecting Fig. 7h and comparing it with Figs 7b and 7e. Finally, in Fig. 7i, both haze and transparency are gone, and are replaced by a lightness illusion involving a surround contrast constellation, similar as in Figs 7c and 7f. The appearance of perceived transparency in Figs 7d, 7e, 7g and 7h is consistent with the classical Metelli rules, in that the luminance contrasts within the transparent portions are smaller than outside of them; however, transparency cannot be directly deduced from the rules because there are no X-junctions in the displays.

In sum, the displays in Fig. 7 demonstrate that luminance gradients may induce qualitative differences in the appearance of uniform regions, such as haze and transparency, if they blend into them (Figs 7a, 7b, 7d, 7e), in which case glow also appears, or show low edge contrast (Figs 7g, 7h), in which case there is no glow, but not when the contrast is high (Figs 7c, 7f, 7i), in which case they may instead induce lightness differences, depending on surround contrast constellations.

For comparison, note that very similar impressions of haze and transparency can be observed in Fig. 8a, which is a replica of a display devised by Adelson (2000, Fig. 24.17), who linked the appearance of these effects to the presence of particular X-junctions. However, such junctions are absent in Fig. 7g, which indicates that they are not strictly needed to generate impressions of transparency and haze in

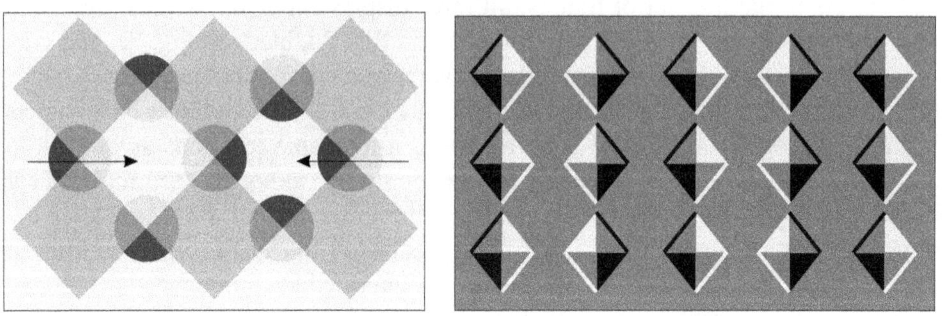

(a) (b)

Figure 8. Two figures by Adelson (1993, 2000), used for comparisons with Figs 7g and 7i. See text for details.

this type of displays, and that simple luminance steps may suffice, combined with decreased figure-ground edge contrasts. It is also interesting to compare the displays involving backgrounds of intermediate luminance to Fig. 8b, which is based on another display by Adelson (1993, Fig. 4c). It contains a set of relatively complex figures, arranged in five columns of three elements each, similar as for the displays in Fig. 7. The background has uniform luminance, but involves a similar lightness illusion within interspaces as in Figs 7c, 7f, and 7i. Adelson (whose original figure shows a somewhat stronger lightness difference, probably because the elements touch vertically) has explained this effect by invoking a variant of the illumination interpretation account, linking it to the impression of transparency in his display. However, a closer analysis of Fig. 8b reveals that it, like Figs 7c, 7f, and 7i, involves a surround contrast constellation (albeit of a somewhat more intricate variety), in that the darker appearing portions of the background are bordered by lighter halves of the figural elements, and *vice versa.*

Figure 9 presents another set of figures demonstrating the qualitative effects of gradients. Consider first Fig. 9a, a display that roughly resembles a window consisting of a wooden frame and three vertical window panes; the frame contains uniform regions, which look planar, and gradients, which convey 3D relief. The uniform regions in the display have the same luminance, but the frame region, which blends into the gradients, looks hazy, whereas the three panes look clear. A similar difference can be noted between the corresponding regions in Fig. 9b, which is the luminance inverse of Fig. 9a. Furthermore, in these displays the lightness difference between the equiluminant hazy (frame) and clear (panes) portions appears more salient than in previous displays, in that the clear white portions look lighter than the hazy white portions (Fig. 9a), whereas the clear black portions look darker than the hazy black portions (Fig. 9b); this is probably due to surround contrast constellations involving the panes. The luminance profiles and the lightness effects in these figures are related to some variants of the Craik-O'Brien-Cornsweet effect (see Todorović, 1987).

Figures 9c and 9d are versions of Figs 9a and 9b that use a very coarse luminance gradient involving only two levels (light gray — dark gray). Note that in these coarse-gradient displays similar differences between the frame and the panes obtain as in the fine-gradient displays, though both the lightness difference and the haze effect appear somewhat reduced. These displays suggest a relation of the presented phenomena to the watercolor effect (Pinna *et al.*, 2001). Figure 9e presents an originally chromatic example of this effect, which was converted to gray scale, and Fig. 9f presents the corresponding luminance inverse. The achromatization has decreased the strength of the original chromatic effect, but a difference between 'frame' and 'panes' regions may still be noticed, especially if viewed on the computer screen. A close inspection of these figures reveals their geometrical and photometrical similarities to Figs 9c and 9d, including the presence and direction of the coarse gradients. The structural similarities between the displays and the phenomenal similarities of the effects presented in Fig. 9 suggest that they may all

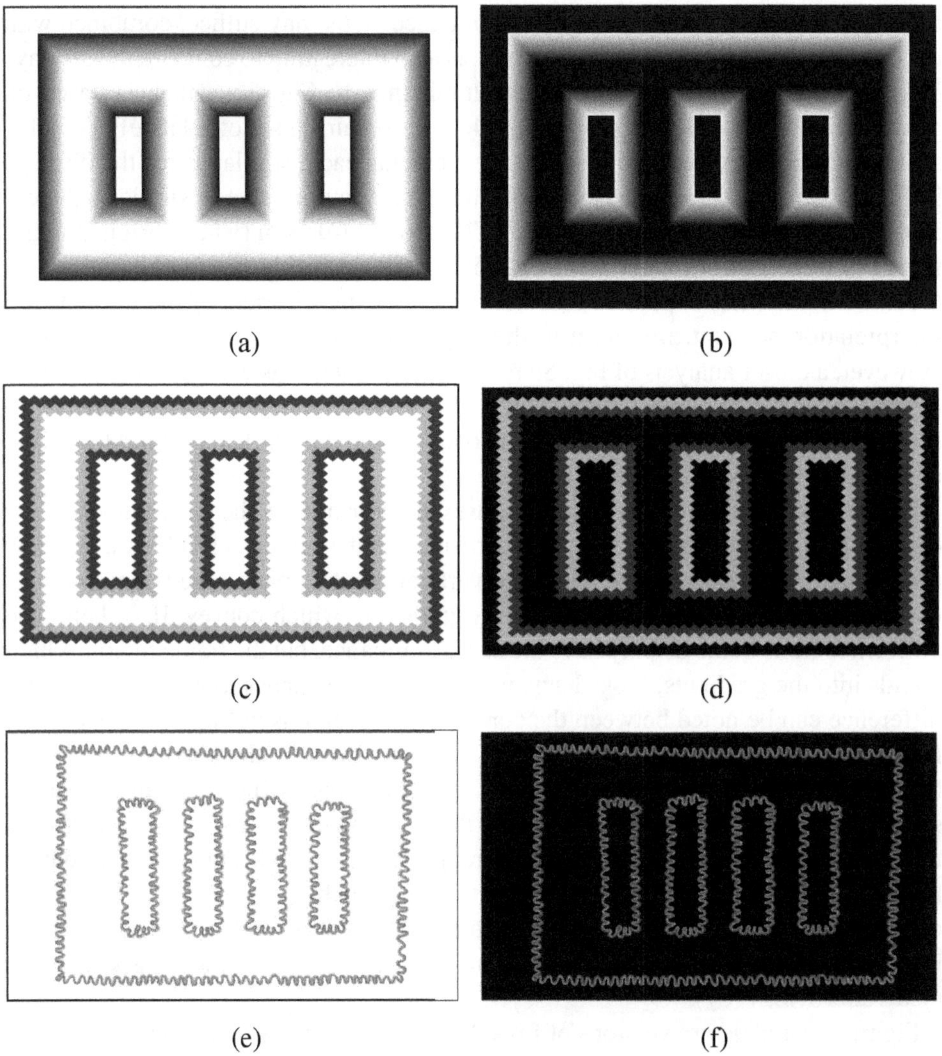

Figure 9. Effects of luminance gradients on qualitative appearance, part III. The haze effect seems strongest for broad and fine gradients (a, b), weaker for broad and coarse gradients (c, d), and weaker still for narrow and coarse gradients (e, f). See text for details.

be caused by common mechanisms based on gradients, either fine-grained or coarse, bordering uniform regions and displaying different luminance contrasts with them.

Figure 10 contains the final set of displays in this section, serving to summarize similarities and differences between many preceding figures, involving a display format useful for a comparative overview of the effects. Consider first Figs 10a, 10b, and 10c. The patches labeled A and B are equiluminant uniform gray targets which appear differently light. The targets are mostly surrounded by gradient patches; the arrows within the gradient patches indicate the directions of luminance increase. These figures have close structural relations to various displays in Figs 1 to 5. Thus

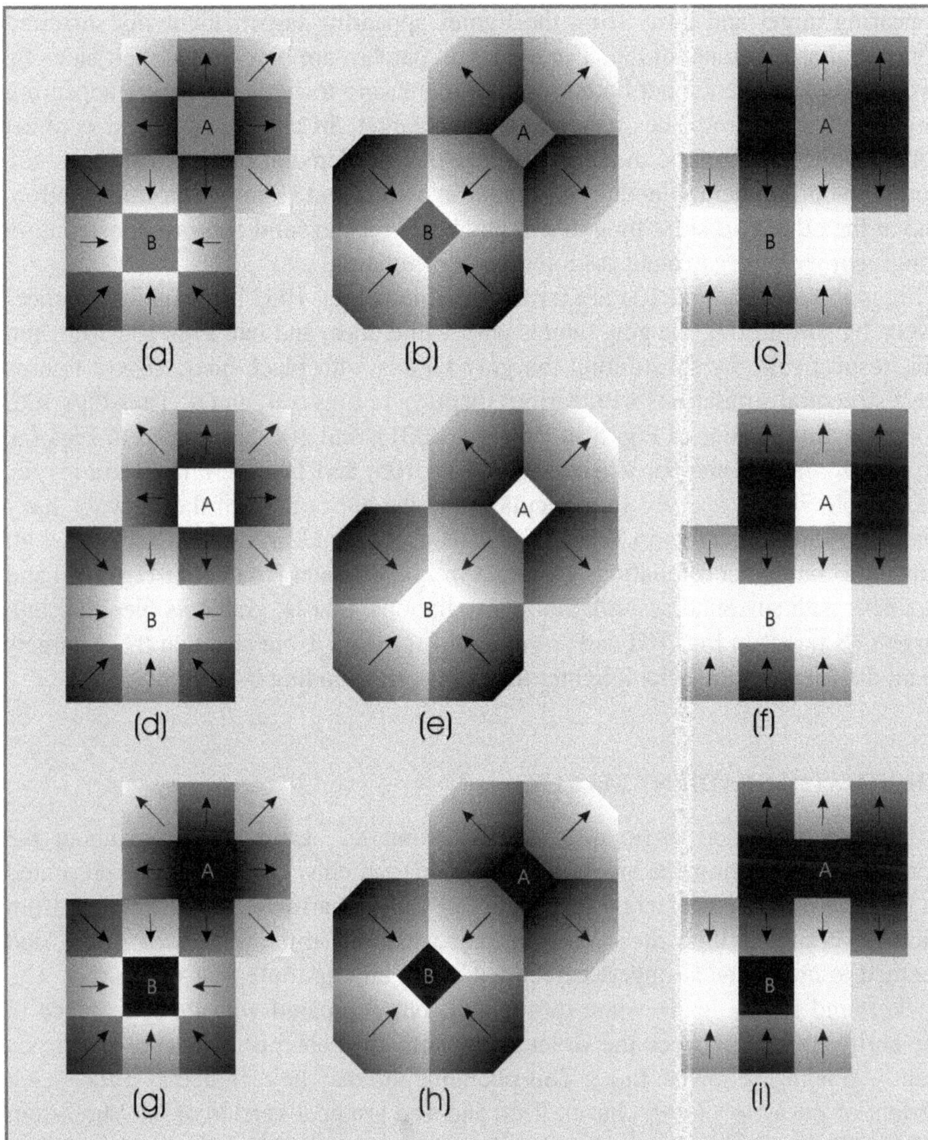

Figure 10. Targets of three luminance levels in different contexts. Gray targets: displays (a), (b), and (c). White targets: displays (d), (e), and (f). Black targets: displays (g), (h), and (i). See text for details.

Fig. 10a is related to Figs 1, 4b, 4e, 5b, and 5d; Fig. 10b is related to displays in Figs 2 and 3; and Fig. 10c is related to Figs 4a, 4c, 5a and 5c. Comparing the three figures brings out the structural similarities and differences between the three main types of lightness illusion displays discussed in this paper. The main differences between them are as follows: in Fig. 10a the target patches are bordered at their four sides by a single type of gradient, with luminance increasing towards the darker appearing target and away from the lighter appearing target, involving surround contrast constellations; in Fig. 10b the target patches are bordered at their sides by two types of gradients, but both types are increasing towards the darker appearing target and away from the lighter appearing target, in another surround contrast constellation; in Fig. 10c the target patches are bordered at their two sides by the same type of gradient, blending into them in a surround assimilation constellation, and at the other two sides by uniform patches, in a surround contrast constellation congruent with the surround assimilation constellation.

Figures 10a, 10b, and 10c are transformed into Figs 10d, 10e, and 10f, respectively, by substituting the gray targets with white ones, and into Figs 10g, 10h, and 10i, respectively, by substituting the gray targets with black ones. These figures have structural similarities with various displays in Figs 6, 7, and 9. Thus Figs 10d, 10e (which is similar to Fig. 8 in Bressan, 2001), and 10f, are related to Figs 6a, 6b, 6e, 6f, 7a, 7d, and 9a, whereas Figs 10g, 10h, and 10i, are related to Figs 6c, 6d, 6g, 6h, 7b, 7e, and 9b. In particular, note the appearance of the clear *vs.* hazy distinction in the targets in Figs 10d and 10g, where clearness is associated with surround contrast constellations (A-target in Fig. 10d and B-target in Fig. 10g) and haziness with surround assimilation constellations, that is, gradients blending into targets (B-target in Fig. 10d and A-target in Fig. 10g). Note also that the A-targets in all displays generally look lighter than the corresponding B-targets.

CHROMATIC GRADIENT EFFECTS

All effects presented up to now were achromatic. Little is known about the corresponding phenomena involving chromatic gradients. One example is presented in Fig. 11a. It consists of a bouquet of curved 'burning arrows', with a gradient from violet to orange at their tips; this color pair was used in the watercolor effect, and seemed to create the strongest perceptual effect among those that were tried. The background is physically white throughout, but compared with its appearance in the periphery and between the violet arrow tails, the center of the figure assumes a faintly glowing yellowish tinge. This phenomenon may be regarded as a chromatic variant of Zavagno's (1999) haze effect, and as a gradient version of the Ehrenstein (1941) illusion, and it is also related to the watercolor effect and displays studied by Da Pos and Bressan (2003). Note that the effect involves color assimilation, in the sense that the induced color in the center is similar to the adjoining color of the tips of the arrows.

Another chromatic assimilation effect is presented in Fig. 11b. It is composed of 'flowers' with 'petals' involving a violet-orange gradient, placed on a physically uniform white background. Note that in the regions of confluence of violet petal

Figure 11. (See color plate IX) Chromatic gradient effects. In both displays the backgrounds are uniform and achromatic, but show chromatic tinges at places of confluence of chromatic gradients. See text for details.

ends the background takes on a faint violet tinge, whereas in the regions of confluence of orange petal ends it takes on a faint orange tinge. An additional effect may be noticed if the display is treated as an auto-stereoscopic image, and the neighboring flowers, involving oppositely directed color gradients, are dichoptically superposed. Using free fusion, the reader may confirm that in the stereo image there is little rivalry and that the colors tend to fuse rather well, such that color variety is reduced and the flowers appear more uniform (though uniformity varies with different computer screens); this is unlike in achromatic versions of the image, not presented here, involving black-white gradients instead of violent-orange ones, which are very hard to fuse and are strongly rivalrous (see Hovis, 1989 for a review of dichoptic color mixing). Furthermore, the color tinges, which are induced into the background in non-stereoscopic viewing, completely disappear in the stereoscopic percept. The reason for this may be either that the induced colors are first generated and then dichoptically fused, or that they are not generated at all at the binocular processing stage.

SIMPLE NEURAL SIMULATIONS

In this paper I have argued against the role of illumination interpretation mechanisms in lightness illusions in flat images. However, regardless whether the presence of luminance gradients does or does not trigger illumination interpretation processes, it is of interest to study the reactions of neurons in the visual system to various configurations of gradients, and compare them to responses to abrupt luminance changes. Figure 12 presents computer simulations of reactions of 1D layers of processing units with on-center off-surround receptive fields to six selected stimuli. Input luminance profiles and corresponding examples of stimuli are depicted in the left-hand column. All stimuli contained a centrally located uniform target region with medium luminance level ($L = 0.8$, indicated by a horizontal dashed line); the position of the targets within the stimuli is indicated as lying between two narrow-dotted vertical lines. The targets were either increments (Inc1, Inc2, and Inc3) or decrements (Dec1, Dec2, and Dec3), that is, they had either higher or lower luminance levels than the immediately adjoining portions of the stimuli. The luminance profiles of the decrements were the same as for the corresponding increments, except for being mirrored upwards about the $L = 0.8$ horizontal line. The targets had inner and outer surrounds. The outer surrounds (lying outside of wide-dotted vertical lines), had low luminance level ($L = 0.5$) for the increments, and high luminance levels ($L = 1$) for the decrements. The inner surrounds (lying between wide-dotted and narrow-dotted vertical lines), had luminance levels that were either equal to those of the outer surrounds (Inc1, Dec1), or involved linear luminance gradients that either connected the outer surrounds and the targets smoothly (Inc2, Dec2), or had inverted profiles, involving abrupt luminance jumps (Inc3, Dec3). The Inc1-Dec1 and the Inc3-Dec3 pair involve surround contrast constellations for

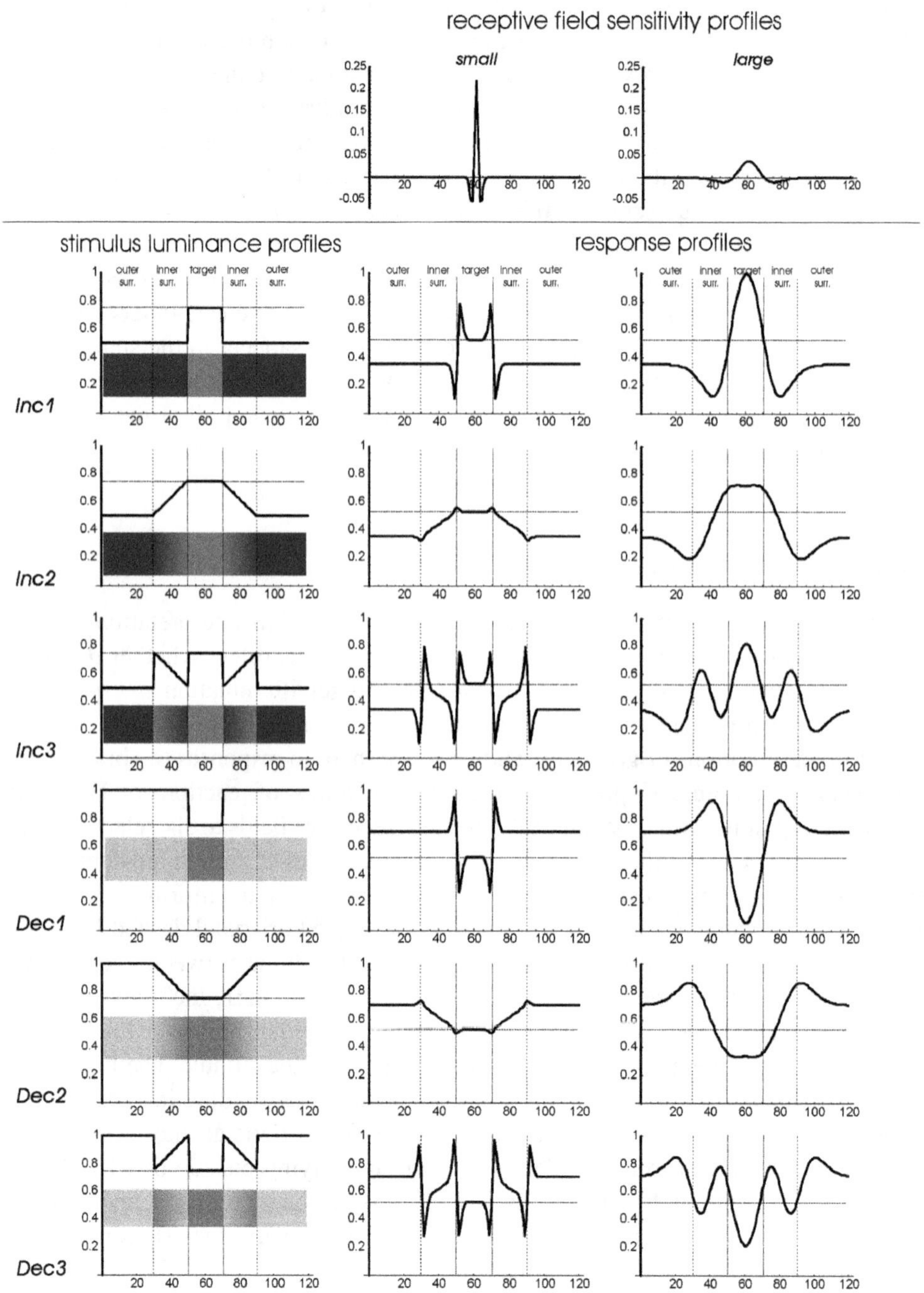

Figure 12. Modeling reactions of layers of two types of on-center neural units to six stimuli. See text for details.

target regions, and the Inc2-Dec2 pair involves a surround assimilation constellation.

The sensitivity profiles of the processing units were defined as differences of two Gaussians, one tall and narrow for the center and the other shorter and wider for the surround. The surround Gaussian was scaled to 90%, making the receptive fields volume-unbalanced (having a positive integral). The consequence of this is that the units exhibited sensitivity to absolute luminance levels, that is, they had positive outputs to uniform stimuli, since the positive reaction of the center was not fully matched by the negative reaction of the surround (see Croner and Kaplan, 1995). Two types of processing units were used, depicted at the top right of Fig. 12, one with small receptive field size (center and surround space constants equal to 1 and 2, respectively) and the other with large receptive field size (six times larger center and surround space constants), very roughly analogous to ganglion P-cells and M-cells. The responses of the unit layers were computed as convolutions of their sensitivity profiles with the stimulus distributions. The 12 resulting output profiles (6 stimulus profiles $\times 2$ receptive field profiles) were subsequently normalized by dividing them by the maximum reaction of all profiles.

Consider first the reactions of small-field units to increments, depicted in the top three graphs in the middle column in Fig. 12. Stimulus Inc1 represents a light patch on darker background (such as in Fig. 6a). The corresponding response profile involves the familiar pattern of overshoots and undershoots associated with the two luminance edges. Note the larger reaction to the target interior than to the background, due to the volume-imbalance of the receptive fields; a completely balanced receptive field would produce zero output for uniform patches of any level. The second stimulus, Inc2, involves linear gradients that blend into uniform regions (such as in Fig. 6b, in portions of striped displays in Figs 4 and 5, and in portions of Figs 9a, 9b and 10). The corresponding regions of the response profile exhibit similar neural gradients. This is another feature due to receptive field imbalance, since volume-balanced units would produce zero outputs for linear gradients (this can be seen by noting that the action of balanced difference-of-Gaussians units resembles taking the second derivative, and the second derivative of any linear function is zero). Note also the rather modest overshoots and undershoots in the response profile. In contrast, the response to stimulus Inc3, involving non-blending gradients (similar to luminance distributions in many portions of displays in Figs 2 to 5 and Figs 9a, 9b and 10), exhibits both neural gradients (as for Inc2) as well as strong overshoots and undershoots (as for Inc1). Turning to decrements, note that Dec1 is similar to Fig. 6c, Dec2 is similar to Fig. 6d and portions of striped displays in Figs 4 and 5 and in Figs 9a, 9b and 10, and Dec3 is similar to portions of displays in Figs 2 to 5 and Figs 9a, 9b and 10. The response profiles of the small-field layer to these stimuli (depicted in the bottom three graphs in the middle column) are mirror images of corresponding profiles for increments.

Are these response profiles relevant for understanding the lightness illusions in images exhibiting similar corresponding luminance profiles? Consider Inc1 and

Dec1, which, regarded together, form the classical simultaneous contrast display. Note that the levels of the response profiles corresponding to target interiors, indicated by dashed horizontal lines, are identical for the two stimuli (and indeed for all six stimuli); however, the presence of the overshoots for Inc1 and undershoots for Dec1 agrees with the direction of the lightness effect in simultaneous contrast (the same patch looks lighter on darker ground than on lighter ground). Thus subsequent processing, roughly corresponding to averaging or smoothing or diffusion confined within the target region, bounded by increments or decrements, would predict the form of the phenomenon (see Hong and Grossberg, 2004; Grossberg and Todorović, 1988). A similar analysis applies to targets in Inc3 and Dec3, but the targets in Inc2 and Dec2 would be predicted to look quite similar.

Consider now the response profiles of the large-field unit layer, depicted in the right-hand column of Fig. 12. Compared to the response profile of small-field units for Inc1, in the response profile of large-field units to the same stimulus the undershoots are much wider, and the two overshoots have merged into a single large overshoot within the target region, strongly exceeding the level of the response within the target interior of the small-field units, indicated by the dashed horizontal line; the response profile for Dec1 is the mirror image of the response to Inc1. Thus the average response within the target region is significantly larger for Inc1 than for Dec1, which again corresponds to the direction of the lightness illusion in the standard simultaneous contrast display. A simple average of the somewhat complementary small-field and large-field response profiles would provide a robust prediction of simultaneous contrast, even without any subsequent diffusion processes within the target regions (though the diffusion would provide more uniform output levels). A similar analysis applies for the Inc3-Dec3 pair. Furthermore, in contrast to small-field simulations, there is a stronger reaction to the target in Inc2 than in Dec2, although the difference is smaller than for the Inc1-Dec1 pair, which is in agreement with MacLeod's (1947) data. The importance of units with large receptive fields for computational models of lightness phenomena was recently stressed and analyzed, in different ways, by Blakeslee and McCourt (2004) and by Hong and Grossberg (2004).

The presented simulations fall far short of a computational model of effects discussed in this paper. Among other shortcomings, they are not 2D, they do not address the clear *vs.* hazy difference, and they do not explain the sometimes surprising strength of the gradient-induced effects. Furthermore, they are unsuitable for the treatment of a class of lightness effects which run counter to predictions based on surround contrast constellations, such as the White effect and related phenomena (see Todorović, 1997). However, most full-fledged models include related types of processing units among their building blocks. Therefore a close analysis of the response profiles and the particular combinations of overshoots/undershoots and gradual reactions present in them may provide insights for future models of effects of luminance gradients on uniform regions.

REFERENCES

Adelson, E. H. (1993). Perceptual organization and the judgment of brightness, *Science* **262**, 2042–2044.

Adelson, E. H. (1995). Checker Shadow Illusion. URL: http://web.mit.edu/persci/people/adelson/checkershadow_illusion.html

Adelson, E. H. (2000). Lightness perception and lightness illusions, in: *The New Cognitive Neurosciences*, Gazzaniga, M. (Ed.), pp. 339–351. MIT Press, Cambridge, MA.

Agostini, T. and Galmonte, A. (2002). A new effect of luminance gradient on achromatic simultaneous contrast, *Psychonomic Bull. Rev.* **9**, 264–269.

Anderson, B. L. (1997). A theory of illusory lightness and transparency in monocular and binocular images: the role of contour junctions, *Perception* **26**, 419–453.

Anderson, B. L. and Winawer, J. (2005). Image segmentation and lightness perception, *Nature* **434**, 79–83.

Beck, J. (1972). *Surface Color Perception*. Cornell University Press, Ithaca, NY.

Blakeslee, B. and McCourt, M. E. (1997). Similar mechanisms underlie simultaneous brightness contrast and grating induction, *Vision Research* **37**, 2849–2869.

Blakeslee, B. and McCourt, M. E. (2003). A multiscale spatial filtering account of brightness phenomena, in: *Levels of Perception*, Harris, L. and Jenkin, M. (Eds). Springer-Verlag, New York.

Blakeslee, B. and McCourt, M. E. (2004). A unified theory of brightness contrast and assimilation incorporating oriented multiscale spatial filtering and contrast normalization, *Vision Research* **44**, 2483–2503.

Bressan, P. (2001). Explaining lightness illusions, *Perception* **30**, 1031–1046.

Bressan, P. (2003a). A fair test of the effect of a shadow-incompatible luminance gradient on the simultaneous lightness contrast. Comment, *Perception* **32**, 721–723.

Bressan, P. (2003b). A fair test of the effect of a shadow-incompatible luminance gradient on the simultaneous lightness contrast. Reply to Logvinenko's reply, *Perception* **32**, 725–730.

Croner, L. J. and Kaplan, E. (1995). Receptive fields of P and M ganglion cells across the primate retina, *Vision Research* **35**, 7–24.

Da Pos, O. and Bressan, P. (2003). Chromatic induction in neon colour spreading, *Vision Research* **43**, 697–706.

Ehrenstein, W. (1941). Über Abwandlungen der L. Hermannschen Helligkeitserscheinung, *Zeitschrift für Psychologie* **150**, 83–91.

Evans, R. M. (1959). *Eye, Film, and Camera in Color Photography*. John Wiley, New York.

Fry, G. A. (1931). The stimulus correlates of bulky color, *Amer. J. Psychol.* **43**, 618–620.

Gilchrist, A. L. (1977). Perceived lightness depends on perceived spatial arrangement, *Science* **195**, 185–187.

Gilchrist, A. (1988). Lightness contrast and failures of constancy: a common explanation, *Perception and Psychophysics* **43**, 415–424.

Gilchrist, A., Kossyfidis, C., Bonato, F., Agostini, T., Cataliotti, J., Li, X., Spehar, B., Annan, V. and Economou, E. (1999). An anchoring theory of lightness perception, *Psychol. Rev.* **106**, 795–834.

Gilchrist, A. and Economou, E. (2003). Dualistic versus monistic accounts of lightness perception, in: *Levels of Perception*. Springer Verlag, New York.

Gillam, B. (1998). Illusions at century's end, in: Hochberg, J. (1998). *Perception and Cognition at Century's End*, Harris, L. and Jenkin, M. (Eds), pp. 95–136. Academic Press, New York.

Grossberg, S. and Todorović, D. (1988). Neural dynamics of 1-D and 2-D brightness perception: A unified model of classical and recent phenomena, *Perception and Psychophysics* **43**, 241–277.

Heinemann, E. G. and Chase, S. (1995). A quantitative model for brightness induction, *Vision Research* **35**, 2007–2020.

Hong, S. and Grossberg, S. (2004). A neuromorphic model for achromatic and chromatic surface representation of natural images, *Neural Networks* **17**, 787–808.

Hovis, J. K. (1989). Review of dichoptic color mixing, *Optomet. Visual Sci.* **66**, 181–190.
Hurvich, L. M. and Jameson, D. (1966). *The Perception of Brightness and Darkness*. Allyn and Bacon, Boston, MA.
Kanizsa, G. (1980). *Grammatica del Vedere*. Il Mulino, Bologna.
Kardos, L. (1934). Ding und Schatten. Eine experimentelle Untersuchung über die Grundlagen des Farbensehens. *Z. Psychol.* **23** (Suppl.), 1–184.
Kennedy, J. M. (1976). Sun figure: an illusory diffuse contour resulting from an arrangement of dots, *Perception* **5**, 479–481.
Kingdom, F. (1997). Simultaneous contrast: the legacies of Hering and Helmholtz, *Perception* **26**, 673–677.
Kingdom, F. (1999). Old wine in new bottles? Some thoughts on Logvinenko's 'Lightness induction revisited', *Perception* **28**, 929–934.
Kingdom, F. (2003a). Levels of brightness perception, in: *Levels of Perception*, Harris, L. and Jenkin, M. (Eds). Springer Verlag, New York.
Kingdom, F. (2003b). Reply to Todorovic's comments on Kingdom's chapter, in: *Levels of Perception*, Harris, L. and Jenkin, M. (Eds) (on accompanying CD-ROM). Springer-Verlag, New York.
Kingdom, F. and Moulden, B. (1992). A multi-channel approach to brightness coding, *Vision Research* **32**, 1565–1582.
Knill, D. C. and Kersten, D. (1991). Apparent surface curvature affects lightness perception, *Nature* **351**, 228–230.
Logvinenko, A. D. (1999). Lightness induction revisited, *Perception* **28**, 803–816.
Logvinenko, A. D. (2003). A fair test of the effect of a shadow-incompatible luminance gradient on the simultaneous lightness contrast (followed by Discussion), *Perception* **32**, 717–730.
Logvinenko, A. D., Adelson, E. H., Ross, D. A. and Somers, D. (2005). Straightness as a cue for luminance edge interpretation, *Perception and Psychophysics* **67**, 120–128.
MacLeod, R. B. (1947). The effects of 'artificial penumbrae' on the brightness of included areas, in: *Miscellanea Psychologica Albert Michotte*, pp. 138–154. Institute Superieur de Philosophie, Louvain.
McCourt, M. E. (1982). A spatial frequency dependent grating induction effect, *Vision Research* **22**, 119–134.
Palmer, S. E. (1999). *Vision Science: Photons to Phenomenology*. The MIT Press, Cambridge, MA.
Pessoa, L. and DeWeerd, P. (2003). *Filling-in: from Perceptual Completion to Cortical Reorganization*. Oxford University Press, Oxford.
Pessoa, L., Mingolla, E. and Neumann, H. (1995). A contrast- and luminance driven multiscale network model of brightness perception, *Vision Researtch* **35**, 2201–2223.
Pinna, B., Brelstaff, G. and Spillmann, L. (2001). Surface color from boundaries: a new watercolor illusion, *Vision Research* **41**, 2669–2676.
Purves, D. and Lotto, R. B. (2003). *Why We See What We Do. An Empirical Theory of Vision*. Sinauer Associates, Inc., Sunderland, MA.
Purves D., Williams, S. M., Nundy, S. and Lotto, R. B. (2004). Perceiving the intensity of light, *Psychol. Rev.* **111**, 142–158.
Ross, W. D. and Pessoa, L. (2000). Lightness from contrast: a selective integration model, *Perception and Psychophysics* **62**, 1160–1181.
Schirillo, J. A. and Shevell, S. K. (1997). An account of brightness in complex scenes based on inferred illumination, *Perception* **26**, 507–518.
Schirillo, J. A. and Shevell, S. K. (2002). Articulation: brightness, apparent illumination, and contrast ratios, *Perception* **31**, 161–169.
Spillmann, L., Fuld, K. and Gerrits, H. J. (1976). Brightness contrast in the Ehrenstein illusion, *Vision Research* **16**, 713–719.
Todorović, D. (1987). The Craik-O'Brien-Cornsweet effect: new varieties and their theoretical implications, *Perception and Psychophysics* **42**, 545–560.

Todorović, D. (1997). Lightness and junctions, *Perception* **26**, 379–394.

Todorović, D. (1999). Achromatic compositions, *Perception* **28** (Suppl.), 71.

Todorović, D. (2002). Comparative overview of perception of distal and proximal visual attributes, in: *Perception and the Physical World*, Heyer, D. and Mausfeld, R. (Eds). John Wiley, Chichester.

Todorović, D. (2003a). Comments on Kingdom's chapter, in: *Levels of Perception*, Harris, L. and Jenkin, M. (Eds) (on accompanying CD-ROM). Springer Verlag, New York.

Todorović, D. (2003b). Comments on Kingdom's reply, in: *Levels of Perception*, Harris, L. and Jenkin, M. (Eds) (on accompanying CD-ROM). Springer Verlag, New York.

Todorović, D. (2003c). Comments on Blakeslee and McCourt's chapter, in: *Levels of Perception*, Harris, L. and Jenkin, M. (Eds) (on CD-ROM). Springer Verlag, New York.

Trunz, E. (1981). *Goethes Werke. Band XIV*. Verlag C. H. Beck, Munich, Germany.

Wade, N. J. (1998). *A Natural History of Vision*. Bradford, Cambridge, MA.

Whittle, P. (1994a). The psychophysics of contrast brightness, in: *Lightness, Brightness and Transparency*, Gilchrist, A. (Ed.), pp. 35–110. Lawrence Erlbaum, Hillsdale, NJ.

Whittle, P. (1994b). Contrast brightness and ordinary seeing, in: *Lightness, Brightness and Transparency*, Gilchrist, A. (Ed.), pp. 111–157. Lawrence Erlbaum, Hillsdale, NJ.

Williams, S. M., McCoy, A. N. and Purves, D. (1998a). The influence of depicted illumination on brightness, *Proc. Nat. Acad. Sci. USA* **95**, 13296–13300.

Williams, S. M., McCoy, A. N. and Purves, D. (1998b). An empirical explanation of brightness, *Proc. Nat. Acad. Sci. USA* **95**, 13301–13306.

Zavagno, D. (1999). Some new luminance-gradient effects, *Perception* **28**, 835–838.

Color, Line, and Space, pp. 161–219 (2006)

A neural model of surface perception: Lightness, anchoring, and filling-in

STEPHEN GROSSBERG * and SIMON HONG

Department of Cognitive and Neural Systems and Center for Adaptive Systems, Boston University, 677 Beacon Street, Boston, MA 02215, USA

Abstract—A neural model is proposed of how the visual system processes natural images under variable illumination conditions to generate surface lightness percepts. Previous models clarify how the brain can compute relative contrast. The anchored Filling-In Lightness Model (aFILM) clarifies how the brain 'anchors' lightness percepts to determine an absolute lightness scale that uses the full dynamic range of neurons. The model quantitatively simulates lightness anchoring properties (Articulation, Insulation, Configuration, Area Effect) and other lightness data (discounting the illuminant, the double brilliant illusion, lightness constancy and contrast, Mondrian contrast constancy, Craik-O'Brien-Cornsweet illusion). The model clarifies how retinal processing stages achieve light adaptation and spatial contrast adaptation, and how cortical processing stages fill-in surface lightness using long-range horizontal connections that are gated by boundary signals. The new filling-in mechanism runs 1000 times faster than diffusion mechanisms of previous filling-in models.

Keywords: Surface perception; lightness; anchoring; filling-in; retinal adaptation; long-range horizontal connections; visual cortex.

1. INTRODUCTION

1.1. From luminance to anchored lightness

The retina receives luminance signals, which are a product of reflectances and illumination levels (Hurlbert, 1986; Lambert, 1760; Wyszecki and Stiles, 1982), from objects in the world. Surface reflectances, or the percentages of light reflected by a surface in each wavelength (also known as albedo), provide information about the material properties of objects. From these luminance signals, the visual system is able to estimate object reflectances by compensating for an immense dynamic

*To whom correspondence should be addressed. E-mail: steve@bu.edu

range of mean illuminations across time, and for a wide dynamic range across a single scene. This process of 'discounting the illuminant' is not sufficient, however, to efficiently see the world because illuminant-discounted signals may represent only the *relative* amounts of light that each object surface reflects to the eyes. For effective perception, the brain also needs to compute an *absolute* lightness scale that can represent the full-range of experience from dim moonlight to dazzling sunlight.

Early neural models of surface lightness perception simulated many classical psychophysical data based upon estimates of relative light levels, including brightness constancy, contrast, and assimilation; Craik-O'Brien-Cornsweet effect; Koffka-Benussi ring; Kanizsa-Minguzzi anomalous brightness differentiation; Hermann grid; Land Mondrians viewed under constant and gradient illumination conditions that could not be explained by Land's Retinex theory; Bergström brightness percepts of step-like and smoothly modulated luminance profiles; Hamada brightness percepts of luminance increments and decrements; Mach bands; low-contrast and high-contrast missing fundamental and nonlinear contrast effects associated with sinusoidal luminance waves; and Ehrenstein brightness enhancement (Cohen and Grossberg, 1984; Gove *et al.*, 1995; Grossberg and Todorovic, 1988; Neumann *et al.*, 1998). Consistent extensions of these models simulated 3D figure–ground brightness percepts, such as Fechner's paradox; binocular brightness summation; Bregman-Kanizsa figure–ground separation; Kanizsa stratification; Munker-White effect; Benary cross; checkerboard percepts; McCollough effect; Necker cube; transparency, and 3D neon color spreading (Grossberg *et al.*, 2002; Grossberg and Kelly, 1999; Grossberg and Swaminathan, 2004; Grossberg and Yazdanbakhsh, 2003, 2005; Kelly and Grossberg, 2000; Ross and Pessoa, 2000).

Given the large amount of already simulated data, the present article developed a model that is consistent with these earlier explanations, while also proposing how an absolute lightness scale may be constructed by the brain. To realize this goal, the new anchored Filling-In Lightness Model (aFILM) provides a more sophisticated account of early visual filtering, lightness filling-in, and lightness anchoring. The model quantitatively simulates, for the first time, key psychophysical data about lightness anchoring, as well as other lightness data to show that it is consistent with earlier model explanations. Neurophysiological and anatomical data that support model hypotheses are also summarized (see Table 1). aFILM can also process complex natural scenes under difficult lighting conditions. Although the present work focuses on achromatic images, a variant of the model has been shown capable of processing chromatic natural images as well (Hong and Grossberg, 2004). The model was briefly reported in Hong and Grossberg (2003).

1.2. Discounting the illuminant by early visual preprocessing

Retinal preprocessing of visual signals contributes to discounting the illuminant and creating a relative lightness scale. These processes include two mechanisms of gain control: *light adaptation* and *contrast adaptation*. Human vision adapts to ten orders of magnitude of daily variations of ambient illumination (Cornsweet, 1970;

Table 1.
Brain data that support model

Anatomy, Experiments	Functional interpretation (equations in model)	Selected references (species)
Outer segment of retinal photoreceptor Inner segment of retinal photoreceptor and HC connection	Automatic gain control (A1–A3)	Koutalos and Yau (1996, frog, salamander, cow, human)
	Shift property by negative feedback (A4–A5)	Verweij *et al.* (1996, goldfish)
HC syncytium	Input dependent gated diffusion (A6–A8)	Kamermans *et al.* (1996, goldfish)
Retinal ganglion cells, LGN cells; Physiology of ON, OFF channels.	Center-surround property of ON, OFF channels (A9–A24)	Jones *et al.* (2000, cat), Roska *et al.* (2000, tiger salamander); Schiller (1992, frog, mudpuppy, goldfish, rabbit, cat, monkey)
V1 simple cells	Polarity-dependent boundary detection (A25–A32)	Alonso *et al.* (2001, cat), Hubel and Wiesel (1962, cat)
V1 complex cells	Polarity-independent boundary detection (A33–A35)	Alonso and Martinez (1998, cat)
Filling-in related cortical activities in V1, V2;	Boundary-gated filling-in (A36–A44)	Hung *et al.* (2001, cat); Paradiso and Nakayama (1991, human)
Psychophysics of filling-in Lightness perception	Anchoring of lightness (A45–A53)	Gilchrist *et al.* (1999, human)

Kaufman, 1974; Martin, 1983; Sakmann and Creutzfeldt, 1969). For example, if the brain gets an input like the one in Fig. 1A, we 'see' it as in Fig. 1B. This *light adaptation* property depends in part on retinal circuitry. Figure 1C shows the model response to varying background illumination. The range of maximal sensitivity shifts with background illumination without undergoing compression, as also occurs in the retina (Rodieck, 1998; Werblin, 1971). Another dimension of adaptation is spatial *contrast adaptation*. For example, if there is a big contrast in the visual field, as in Fig. 1D, the brain can rescale input signals to see the dark side as well as the bright side of the scene, as in the model simulation of Fig. 1E. Since retinal ganglion cells, which are the sole output units of the retina, have firing rates that vary over less than three orders of magnitude, the visual system needs to compress the dynamic range of input at the retinal level, without a loss of sensitivity. Mechanisms of contrast adaptation are still undergoing intensive experimental study (Baccus and Meister, 2002; Demb, 2002). Some retinal gain control mechanisms contributing to adaptation include: (1) Ca^{2+} ion-mediated negative feedback at

(A) INPUT

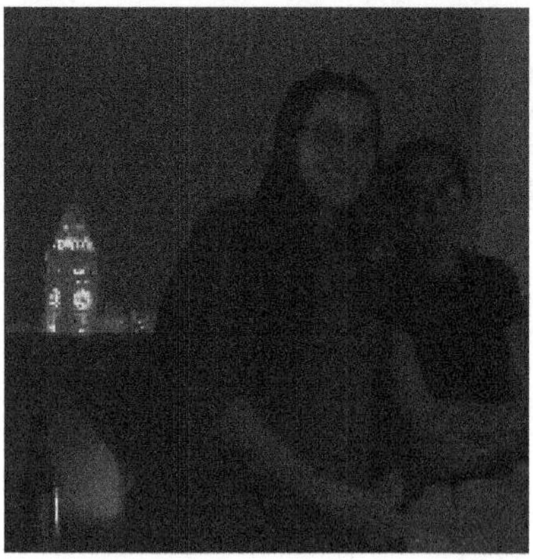

(B) SIMULATION

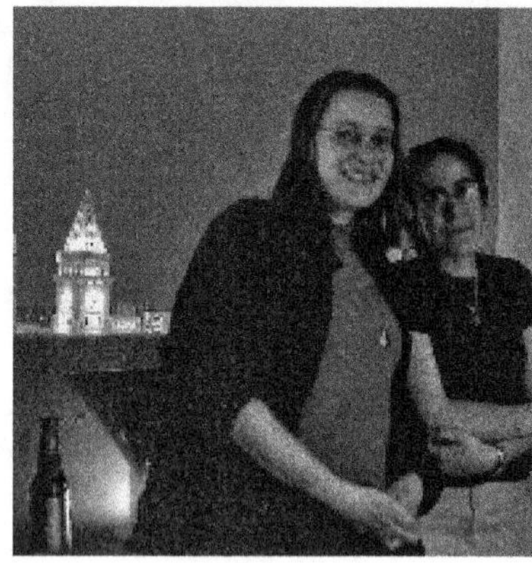

(C) SHIFT PROPERTY

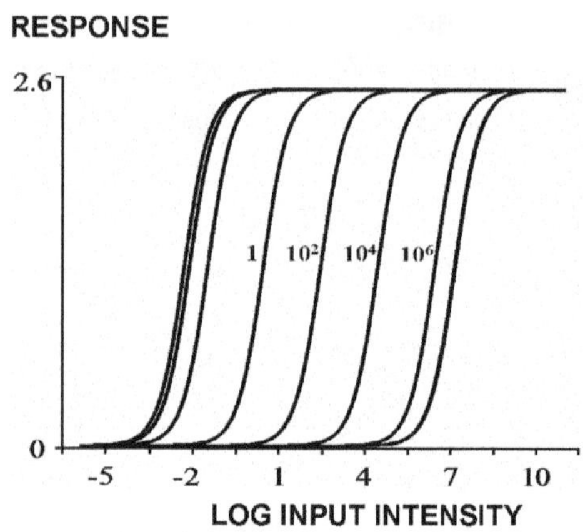

(D) INPUT

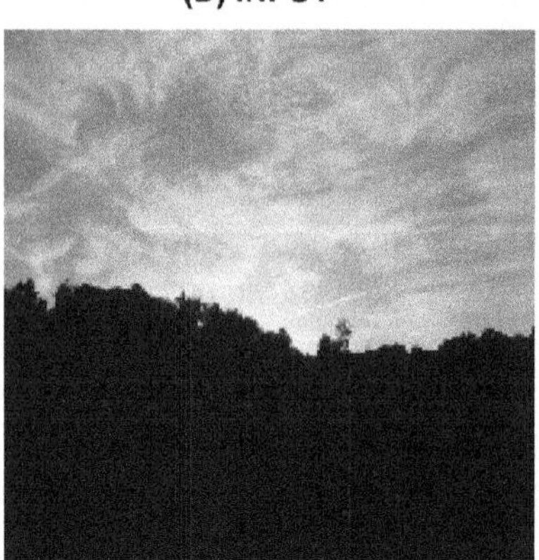

(E) SIMULATION

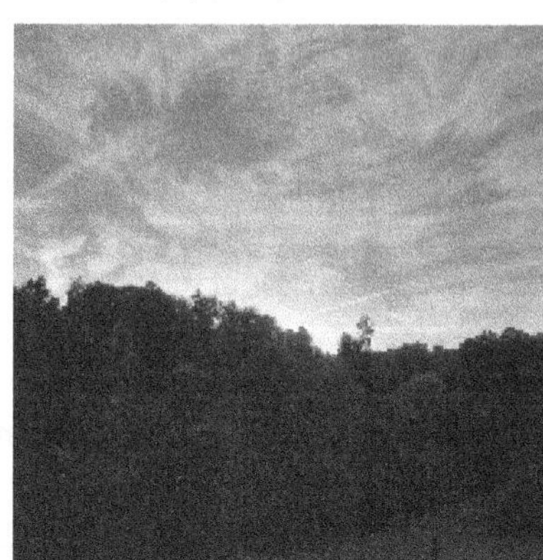

the photoreceptors (Koutalos and Yau, 1996) and bipolar cells (Nawy, 2000); (2) bleaching of photopigments (Dowling, 1987; Fain, 2001); (3) surround negative feedback by horizontal cells (HC) (McMahon *et al.*, 2001; Thibos and Werblin, 1978; Werblin, 1974); and (4) a circuitry switch from cones to rods (Mills and Massey, 1995; Ribelayga *et al.*, 2002).

1.3. Some anchoring hypotheses

Surface lightness percepts cannot fully be explained by such low-level mechanisms. For starters, visual percepts depend upon appropriate interactions between both ON and OFF channel signals that are largely segregated up until cortical area V1 (Schiller, 1992; Schiller *et al.*, 1986). Attempts to explain surface lightness range from the classic *inference* theory of Helmholtz (1866) to recent theories that Gilchrist and his colleagues classify as *intrinsic image* theories (Arend, 1994; Gilchrist *et al.*, 1999). Several theories propose that lightness is derived from luminance ratios among surfaces in a display, but these computations can only recover relative lightness values. There remains the problem of mapping relative lightness values to the absolute lightness values that are experienced. One proposed solution is the *average luminance rule* of Helson (1943), which postulates that the average luminance of the display, defined as middle gray, is an 'anchor' for other luminances; higher luminances than average luminance will then have higher lightness values than middle gray. Figures 2A and 2B show an example where this rule makes a correct prediction. However, in response to Fig. 2C, it makes the error shown in Fig. 2D: the whiteboard becomes middle gray. The *average luminance rule* thus does not explain lightness data quantitatively.

Wallach introduced an anchoring hypothesis which became known as *highest-luminance-as-white* (HLAW) rule (Horn 1977; Land and McCann, 1971; Wallach, 1948, 1976). This rule assumes that the perceptual quantity 'white' is assigned to the highest luminance in a given scene as the standard, and that lower luminant surfaces are assigned to gray values relative to it. According to this rule, the whiteboard in Fig. 2E should look white, as in Fig. 2F. In cases like Fig. 2G, however, the HLAW rule makes a wrong prediction, as shown in Fig. 2H. The white curve in Fig. 2H that is superimposed on the image shows the profile of the predicted

---

Figure 1. Retinal adaptation. (A–B) Input and the simulation of the model reflecting the result of light adaptation of the retina. (C) Shift property of sensitivity of the model retinal units. The model retina simulates the light adaptation property by automatically shifting its operating range to adapt to the ambient luminance of the visual field. When the luminance is too low, it simulates the physical limit of adaptation (the saturation of shifting on the left end of the graph). For clarity only four mean input intensities are shown besides the corresponding curves. The visible 3 leftmost curves have mean luminances of 10^{-4}, 10^{-3}, and 10^{-2} from the left-end, respectively. The visible rightmost curve has a mean luminance of 10^{7}. (D–E) Input with high spatial contrast and the model simulation of the input. It is assumed that the retinal circuit is responsible for this kind of non-linear rescaling of contrast that makes the brain 'see' the dark part as well as the bright part. Photo courtesy of Arash Fazl and Bob Wagner.

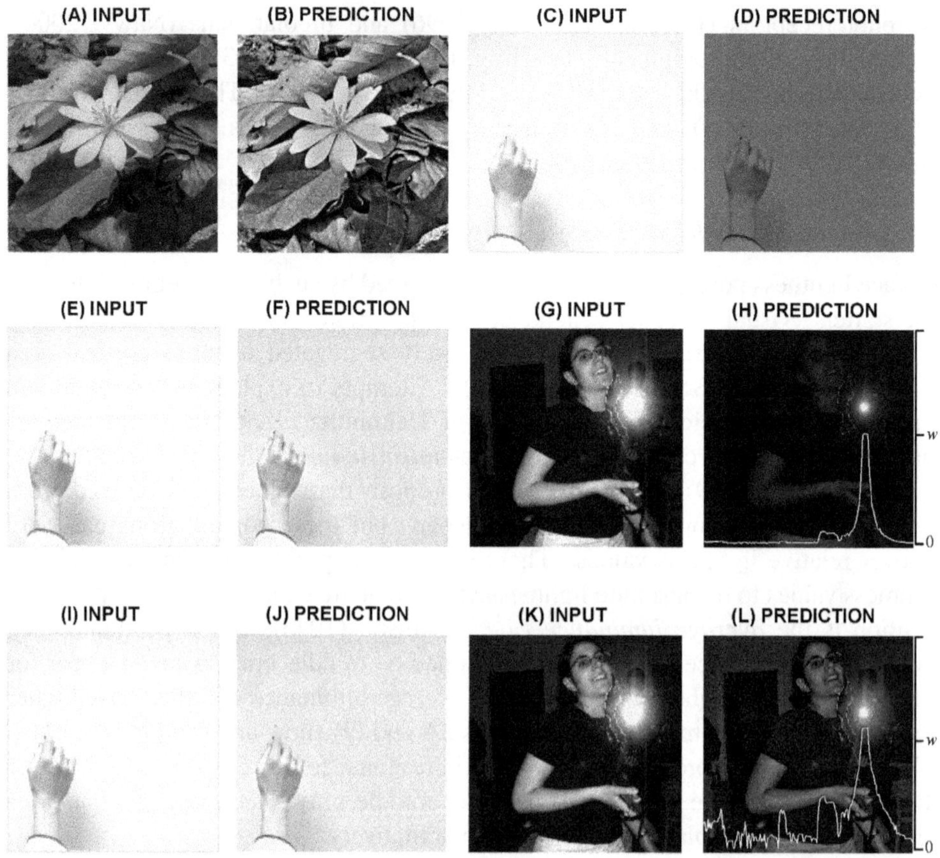

Figure 2. Predictions by various lightness theories. (A–B) Input and correct prediction of lightness by *average luminance rule*. (C–D) Input and wrong prediction by *average luminance rule*. According to the *average luminance rule*, the whiteboard should look middle gray. (E–F) Input and correct prediction of lightness by *highest-luminance-as-white* (HLAW) rule. (G–H) Example where HLAW rule makes an error. HLAW rule makes an error due to a prominent highest luminance like the one in (G). (I–L) Inputs and corresponding predictions made by the model developed here. The model with a new rule called *blurred-highest-contrast-as-white* (BHCAW) rule correctly predicts percepts. See the text for further explanations.

lightness along the horizontal section of the image that crosses the light source. The value 'w' in Fig. 2H marks the lightness value 'white' along the vertical axis. By converting the intense illumination source into 'white', the HLAW rule drives other lightness values to dark levels.

To overcome these shortcomings, the current model proposes how brain dynamics may instantiate a new anchoring rule called the *blurred-highest-contrast-as-white* (BHCAW) rule. Blurring, or spatial integration, is sensitive to the area subtended by the highest luminance, thus introducing spatial scale into the lightness computation (Fig. 3A). Blurring also predicts why some surface regions appear self-luminous (Fig. 3B): see Section 2.5 for further explanation.

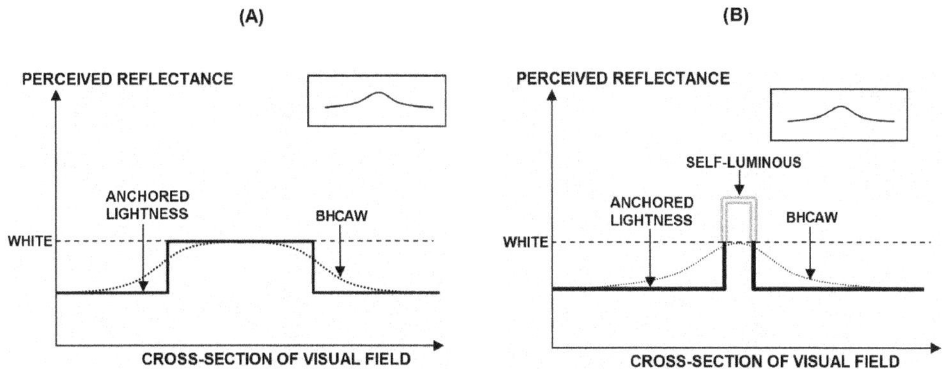

Figure 3. *Blurred-highest-contrast-as-white* (BHCAW) rule and spatial scale. (A) BHCAW rule with a large area of highest luminance. The dashed line indicates the value of WHITE which the blurred highest luminance attains. The thick line (anchored lightness) illustrates a 1D profile of the anchored lightness. In this example, the blurred highest luminance equals white because the area of the highest luminance is at least as large as the kernel used for blurring in the BHCAW module (the inset). (B) BHCAW rule with a small area of highest luminance. Here the blurred highest luminance is smaller than the highest luminance because of the small size of the area of highest luminance relative to the blurring kernel. When the blurred highest luminance is anchored at white, the highest luminance gets pushed above white, becoming self-luminous.

Figures 2I and 2J show a model simulation that provides a result like the HLAW rule in Figs 2E and 2F. Figures 2K and 2L, in contrast, show how the BHCAW rule can correctly predict lightness when a bright light source occurs, which was not done by HLAW. The curve on Fig. 2L shows the simulated lightness profile of the image horizontal section that crosses the light source. The peak of the curve going above white 'w' predicts that the light source will look self-luminous. The model can also quantitatively simulate the four sets of data (Articulation, Configuration, Insulation, and Area Effect) that Gilchrist and his colleagues (1999) have proposed should be explained by any quantitative lightness theory, as is shown next.

1.3.1. Articulation Effect. Figures 4A to 4E illustrate the procedure and the percepts of the Articulation Effect: A black patch (reflectance 3%) is fixed in front of a homogenous dark background (Fig. 4A). When the patch gets illumination 30 times that of the dark background resulting in the luminance of 1.4 ftL (foot Lambert), it looks white (Fig. 4B). (This 30-to-1 foreground-background illumination setting is also used in the Configuration, Insulation, and Area Effect.) When a real white patch (reflectance 90%) appears near the white-looking black patch, the black patch appears gray (Fig. 4C). In the experiment, the subjects indicated perceived reflectance by selecting a match from a Munsell chart of 16 examples. The Munsell chart was illuminated with a different light source so that the luminance of the whitest white, Munsell 9.5, was 160 ftL. That a black surface can look white when intensely illuminated, as illustrated in Fig. 4B, is called the Gelb effect (see Cataliotti and Gilchrist, 1995). As more gray patches are added, the less luminant ones ones look increasingly dark (Figs 4C, 4D and 4E). This darkening effect

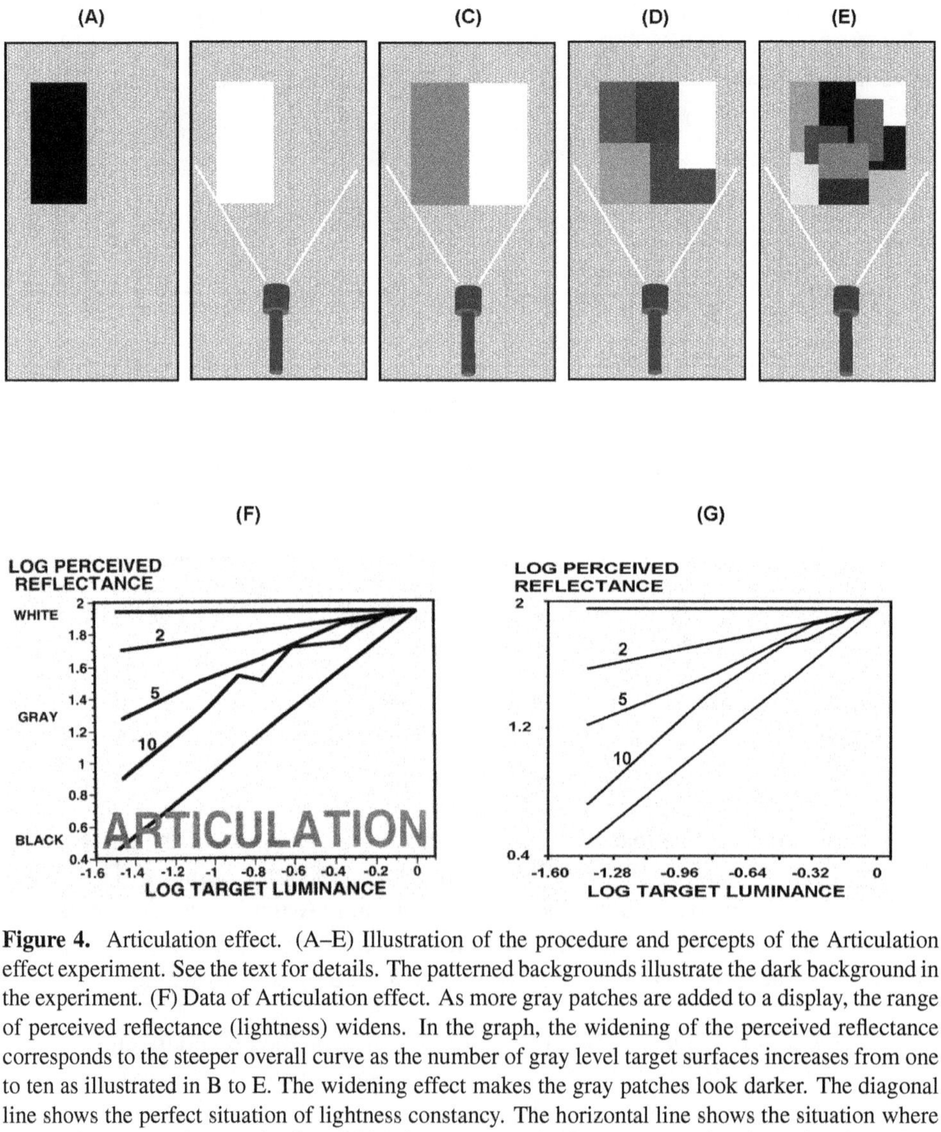

Figure 4. Articulation effect. (A–E) Illustration of the procedure and percepts of the Articulation effect experiment. See the text for details. The patterned backgrounds illustrate the dark background in the experiment. (F) Data of Articulation effect. As more gray patches are added to a display, the range of perceived reflectance (lightness) widens. In the graph, the widening of the perceived reflectance corresponds to the steeper overall curve as the number of gray level target surfaces increases from one to ten as illustrated in B to E. The widening effect makes the gray patches look darker. The diagonal line shows the perfect situation of lightness constancy. The horizontal line shows the situation where there is just one surface on the Ganzfeld (a homogeneous background covering the entire visual field with no other visual cues). (G) Simulation results. Perceived reflectance in the model is anchored lightness of the simulation. See the text for details. Figure F is from Gilchrist *et al.* (1999).

does not affect the highest luminance surface, which remains 'anchored' to white (Fig. 4F). Figure 4G shows the that the model can quantitatively simulate these data. (No effort was made to show error bars in the simulations since none appeared with the data.) Even in the two-Mondrian case in Fig. 4C, the reflectances of the patches range from black to white covering the full span of reflectance used in the experiment. Adding different luminance patches is thus just a process of 'articulation'.

1.3.2. Configuration Effect. Figures 5A and 5B illustrate the procedure and the percepts of the Configuration Effect: a Mondrian display in Fig. 5B, namely, a 2D arrangement of juxtaposed gray patches, widens the range of perceived reflectance compared to the linear arrangement of patches shown in Fig. 5A. That is, the dark patches in Fig. 5A appear lighter than the corresponding dark patches in Fig. 5B. Figures 5C and 5D show the data. Comparison of the 5C and 5D shows that this effect becomes greater with more local articulation. Figures 5E and 5F show that the model can quantitatively simulate this effect.

1.3.3. Insulation Effect. Figures 6A to 6C show the procedure and the percepts of the Insulation Effect. When the staircase arrangement is surrounded by a white insulating region, the range of perceived reflectance widens (Fig. 6B). This does not occur when the staircase is insulated by a black border (Fig. 6C). Figure 6D summarizes the data and Fig. 6E the simulation.

1.3.4. Area Effect. The lower part of Fig. 7A shows the Area Effect set-up. The subject's head is covered by a dome that is divided into two regions. The upper part of Fig. 7A illustrates the stimuli and corresponding percepts. When the highest luminance area occupies more than half of the visual field, it appears white while the darker part looks gray. As the darker area occupies more than half of the visual field, it approaches white, while the lighter area appears self-luminous. Figure 7B shows the data and Fig. 7C the model simulation.

No published models have yet simulated these data using an anchoring process, among other stages in processing lightness information. How the aFILM simulates these and other lightness data is explained below. The reader who prefers can skip to the Results, Section 3, for model simulation explanations before reading about the model in Section 2.

2. MODEL DESCRIPTION

Figure 8 provides a model macrocircuit. Variants of its processing stages, except the Anchored Lightness and BHCAW stages, have earlier been used to simulate lightness data. The present model refines the operations in these stages as well, as indicated below. The Retinal Adaptation stage adapts to ambient luminance and spatial contrasts. Using the adapted signal, the Contrast stage generates contrast signals using multiple-scales of antagonistic ON-center OFF-surround and OFF-center ON-surround processes. The light-adapted signal also goes without change via a parallel pathway to the Luminance stage. The model then branches into two cortical streams, the boundary and surface streams, which have previously been modeled as a Boundary Contour System (BCS) and Feature Contour System (FCS), respectively (Grossberg, 1994, 1997; Grossberg and Kelly, 1999; Grossberg and Mingolla, 1985a, b; Grossberg and Howe, 2003; Grossberg and Swaminathan,

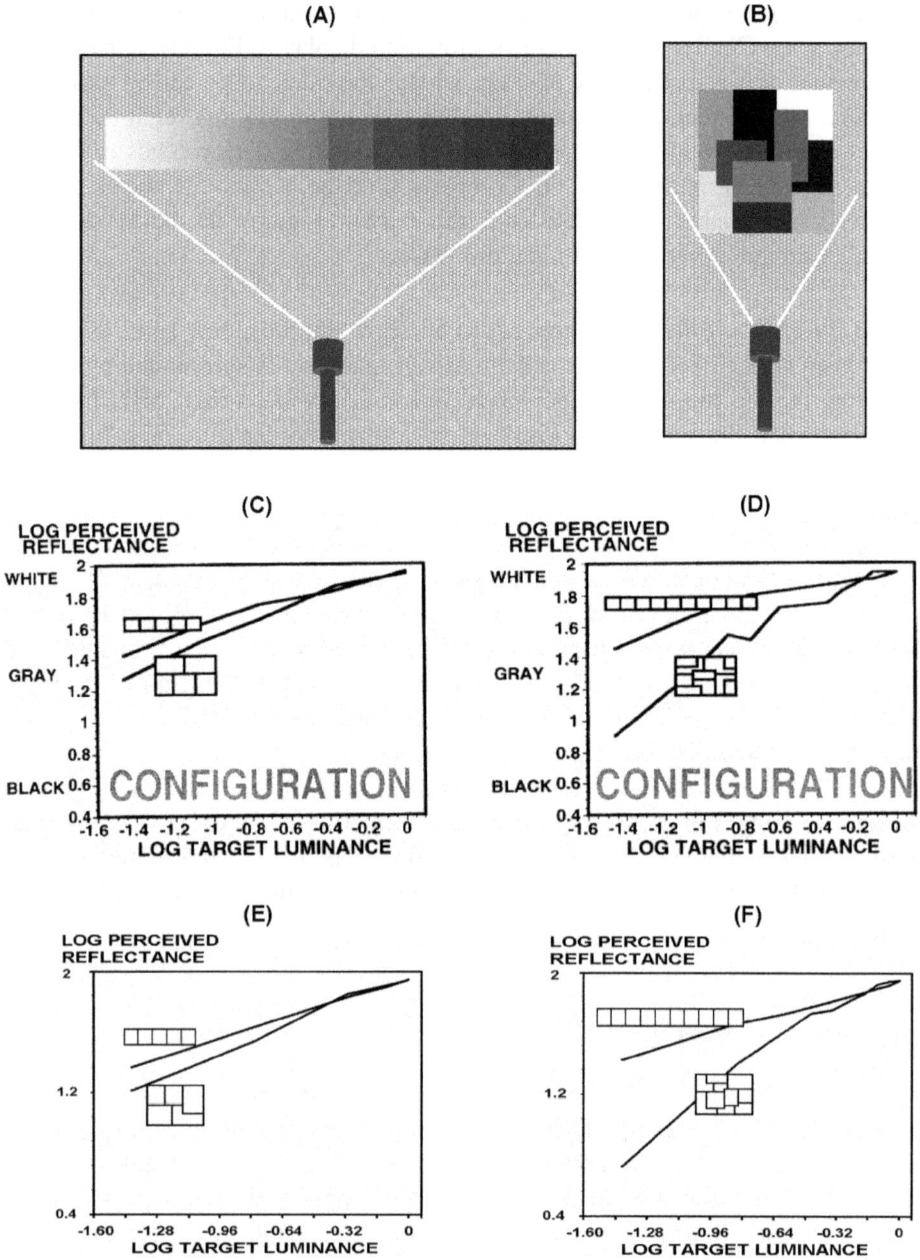

Figure 5. Configuration effect. (A–B) Illustration of the experimental settings and the percepts of Configuration effect. See the text for details. (C–D) Data of Configuration effect. The lower inset of each figure shows the Mondrian arrangement; the upper inset, staircase arrangement. Mondrian arrangement of gray target surfaces widens the range of lightness compared to the staircase arrangement. Comparison of C and D shows that articulation makes the effect bigger. (E–F) Simulation results, corresponding C and D, respectively. The model fits the data of configuration effect in the anchoring theory. See the text for details. The figures C, D are from Gilchrist *et al.* (1999).

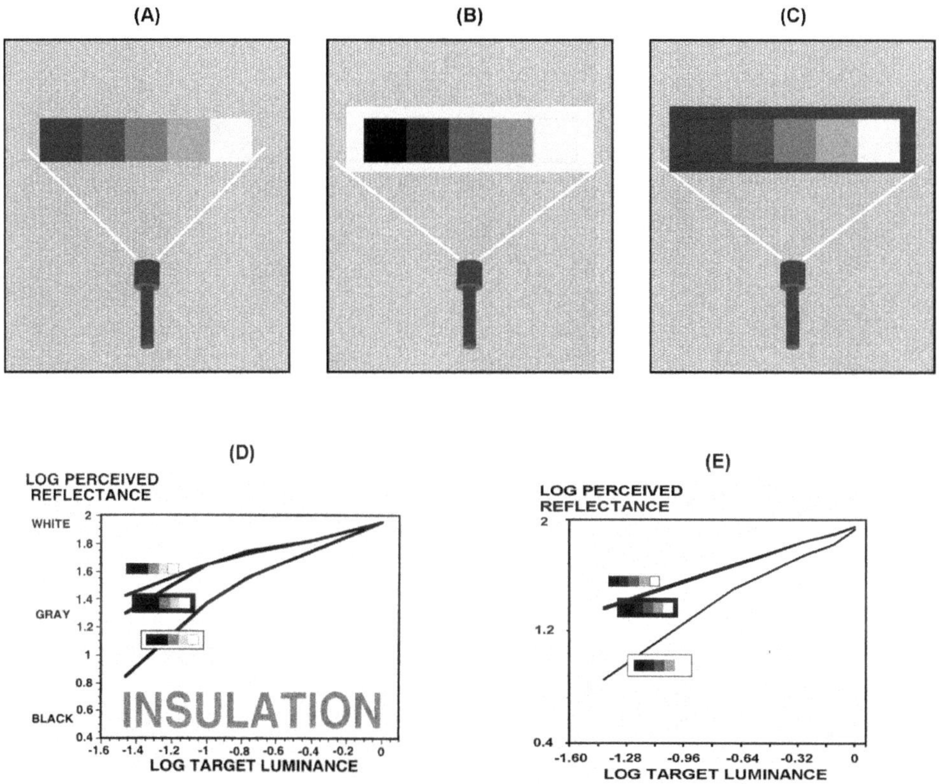

Figure 6. Insulation effect. (A–C) Illustration of the experimental settings and the percepts of Insulation effect. See the text for details. (D) Data of Insulation effect. Insulation by a white surrounding widens the range of perceived reflectance. This effect does not seem to happen when a black surrounding is used for insulation. (E) Simulation results. The model fits the data of configuration effect in the anchoring theory. See the text for further explanation. Figure D is from Gilchrist *et al.* (1999).

2004; Grossberg and Todorovic, 1988; Kelly and Grossberg, 2000). The luminance and contrast signals are pooled at the Surface Filling-In stage, where their spread via long-range horizontal connections is gated, or blocked, by contrast-sensitive signals from the Boundary stage. The filled-in signals are then rescaled at the Anchored Lightness stage. These anchored signals represent perceived lightness in the model.

2.1. Retinal adaptation

The model retina calculates the steady-state of retinal adaptation (light adaptation and spatial contrast adaptation) to a given input image. It adapts the response of photoreceptors to varying levels of incoming light, since otherwise the visual process could be desensitized by saturation right at the photoreceptor. Light adaptation, at the photoreceptor outer segment, protects each photoreceptor from saturation by using intracellular temporal adaptation that shifts the photoreceptor

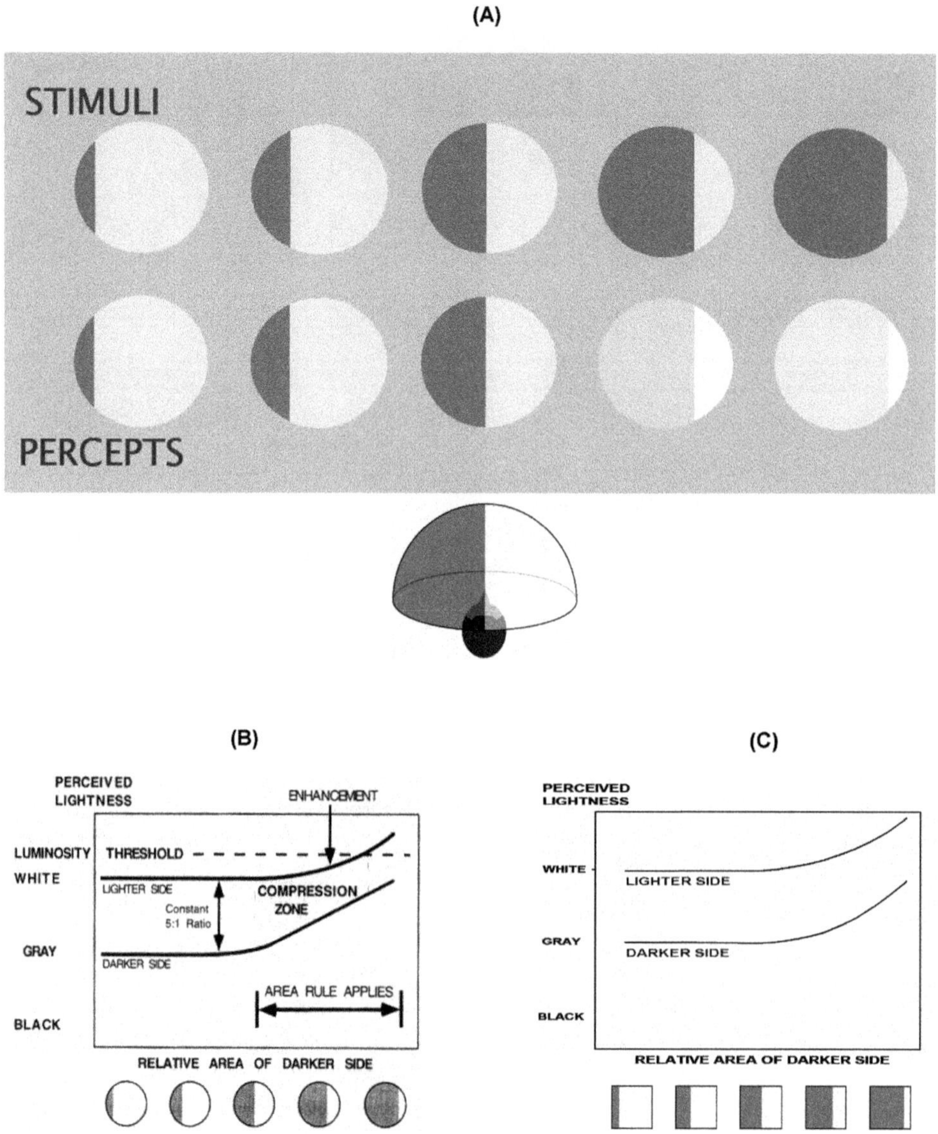

Figure 7. Area effect in divided Ganzfeld situation. (A) Illustration of the experimental settings and the percepts of Area effect. See the text for details. (B) Qualitative illustration of the area effect. As the non-highest luminance area becomes bigger than the half of the visual field, it approach to white, while the smaller area of highest luminance becomes luminous. The divided discs along the abscissa with light and dark surfaces show the configurations of the stimuli. (C) Simulation result of Area effect. The model simulates the concept of the effect quantitatively. The squares along the abscissa with light and dark surfaces show the configurations of the stimuli. See the text for details. Figure B is from Gilchrist *et al.* (1999).

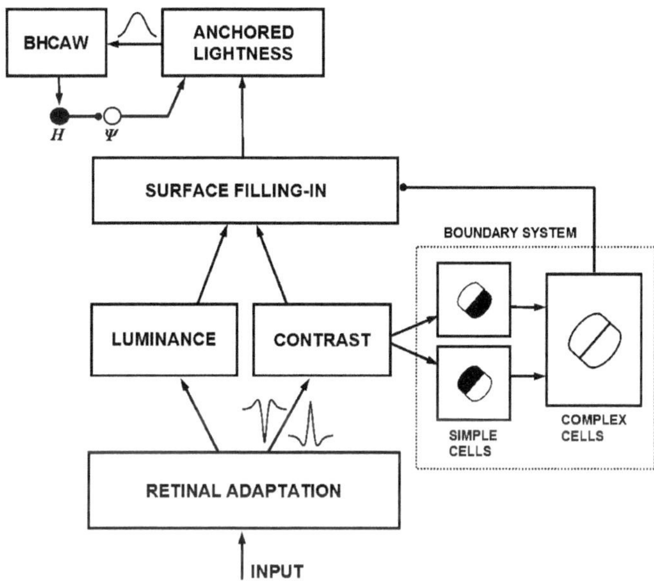

Figure 8. Illustration of the model. See the text for details. Each box indicates an array of cells doing a similar task. Arrow-heads indicate excitatory signals for the post-synaptic units; round-heads indicate inhibitory signals. The Mexican-hat shape and the up-side-down shape of it between the retinal adaptation and contrast modules illustrate the one-dimensional shapes of the on-center off-surround and off-center on-surround antagonistic filters for contrast calculation. The bell-shaped curve between the anchored lightness and BHCAW modules illustrates the one-dimensional shape of the blurring kernel for anchoring. For clarity, the boundary system shows just one orientation. In the simulation, four orientations are used.

sensitivity curve (Gated input in Fig. 9; Baylor *et al.*, 1974a, b; Carpenter and Grossberg, 1981; Koutalos and Yau, 1996); see Appendix A, equations (A1)–(A3) for this mechanism.

The light-adapted signal is further processed at the photoreceptor inner segment where it gets feedback from a horizontal cell (HC) that is connected with other HCs by gap junctions, forming a syncytium that is sensitive to spatial contrast (Fig. 9). HC inhibition further adjusts the sensitivity curve to realize spatial contrast adaptation. It is assumed that the permeability of gap junctions between HCs decreases as the difference of the inputs to HCs from coupled photoreceptors increases. In Fig. 9, for example, where the input pattern has a steep difference (the thick and thin input arrows), the permeability between the left and right HCs decreases. When there is not much difference in inputs, the permeability between the HCs remains large. The model retina hereby segregates and selectively suppresses signals in regions that have strong contrasts, such as a light source. Figure 9 shows only connections between nearest neighbors. In simulations, connections reaching farther than nearest neighbors are also used that model the connectivity and cell types in the retina (Masland, 2001; Sterling, 1998). Inhibition

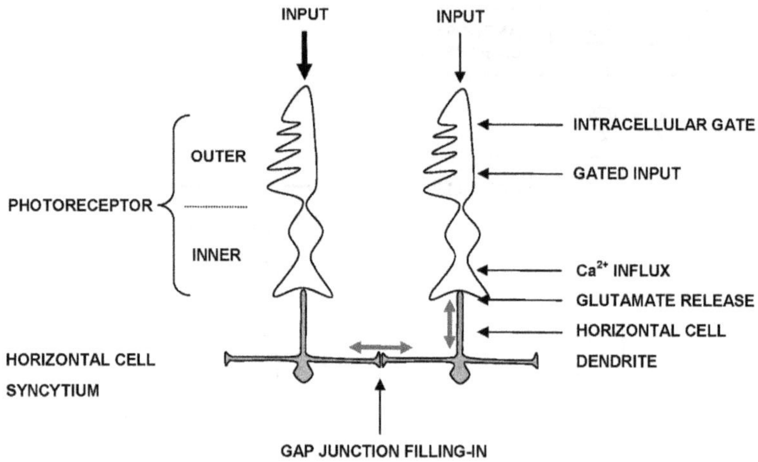

Figure 9. Circuit of retinal adaptation. Two stages of Retinal adaptation are implemented: One, a light adaptation stage at the outer segment of the photoreceptors, the other, a spatial contrast adaptation stage at the negative feedback circuit between the inner-segments of photoreceptors and a syncytium of HCs. It is assumed that the permeability of the gap junctions between HCs decreases as the difference of the inputs to the HCs from the coupled photoreceptors increases. For simplicity only the connections between nearest neighbors are shown. In simulations, long-range connections are also allowed. The gray bidirectional arrows show the mutual influence between connected units. See the text and Appendix for further details.

of the HC on the photoreceptor controls the output of the photoreceptor (Glutamate release in Fig. 9) by modulating Ca^{2+} influx at the photoreceptor inner segment. This feedback prevents the output from being saturated by localized high-contrast input signals. It hereby helps us see a room lit by a light bulb, the light bulb itself, and the label on it. See Appendix A, equations (A4)–(A8).

2.2. Multiple-scale Contrast and Luminance Stage

The retinally-adapted signal is processed by the Contrast stage, which is realized by center-surround networks. The Contrast stage thus receives normalized input signals. The Contrast stage carries out spatial frequency-specific processing in multiple scales using cell types with on-center off-surround (ON cells) or off-center on-surround (OFF cells) found in the retina (Barlow, 1953; Cook and McReynolds, 1998; Kuffler, 1953; Werblin and Dowling, 1969) and the lateral geniculate nucleus (LGN) (Dubin and Cleland, 1977; Hubel and Wiesel, 1961; Jones *et al.*, 2000; Schiller, 1992). Retinal and LGN mechanisms are lumped together to provide the simplest realization of mechanisms that are needed to derive the desired results: see Appendix A, equations (A9)–(A24).

The 1D cross-sections of the contrast operators are illustrated in Fig. 10 between the retinal adaptation and contrast stages. For example, a feedforward shunting on-center off-surround network (Grossberg, 1980, 1983) extracts local contrasts and

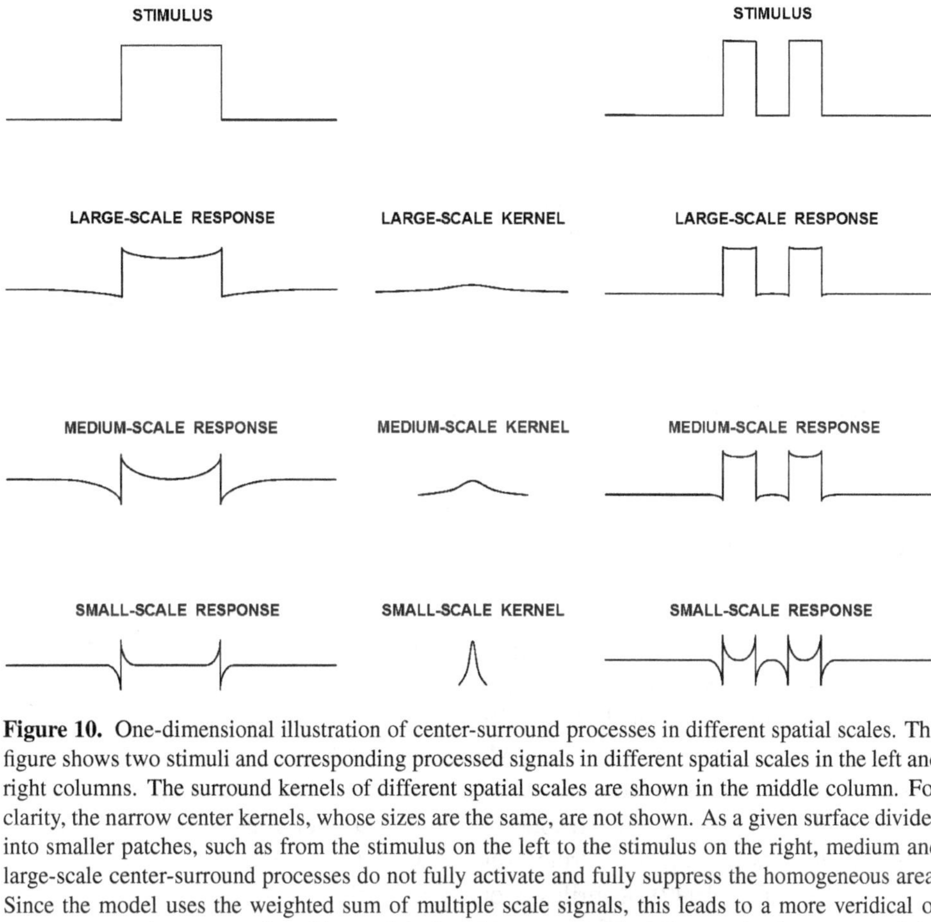

Figure 10. One-dimensional illustration of center-surround processes in different spatial scales. The figure shows two stimuli and corresponding processed signals in different spatial scales in the left and right columns. The surround kernels of different spatial scales are shown in the middle column. For clarity, the narrow center kernels, whose sizes are the same, are not shown. As a given surface divides into smaller patches, such as from the stimulus on the left to the stimulus on the right, medium and large-scale center-surround processes do not fully activate and fully suppress the homogeneous area. Since the model uses the weighted sum of multiple scale signals, this leads to a more veridical or non-compressed representation. The model takes this contrast calculation mechanism as part of the Articulation, Configuration, and Insulation effects of lightness anchoring. See Lightness Anchoring section for further details.

significantly attenuates illumination gradients and background levels in a scale-specific manner. It does so using automatic gain control that divides the input of the center by the local average represented by the surround, thus estimating the local contrast:

$$\text{Contrast} = \frac{L_{spot} - L_{background}}{L_{spot} + L_{background}},$$

where L_{spot}, and $L_{background}$ are the luminances of the probe stimulus and background, respectively. Using different sizes of surround, the system extracts small-scale to large-scale contrasts. These various surround sizes simulate the different sizes of lateral inhibition cell types in the retina (for a review, see Masland, 2001). The model uses a fixed narrow center kernel with the different surround scales (Grossberg *et al.*, 1995; Mingolla *et al.*, 1999) and thereby also simulates the output of a sharp center at the ganglion cells due to interactions in the retinal network (Cook and McReynolds, 1998; Roska *et al.*, 2000). This is sufficient to simulate the targeted lightness data. Although the shunting center-surround network is presented separately from the retinal adaptation stage, it also contributes to background adaptation. Likewise, the adaptation by the HC network is a type of center-surround process with a large surround scale.

Figure 10 illustrates how three different scales, that are defined by different Gaussian off-surround filter widths, respond to luminance inputs. Since the large-scale signal represents the luminance signal better (see large-scale response in Fig. 10), it plays the role of a *luminance signal.* In order to preserve image resolution, single-scale models typically use a small scale. However, a small scale may exhibit *brightness bowing* (see small-scale response in Fig. 10) because a small-scale (high frequency) center-surround unit acts like an edge detector, and thereby suppresses information from large homogeneous surface regions.

The model postulate of multiple scales is consistent with electrophysiological observations in V1 of alert primates (Bartlett and Doty, 1974; Kayama *et al.*, 1979; Kinoshita and Komatsu, 2001; Komatsu *et al.*, 1996) and anesthetized cats (MacEvoy *et al.*, 1998), where cells code not only edge signals but also uniform surface luminance as well. The electrophysiological study with alert monkeys by Friedman *et al.* (2003) also shows that cells in V1 and V2 code uniform color surface information. In the LGN, uniform surface luminance coding units have been found in anesthetized primates (Marrocco, 1972) and cats (Papaioannou and White, 1972) as well as in alert primates (Barlow et al., 1978; Kayama *et al.*, 1979). When surface luminance was temporally modulated, cells in the LGN and V1 of anesthetized cats coding the surface region were modulated (Rossi *et al.*, 1996; Rossi and Paradiso, 1999). Kahrilas, Doty, and Bartlett (1980) failed to detect such neurons in the visual cortex of the awake rabbit. Their data suggest that there may be some differences between species, and techniques of anesthesia also seem to play an important role.

2.3. Boundary formation

Boundary-gated filling-in has helped to simulate many psychophysical and neural data about surface perception (Arrington, 1994; Cohen and Grossberg, 1984; Grossberg and Mingolla, 1985a, b; Grossberg and Kelly, 1999; Grossberg *et al.*, 2002; Kelly and Grossberg, 2000; Pessoa *et al.*, 1995). Grossberg and Todorovic (1988) developed this concept to simulate psychophysical data about brightness (perceived luminance). In their model, the illuminant is discounted by a contrast-detecting on-center off-surround network among cells obeying shunting equations. The surviving contrast signals are used to fill-in a surface brightness estimate within a region surrounded by boundaries that are themselves derived from the illuminant-discounted contrast signals. The anchored Filling-In Lightness Model (aFILM) adopts this hypothesis to explain lightness data using contrast and luminance signals together to trigger filling-in of lightness in regions that are defined by surrounding boundaries (Fig. 8). The filling-in mechanism is, however, assumed not to be diffusion, but rather a much more rapid propagation of signals using long-range horizontal connections.

2.3.1. Simple cells. Boundary formation begins at model simple cells that simulate orientationally tuned simple cells in layer 4 of cortical area V1 (Fig. 8), which have contrast-polarized and oriented ON (excitatory for luminance) and OFF (excitatory for darkness) regions in their receptive fields (Bullier and Henry, 1979; Gilbert, 1977; Hubel and Wiesel, 1962). Model simple cells pool model ON cell LGN outputs in their ON region and OFF cell LGN outputs in their OFF region, consistent with data showing that the ON and OFF subfields of simple cells originate from projections of ON and OFF cells in the LGN, respectively (Alonso *et al.*, 2001; Lee *et al.*, 2000; Reid and Alonso, 1995). Receptive fields of a simple cell are modeled by a pair of elongated Gaussian kernels with shifted centers (Grossberg *et al.*, 1995; Mingolla *et al.*, 1999; Pessoa *et al.*, 1995). The model simple cell is maximally active when a luminance edge is aligned with the oriented border between the ON and OFF regions due to a push-pull design of the ON and OFF regions. For example, a simple cell with a vertical orientation and a light-dark polarity from left to right pools excitatory inputs from on-center off-surround contrast signals on the left side of the kernel and off-center on-surround signals from the right side of the kernel, and also pools inhibitory inputs from on-center off-surround contrast signals on the right side of the kernel and off-center on-surround signals from the left side of the kernel. Since the output is a rectified sum of the filtered signals, the simple cell is active only when a luminance imbalance occurs with the correct polarity across the oriented axis: see Appendix A, equations (A25)–(A32).

2.3.2. Complex cells. The boundary signals of an object need to be joined together even in cases where a scene's contrast polarity reverses along the border of the object, such as at the edge of a middle gray object on a white-and-black checkerboard background (Grossberg, 1994). The model does this using model complex cells that pool a pair of light-dark and dark-light simple cell signals of the same orientation at each position. Pooling simulates the known complex cell property of responding to oriented luminance edges without clear ON/OFF subfield zones (Mechler and Ringach, 2002): see Appendix A, equation (A33). A hierarchical combination of simple cell outputs at complex cells (Hubel and Wiesel, 1962; Schiller *et al.*, 1976) is supported by recent experimental data (Alonso and Martinez, 1998; Dresp and Grossberg, 1997; Martinez and Alonso, 2001), theoretical analysis (Sakai and Tanaka, 2000) and modeling studies (e.g. Gove *et al.*, 1995; Grossberg and Mingolla, 1985a). Additional feedback interactions exist (see Raizada and Grossberg, 2003), but are not needed for present purposes.

2.4. Surface Filling-in

At the Surface Filling-In stage, cells pool signals from both the Contrast and the Luminance stages. Three scales are pooled: small-scale and medium-scale contrast signals and large-scale luminance signals: see Appendix A, equation (A38). These pooled signals fill-in along long-range horizontal connections, a new model feature: see Appendix A, equations (A40)–(A44). Filling-in is blocked by signals from the Boundary System stage.

As noted above, boundary-gated surface filling-in has been used to explain many psychophysical data about brightness and color perception and 3D figure-ground perception. The filling-in mechanism utilizes two streams of the What cortical visual pathway: the surface stream has been predicted to run through the blobs and V2 thin strips to V4, whereas the boundary stream runs through V1 interblobs and V2 pale stripes to V4 (Grossberg, 1994). These two streams have been proposed to compute complementary properties during visual information processing (Grossberg, 2000). A growing list of experiments support the existence of such a filling-in process. The important psychophysical study of filling-in by Paradiso and Nakayama (1991) attempted to catch filling-in 'on the fly'. Arrington (1994) used the Grossberg and Todorovic (1988) model to fit the Paradiso and Nakayama (1991) data about the dynamics of filling-in. The dynamics of aFILM are consistent with the Arrington (1994) explanation.

Filling-in in the blind spot is another example of surface filling-in (Komatsu *et al.*, 1996, 2000). Surface representations can be formed early in visual processing even without top-down cognitive signals (Kamitani and Shimojo, 2003). Sasaki *et al.* (2001) showed using fMRI that when a human subject perceives a transparent illusory region bounded by illusory contours, the V1 region corresponding to the illusory visual field became active. These data, combined with data about illusory contour representations known to exist in V2 (von der Heydt *et al.*, 1984), are consistent with the possibility that surface representations may start to form in V2. More direct evidence comes from electrophysiology combined with cortical imaging: Hung *et al.* (2001) reported that the Craik-O'Brien-Cornsweet Effect can be detected in V1 and is prevalent in V2. In their experiment, the activities of cells having receptive fields inside the homogeneous surfaces were modulated with cusps at the edge of the surfaces. The large spatial scale needed to fully integrate information across visual space (Angelucci *et al.*, 2002) also marks V2 as a processing stage where surface representations start to get formed. Data concerning border ownership representations in V2 and V4 (Zhou *et al.*, 2000) are also consistent with this conclusion. See Grossberg (1994, 1997) for further discussion.

Figure 11 shows a 1D illustration of the model filling-in network. The round units on top represent the cells in the Surface Filling-In stage. The cells in the Contrast and Luminance stage feed their signals topographically to the corresponding filling-in cells. These signals spread between filling-in cells along long-range horizontal connections with Gaussian receptive fields. Signal propagation is gated

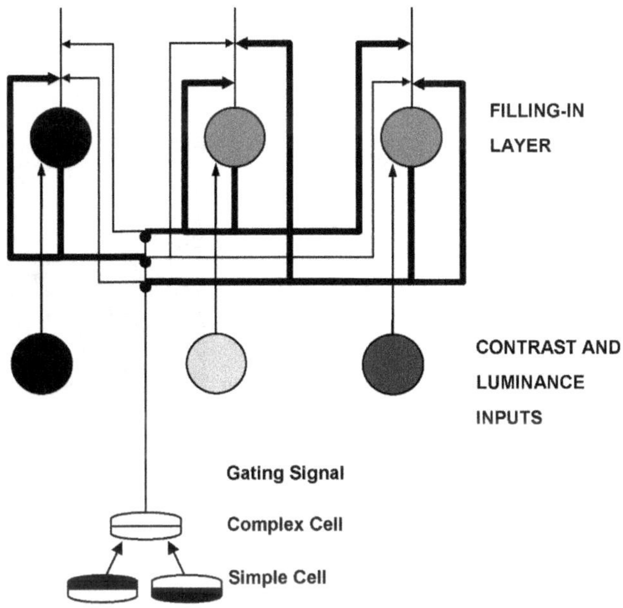

Figure 11. Schematic representation of the parallel filling-in mechanism. In the filling layer, the signals coming from the contrast and luminance units are homogenized within areas defined by luminance boundary signals. The horizontal connections that propagate signals are gated by the gating signals coming from the complex cells. The gated lateral connections have smaller conductances (thin lines) than the other ones (thick ones). The gray levels in the round units represent the activities of them. Just one set of gating signal is shown for clarity.

by Boundary System signals (represented by a vertical line in Fig. 11) coming from the complex cells (for simplicity just one set of gating cells is shown). The gated horizontal connections have smaller conductances (thin horizontal lines) than the other ones (thick horizontal lines). The gray levels of the filling-in cells and of the contrast and luminance input cells represent the level of cell activation. The Filling-In Layer illustrates two homogeneous filled-in regions, black and middle gray, separated by a boundary signal.

2.5. Anchored Lightness

Anchored Lightness stage cells receive inputs from the Surface Filling-In stage. These cells are modulated by a feedback signal originating from the anchoring cells themselves that renormalizes their activities to realize an absolute lightness scale (Figs 8 and 12A). As noted in Section 1.3, the new Blurred-Highest-Contrast-As-White (BHCAW) anchoring rule overcomes problems of the classical HLAW rule: see Appendix A, equations (A45)–(A53), for the anchoring equations.

Figures 3A and 3B illustrate the model's explanation of the area effect for a two-field Ganzfeld configuration, as in Fig. 7. In such a display, there are just two homogeneous surfaces with different luminances, one the target surface, the other

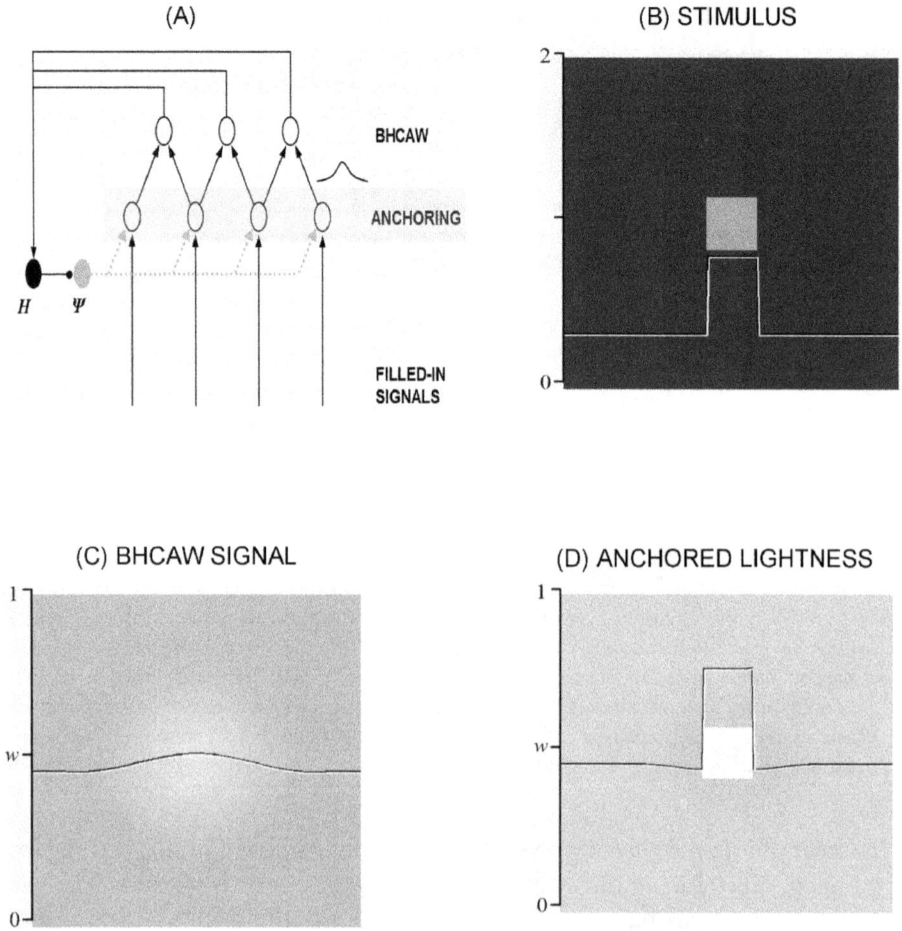

Figure 12. BHCAW rule and Area Effect in a two-field Ganzfeld configuration. (A) Model circuit of lightness anchoring. The activities of the anchoring units are locally pooled by BHCAW units to form a blurred version of the anchoring signals. The filter used to generate the blurred signals is shown as a bell-shaped figure between the anchoring and BHCAW modules. The BHCAW signals are fed to an inhibitory unit H. The unit H becomes active when any of the BHCAW unit exceeds its threshold set to WHITE and fires. When active, H inhibits the tonically active unit Ψ that modulates the activities of anchoring units. This circuit allows the activities of anchoring units to grow until at least one of the blurred version of anchoring signals, BHCAW, meets the criterion of WHITE. See Appendix A for mathematical details. One thing to notice is that the inhibition by H on Ψ lowers but does not completely shut off the activity of Ψ, leaving a chance to the BHCAW signals to go beyond WHITE when the bottom up signal is strong enough, for example, a bright light source of some size. In such a case, even the BHCAW rule will be violated. (B–D), Two-dimensional simulation of two-field Ganzfeld configuration. The curve on each figure represents the activities of the units along the horizontal midline. This convention applies to all the following figures. The scale for the curve is denoted on the left side of each figure. B.stimulus shows the input configuration. D.anchored lightness shows the area effect corresponding to the one in Fig. 4B. Note that the highest activity of the BHCAW module in Fig. C is anchored to white (w).

the Ganzfeld. To achieve anchoring, the model first computes a blurred version of the filled-in surface activity, called the BHCAW signal (Fig. 12A). The highest value of this blurred pattern is anchored to white by using a feedback signal, labeled ψ (Figs 8 and 12A), that multiplicatively rescales the filled-in surface signals by automatic gain control. The process H, which inhibits ψ, becomes activated whenever any BHCAW signal exceeds a threshold that determines the value of white (White in Figs 3A and 3B). Since the highest value of the *blurred* filled-in activity is used for anchoring, the anchored lightness (Anchored Lightness), or unblurred pattern, will look self-luminous (Fig. 3B) when the area of the highest filled-in activity is not broad enough to span the blurring kernel. This happens because the blurring kernel averages lower activities as well at this position, so White will be a smaller value than the maximum filled-in activity. As the area of the highest filled-in activity becomes larger, this mechanism predicts that the background will approach White because of the small difference between the highest and background signals that are averaged by the blurring kernel (Fig. 3A). Figures 12B to 12D show a 2D simulation of the two-field Ganzfeld configuration. The curve in each figure shows the activities of the cells along the horizontal middle section of the 2D image. The labels on the left side of each figure indicate the scale of vertical axis for the curve; in particular, w denotes white.

3. SIMULATION RESULTS

The first results demonstrate that the model can simulate various classical lightness properties, while also anchoring the results.

3.1. Light adaptation

As noted in Section 1.1, the model retina can realize light adaptation properties: see Figs 1A and 1B. Figure 1C illustrates the shift of retinal sensitivity with background illumination that makes model light adaptation possible. This shift property emulates neurophysiological data of Werblin (1971), among others. The leftmost curve of the shift property at lower values of background luminance corresponds to the physical limit of light adaptation observed in retinal ganglion cells (Barlow and Levick, 1969; Enroth-Cugell and Shapley, 1973a). Over a wide range of background luminances, the model obeys the classical Weber law (Grossberg, 1983). Intracellular temporal adaptation in each photoreceptor acts like a divisive negative feedback signal: see Appendix B for stimuli used in this simulation.

3.2. Lightness constancy and discounting illumination gradients

Figures 13A and 13B illustrate how the model discounts the illuminant in response to two light patches on a dark gradient of illumination in the background. To generate the input, light patches with the same reflectance and a background with

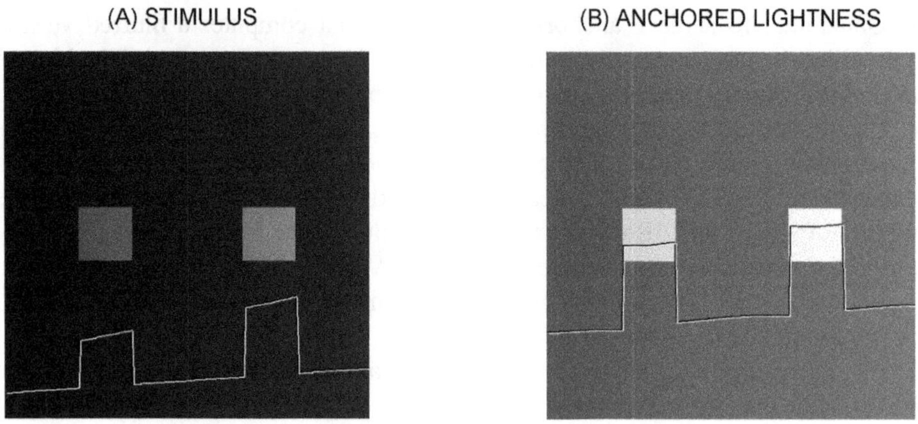

Figure 13. Discounting illuminant. Unevenly illuminated two light patches with identical reflectance (A.stimulus generate a percept that discounts the illumination (B.anchored lightness). However, the model also predicts a bit of bias introduced by the illumination gradient. The light patch on the right looks a bit lighter than the left one. The model also picks up the illumination gradient itself using the large scale.

a smaller constant reflectance were multiplied by a gradient of illumination. The curves on Figs 13A and 13B show the input intensities and anchored lightnesses along the horizontal midline, respectively. Figure 13B shows the property of lightness constancy and illumination discounting: The light patch on the left is almost as light as the one on the right, unlike the one in Fig. 13A. This property comes from the ratio-calculating property of the local contrast units. Figure 13B also shows that, when the gradient of illumination is big enough, the model exhibits a lightness bias where the square patch with higher illumination looks slightly lighter than the one with lower illumination This property is due to the large scale that computes a more veridical representation of the stimuli. This prediction is supported by the observation that, when subjects are asked to decide the perceived reflectance of surfaces, they always give a higher value to the highly illuminated one than the same one with low *illumination* (Gilchrist *et al.*, 1999, p. 826).

3.3. Lightness contrast

Figure 14 shows a simulation of lightness contrast. The two middle gray patches in Fig. 14A have identical luminance. Small and medium scales calculate local ratio contrasts, and their contribution makes the light square on the dark background look lighter than the one on the bright background, even though they have identical luminance (Fig. 14B).

3.4. Contrast constancy: evenly and unevenly illuminated Mondrians

Figures 15A and 15B show an evenly illuminated Mondrian and the corresponding model percept, respectively. A part of the square on the upper left of each figure is

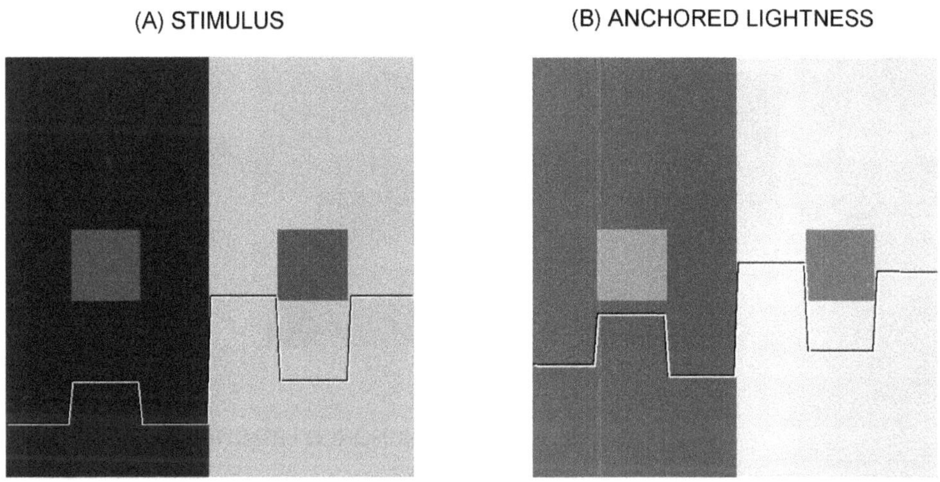

Figure 14. Simultaneous contrast. Two identical square patches on different backgrounds (A.stimulus) are perceived differently (B.anchored lightness). The one on the dark background looks lighter. Local contrast signals provide the source of this difference.

cut and pasted into the square on the bottom right. Since both squares have the same luminance, Fig. 15A shows no cut patch in the bottom right square. Figure 15B shows that the square on the top left is perceived to be lighter than the bottom right square, because the square on the right bottom is surrounded by lighter surfaces than the square on the left top. The square on the right bottom thus receives more surround suppression than the square on the upper left.

Figures 15C and 15D show that the model generates the same result when a light source at the bottom right corner creates a gradient of illumination from bottom right to top left. The model output (Fig. 15D) shows that the upper left square still looks lighter than the bottom right square despite the fact that the luminance at the bottom right is higher. This happens because the small and medium scales calculate the local contrasts and ignore the large scale illumination gradient. This 'contrast constancy' calculation by the two smaller scales overrides the prediction by the big scale that picks up the gradient. Grossberg and Todorovic (1988) first simulated this effect with a single contrast scale.

3.5. Craik-O'Brien-Cornsweet effect

The Craik-O'Brien-Cornsweet effect (Cornsweet, 1970) is challenging because filtering alone cannot explain the percept. As with the percepts above, the model explains this illusion using a combination of illuminant-discounting filtering followed by boundary-gated surface filling-in, as first simulated in Grossberg and Todorovic (1988). Figure 16A shows the stimulus with a uniform background luminance with a luminance cusp in the middle. Figure 16B shows the anchored lightness percept in which the left half of the image looks uniformly lighter than the right half. At the Surface Filling-In stage (Fig. 8), the cusp-like inputs from

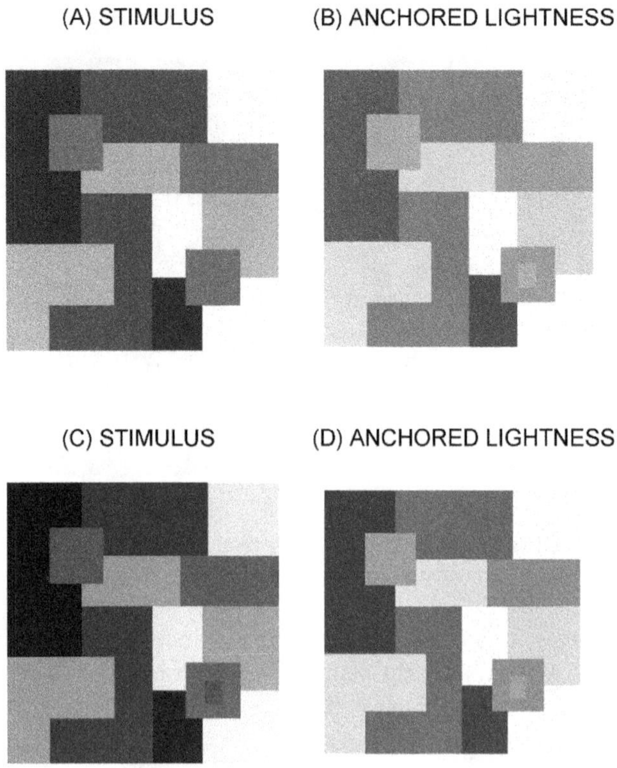

Figure 15. Evenly and unevenly illuminated Mondrians. To facilitate the comparison, a part of the square on the upper left of each figure has been cut and pasted to the square on the bottom right of the figure. (A–B) Evenly illuminated Mondrian. Stimulus and anchored lightness panels show the configuration of an evenly illuminated Mondrian stimulus and the output of the model, respectively. (C–D) Unevenly illuminated Mondrian. The stimulus shows the differently illuminated target surfaces because of the illumination gradient. The gradient of illumination is made by a light source located at the bottom-right corner. Anchored lightness shows the final output of the model. See the text for details.

the Contrast stage (Fig. 16D) are smoothed within areas defined by Boundary stage signals (Fig. 16C). This smoothing makes the surface on the left lighter than the one on the right because of the larger contrast activities within this region (Fig. 16D). The area-defining boundaries play a critical role. When no boundaries surround the luminance cusp, the illusion is not seen because the contrasts can spread via filling-in around both sides of the cusp region and smooth one another; see Grossberg and Todorovic (1988).

3.6. Double Brilliant illusion

Bressan (2001) described a Double-Brilliant illusion wherein the diamond that has less contrast around it (Fig. 17B) looks more brilliant than the one having more contrast around it (Fig. 17A), even though both diamonds have the same luminance (Fig. 17C). How this brilliant appearance gives rise to a judgment of 'lightness' may

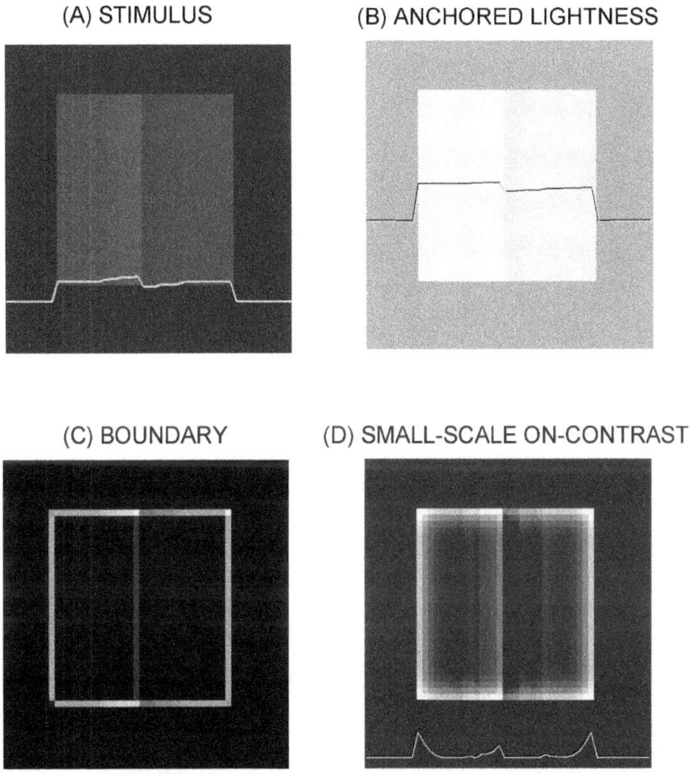

Figure 16. Craik-O'Brien-Cornsweet effect. (A) Two divided identical surfaces with a luminance cusp in the middle. (B) Simulated lightness of the model. The two surfaces are perceived differently. The boundary-gated homogenization of surface signals through a filling-in process makes the surface on the left look slightly lighter than the one on the right. (C) Boundary. (D) Small-scale contrast signals for the two surfaces. Left surface has more activities than the right one explaining the difference at the filled-in surface lightnesses in B.

depend on subtle stimulus matching properties in a particular experiment. Bressan (2001, p. 1042, caption of Fig. 8) writes: 'Both diamonds are white, and mounted on luminance-ramp settings. But the diamond sitting on the dark end of the ramp (left) looks whiter than the one sitting on the light end of the ramp (right)'. There do not seem to be any parametric psychophysical data, however, to characterize the nature of the judgment, which may include a variety of factors, including cognitive ones.

The model simulates the more brilliant appearance of Fig. 17B as a manifestation of spatial contrast adaptation, as in Fig. 1D and 1E (see Appendix A, equations (A4)–(A8)), and thereby raises the issue of what additional perceptual processes or criteria might lead to the claim that Fig. 17A is lighter. In particular, the permeability of gap junctions in the horizontal cell (HC) syncytium decreases only where there is a sharp luminance edge in the input (Fig. 9). The gradual change of luminance around the diamond in Fig. 17B does not block the diffusion of

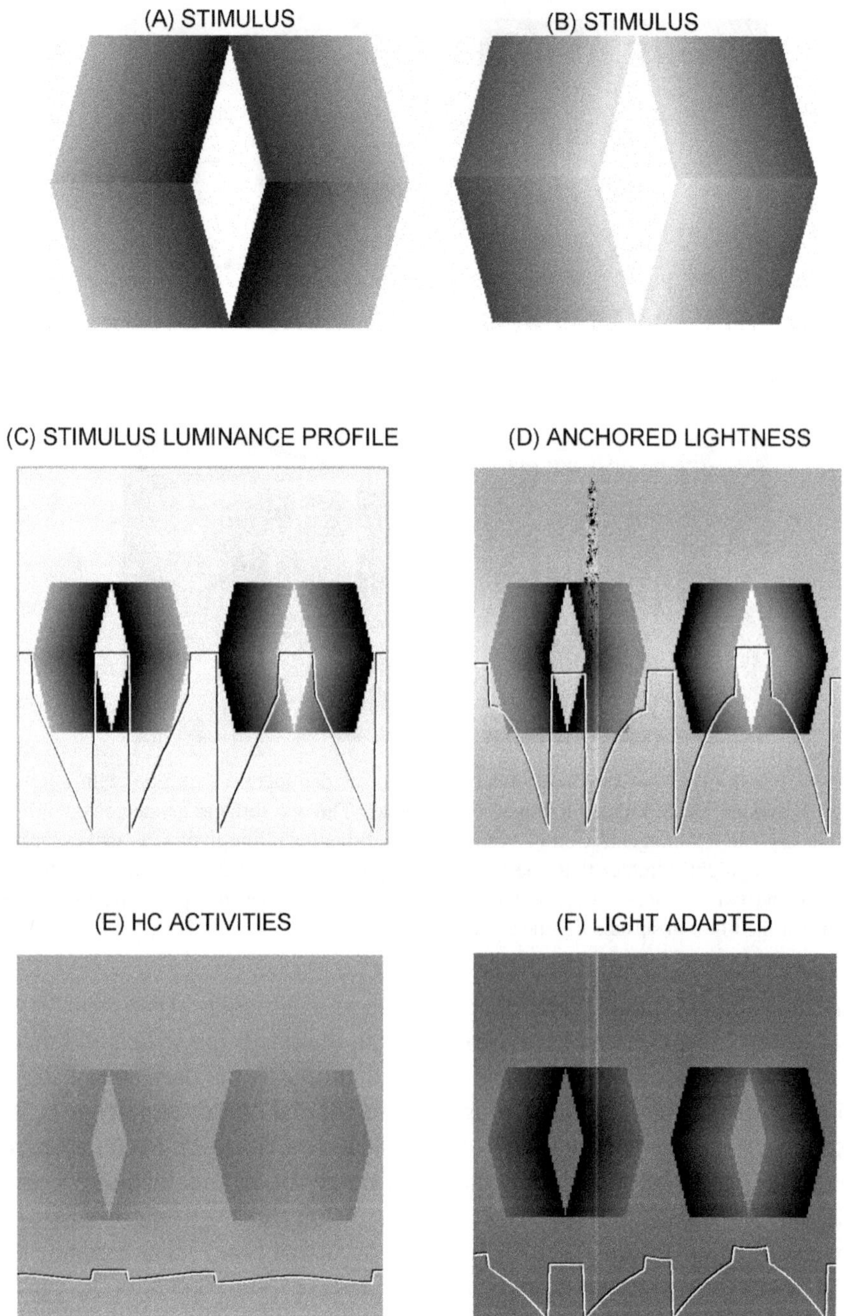

Figure 17. Double Brilliant Illusion. (A–B) Stimuli. (C–D) Stimulus and the output of simulation, respectively. (E–F) Simulated activities of HCs and the steady outputs of photoreceptor inner segments, respectively. See the text for more details. The Figs A and B are from Bressan (2001).

signals across the HC syncytium. The luminance edges around the diamond in Fig. 17A do block diffusion and segregate the diamond region from the rest of the figure. This gated-diffusion process is simulated in Fig. 17E. The segregated large signals shown in the diamond region on the left of 17E suppress the corresponding region of photoreceptor outputs. This results in a less active diamond region on the left in Fig. 18F compared to the diamond region on the right. The anchored lightness of the model in Fig. 18D reflects this difference. The model interprets the contrast adaptation mechanism to be retinal, hence monocular. If this is a correct assumption, then dichoptic presentations of different stimulus parts to each eye may yield different lightness effects. Figures 1D and 1E and Section 4.1 around Fig. 18 show how spatial contrast adaptation can greatly facilitate perception of scenes with large contrast differences.

3.7. Anchoring properties

In addition to simulating classical lightness effects, the model also simulates, for the first time, the four major effects of lightness anchoring (Articulation, Configuration, Insulation, Area Effect) that have been described by Gilchrist and his colleagues.

3.7.1. Articulation Effect. The Articulation effect says that, as the display contains more gray surfaces, the range of perceived lightness widens (Figs 4A–4F). Even in the two-Mondrian case in Fig. 4C, the reflectances of these patches range from black to white, covering the full span of reflectance. Adding more gray patches does not result in a wider range of reflectances. Figure 4G summarizes the model simulation. As the number of surface patches having different luminances increases, the image contains more high spatial frequencies. The medium and large spatial scale kernels thus have less chance to fully activate and suppress the homogeneous area of each patch. Figure 10 illustrates the situation: The divided square luminances on the right cause higher contrast signals in the medium and large scales compared to the corresponding contrast signals on the left column with a larger square stimulus. The loss of suppression by each spatial scale results from the mismatch between the size of the filters and of the scenic patches. Mismatch means less suppression, thus a more veridical representation for that scale, and thus a more veridical percept. The BHCAW process assures that the data remain anchored at white.

3.7.2. Configuration Effect. The Configuration effect says that, when a display contains gray surfaces arranged in a Mondrian, a wider range of lightnesses is perceived than when the same gray surfaces are arranged in a luminance staircase. Figures 5E and 5F summarize model simulations. The model explains this effect much as it does the Articulation effect: in the Mondrian, the intermingled luminance patches are arranged in a more radially compact way. The round-shaped surround kernels in the Contrast stage are thus influenced by more luminances of surrounding surfaces compared with the staircase arrangement. This lets the surround kernels set

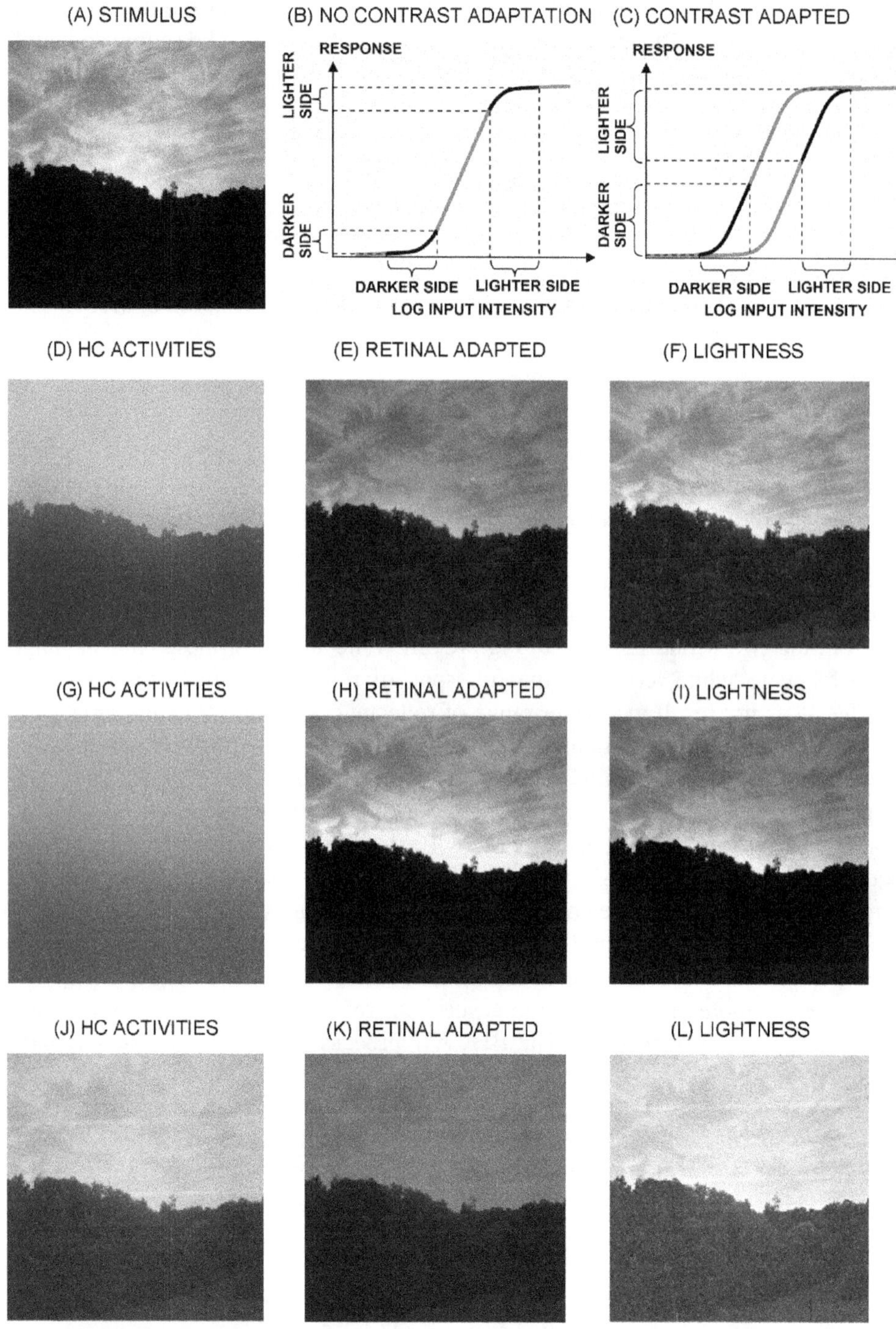
(A) STIMULUS
(B) NO CONTRAST ADAPTATION
RESPONSE
LIGHTER SIDE
DARKER SIDE
DARKER SIDE
LIGHTER SIDE
LOG INPUT INTENSITY
(C) CONTRAST ADAPTED
RESPONSE
LIGHTER SIDE
DARKER SIDE
DARKER SIDE
LIGHTER SIDE
LOG INPUT INTENSITY
(D) HC ACTIVITIES
(E) RETINAL ADAPTED
(F) LIGHTNESS
(G) HC ACTIVITIES
(H) RETINAL ADAPTED
(I) LIGHTNESS
(J) HC ACTIVITIES
(K) RETINAL ADAPTED
(L) LIGHTNESS

the local means (surround activities) to be more different from the corresponding center activities, resulting in a bigger range of perceived reflectances. In other words, if all the adaptation and contrast stage surround activities were the same as their center activities, surround inhibition would drive them all to zero. The radially compact arrangement decreases the distance between different levels of gray patches, thereby inducing stronger lateral inhibition. The influence of the distance between an inducer and test surfaces has been observed in lightness (Newson, 1958) and brightness experiments (Cole and Diamond, 1971; Fry and Alpern, 1953; Leibowitz *et al.*, 1953), where the darker test surface became lighter with increasing distance from the inducer due to surround inhibition. Again, BHCAW process anchors the perception of white.

3.7.3. Insulation Effect. The Insulation effect of Fig. 6 shows that, when the staircase display is insulated by a white surround, the range of its perceived reflectance widens. Figures 6D and 6E show the data and simulation results, respectively. Spatial contrast also helps to explain this effect: insulation of gray surfaces with a white surround results in an expansion of the range of lightness due to the newly added suppression on dark patches by the surround. Insulation by a black surround, however, may not cause much difference because the gray surfaces are illuminated 30 times more than the background. Since the gray patches are already not getting much background inhibition, introducing black insulation does not significantly change the amount of inhibition, thus hardly changing the percept. The BHCAW process again anchors the lightness.

3.7.4. Area Effect. The Area effect in Fig. 7 shows that, in a two-field Ganzfeld situation, as an area other than the area of highest luminance becomes larger than the half of the visual field, its lightness approaches white while the highest luminance area is pushed above white. Figure 7C shows the model simulation, which closely fits the data. Section 2.5 predicts how self-luminosity of a small highest luminance area may be explained by the BHCAW rule: when the highest luminance area is smaller than the blurring kernel at the anchoring stage, the blurred filled-in surface signals will have lower highest activities compared to the un-blurred image

Figure 18. Effect of gated diffusion for spatial contrast adaptation of the model. (A) Stimulus. (B) Illustration of retinal sensitivity curve with no contrast adaptation. The signals for the dark part (darker side along the abscissa) have been mapped to the very low part of the response axis (darker side along the ordinate). The signals for the light part (lighter side along the abscissa) have been mapped to the saturating portion of the curve (lighter side along the ordinate). (C) Illustration of contrast adaptation. With the two different sensitivity curves, the model retina has mapped the widely separated input signals (darker side and lighter side along the abscissa) to quite 'visible' portions of the response (darker side and lighter side along the ordinate). (D) The steady-state activities of HCs at the input A. (E) Retinally adapted signals. Retinal adaptation with gated diffusion at the HC syncytium gives the properly rescaled steady-state output at the inner-segment of the photoreceptor. (F) The final output of the model. (G–I) Simulation with free diffusion among connected HCs. (J–L) Simulation with no diffusion among HCs. See the text for further details.

(Fig. 3B). Since blurred signals anchor lightness, the highest luminance area will look lighter than white. Figure 3B corresponds to the increasing portions of curves in Figs 7B and 7C, whereas Fig. 3A corresponds to the flat regions of these curves.

4. DISCUSSION

The aFILM model integrates known neuroanatomy, electrophysiology, and psychophysics (e.g. Table 1) to clarify how the brain generates a representation of surface lightness. The following discussion analyzes the model's assumptions and limitations.

4.1. Retinal adaptation

The model simulates retinal adaptation using two mechanisms: see Fig. 9. First, at the outer-segment of the photoreceptors, the model realizes an intracellular mechanism of light adaptation whose sensitivity to light is controlled by concentrations of chemicals, such as Ca^{2+} ions, that temporally average visual stimuli (Koutalos and Yau, 1996). Carpenter and Grossberg (1981) used such a mechanism to quantitatively simulate the Baylor *et al.* (1974a, b) data about turtle cone adaptation. Second, at the inner-segment of the photoreceptors, the model simulates spatial contrast adaptation by modulating gap junction communication between horizontal cells (HCs) and thus how HCs influence glutamate release from the photoreceptor (Fahrenfort *et al.*, 1999; Verweij *et al.*, 1996). The permeability of HC gap junctions is affected by various mechanisms, including neurotransmitters (DeVries and Schwartz, 1989, 1992; McMahon, 1994; Xin and Bloomfield, 2000) and transjunctional voltage (Lu *et al.*, 1999; Spray *et al.*, 1979).

For example, for two HCs connected by a gap junction, the permeability of the junction decreases as the difference increases between the inputs that the HCs receive from the photoreceptors (Fig. 9). Such a model retina can properly rescale inputs that have too much contrast, such as the one in Fig. 18A. Figure 18D shows the steady-state HC activities; the dark and light image regions deliver different suppressive feedback signals to the photoreceptors. Figure 18B shows how the retina would respond without spatial contrast adaptation; note that the responses at both the darker and lighter sides of the image are highly compressed, hence insensitive to input contrast differences. Figure 18C shows how spatial contrast adaptation generates two distinct sensitivity curves at photoreceptor inner-segments due to the two different negative feedback levels of the HC network. The network hereby rescales responses at the inner-segment of the photoreceptor to be more sensitive to input contrast differences at both darker and lighter image regions. The rescaled steady-state output of the photoreceptor inner-segments are shown in Fig. 18E. The output of the model photoreceptor in Fig. 18F shows visible dark and light image regions. Figures 18G–18I show a simulation without the HC gating mechanism. The adapted signals in Fig. 18H and the output in Fig. 18I show signal distortion (a halo) along the border of the dark and light parts, and the dark part is less visible. Figures 18J–18L show a simulation with no diffusion among HCs. The results show a prominent compression of signals.

In summary, in addition to the light adaptation at the outer-segment of the photoreceptor, which shifts its sensitivity curve, HC negative feedback further shifts photoreceptor sensitivity in response to the spatial context of input contrasts: see Appendix A, equations (A1)–(A8).

Kamermans *et al.* (1996) discussed how negative feedback between the photoreceptor and the HC network might influence the length constant of the HCs. The present analysis does not need this refinement to quantitatively explain the targeted data. For the same reason, the model also does not simulate the cone-rod circuitry switch (Mills and Massey, 1995; Ribelayga *et al.*, 2002) and the pupillary light reflex (Dowling, 1987), which also influence adaptation.

4.2. Luminance and contrast: multiple-scale filtering

The neuroanatomy of the retina already includes multiple-scale representations whereby center-surround processes shape the outputs of ganglion cells having diverse receptive field sizes (Masland, 2001; Roska *et al.*, 2000; Werblin *et al.*, 2001). The V1 map includes cells whose optimal spatial frequencies increase as their positions move away from blob centers (De Valois and De Valois, 1988; Edwards *et al.*, 1995). Issa *et al.* (2000) also show gradual changes in cell spatial frequency preferences that conform to the hypercolumn cortical organization in V1. Data about cell responses in LGN, V1 and V2 to uniform surface luminance and color also support the existence of large spatial scales (Bartlett and Doty, 1974; Friedman *et al.*, 2003; Komatsu *et al.*, 2000; Marrocco, 1972; Papaioannou and White, 1972). The model hypothesis that achromatic cells in the blob stream pool their multiple-scale representations has not yet been directly tested. The pooling of ON and OFF signals, however, is consistent with the finding that the segregation of ON and OFF channels from the retina and LGN, and their projection to layer 4 in V1 is largely lost in the cortex of the ferret (Chapman and Gödecke, 2002; see Schiller (1992) for a review).

Multiple-scale filtering mechanism has also been proposed as part of a recent evolution of the Retinex model (Jobson *et al.*, 1997a, b; Rahman *et al.*, 1996, 1997). Instead of using predetermined multiple-scale filters all the time and everywhere on the image, the current model reconfigures the shape of the filters to meet the demanding contexts for a given image using the horizontal gating system (see Appendix A: Retinal adaptation). For example, when there is a bright spotlight like the one in Fig. 2G, the HC filters around the edge of the spot light would shrink by disconnecting the communication between the light and the surrounding dark areas using the diffusion gating mechanism illustrated in Fig. 2. The result of this dynamic control of the shape of the HC filters is a natural-looking compression of input signals as illustrated in Fig. 18 where the two extreme input areas (Fig. 18, Darker Side, Lighter Side) are pulled in to the visible area (Fig. 18C, Fig. 18E).

The competence of this mechanism has been demonstrated and compared with Retinex in Hong and Grossberg (2004, Section 3.5), where enhanced visibility is demonstrated without sacrificing a natural appearance.

4.3. Boundary representation

The model's Boundary System stage does not, for simplicity, implement the boundary completion property whereby the visual system forms perceptual groupings from disjoint image inducers (Field *et al.*, 1993; Grossberg and Mingolla, 1985a, b; Grossberg and Raizada, 2000; Kellman and Shipley, 1991; Raizada and Grossberg, 2001; von der Heydt *et al.*, 1984). Incorporation of this property would explain more psychophysical data about such properties as illusory contours (Gove et al., 1995; Grossberg and Mingolla, 1985a), 3D figure-ground separation (Grossberg, 1994; Kelly and Grossberg, 2000), and surface noise suppression in response to noisy images (Grossberg *et al.*, 1995; Mingolla *et al.*, 1999). Surface noise suppression in response to noisy images is necessary because visual signals that reach the retina are distorted and occluded by the retinal layers, blind spot, and veins. Boundary completion by bipole cells can group pixels from such noisy images into coherent boundaries that define object contours. Surface filling-in smoothes pixel values that are surrounded by the same boundaries and thus allows the brain to distinguish between noisy signals due to preprocessing artifacts and those that represent object properties. For figures that do not require significant boundary completion, surface noise suppression can be achieved by the present model's simplified boundary and surface filling-in properties, as illustrated by Fig. 19D. Figure 19C uses a smaller boundary-gating parameter than Fig. 19D, hence smoothes the Lena image less.

4.4. Surface filling-in

A new mechanism of surface filling-in, called Gated Blurring, uses propagation via long-range horizontal connections that are predicted to be within cortical layer 2/3 of V2, among other places. Previous models used nearest-neighbor diffusion for filling-in. Both types of model predict that filling-in is gated by boundary signals that block signal propagation across positions at which luminance contrasts are registered (Fig. 11). This prediction of the model is consistent with the known horizontal connections in the visual cortices (Angelucci *et al.*, 2002; Gilbert and Wiesel, 1979; Rockland and Lund, 1982; Stettler *et al.*, 2002; Yabuta and Callaway, 1998). Long-range horizontal connections have previously been predicted to carry out boundary completion via cells in layer 2/3 of cortical areas V1 and V2 (Grossberg, 1999; Grossberg and Mingolla, 1985a, b; Grossberg and Raizada, 2000; Grossberg and Swaminathan, 2004; Raizada and Grossberg, 2003). The new proposal implies that both boundary completion and surface filling-in may be accomplished by long-range connections, presumably within the parallel boundary and surface processing streams that are projections of interblobs and

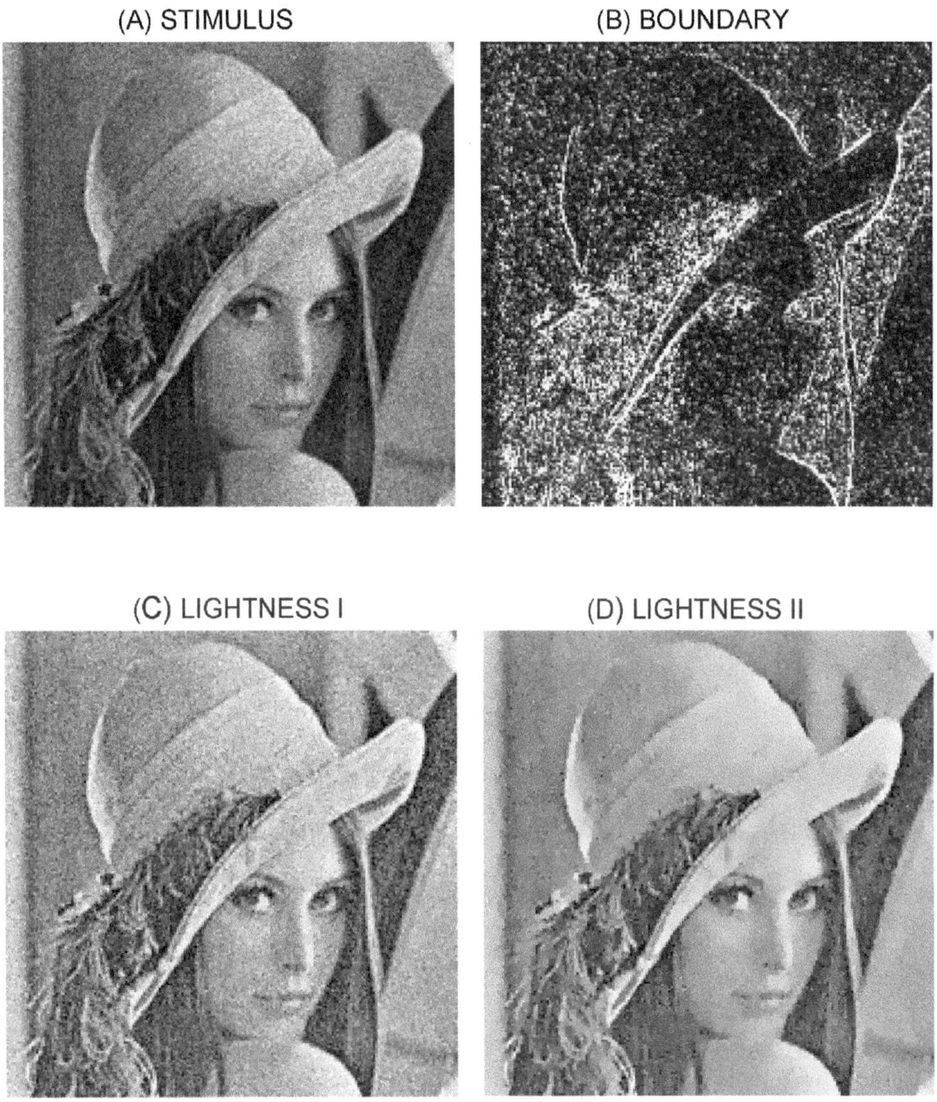

Figure 19. Noise suppression property of the model. (A) Input with Gaussian noise of signal-to-noise ratio of 10 dB. (B) Boundary signals at input A. Despite the disruptive noise, boundary signals still show coherent representation of the edge signals. (C) Output of the model without parameter change. The model does not show much of noise suppression property. (D) Output of the model with a smaller gating parameter ($\varepsilon = 20$). With a smaller gating parameter, the model shows the noise suppression property. See the text for further discussion.

blobs, respectively. The boundary-gating mechanism, by selectively allowing communication between only certain connections, dynamically restructures cell receptive fields. One possible mechanism is axo-axonal gating of horizontal connections, which is consistent with the report by Kobayashi *et al.* (2000) of norepinephrine-mediated suppression of horizontal propagation in V1. The new model mechanism runs at least 1000 times faster than previous nearest-neighbor-based diffusion models, and thus clarifies how filling-in can occur with realistic delays. For example, 10 iterations of the non-diffusive filling-in process were used to generate the filled-in image of the Craik-O'Brien-Cornsweet effect in Fig. 16B. With nearest-neighbor diffusion, about 10 000 iterations were needed. A simulation using long-range diffusion with interactions beyond nearest-neighbors needed 100 iterations.

A long-range filling-in mechanism has also been proposed by Sepp and Neumann (1999). However, instead of the inter-scale competition used in their model, where different scales compete to gain control to fill a given area, the current mechanism uses boundary signals to dynamically reconfigure the shape of the filling-in areas. Our current approach is intrinsically faster because it does not use inter-scale competition, and eliminates the possibility of blurring caused by imperfect scale selection for a given area in the Sepp-Neumann model. Another approach to parallel filling-in was made by Fischl and Schwartz (1997) who used an approximate solution of a nonlinear convolution. The current mechanism distinguishes itself by combining convolution type filling-in with a direct gating mechanism by boundary signals. The parallel diffusion mechanism of the model is a generalization of the boundary-gated diffusive filling-in model of Grossberg and Todorovic (1988).

4.5. Lightness anchoring

The model assumes that lightness anchoring happens after surface filling-in occurs in V2 or V4. Within the larger FACADE theory of 3D vision and figure-ground separation (Grossberg, 1994), this allows surface representations at different depths to have their own anchors. Within FACADE, V4 is the area where modal, or visible, 3D surfaces are represented, whereas V2 can fill-in amodal surface representations. The electrophysiological experiment by MacEvoy and Paradiso (2001) reported lightness constancy in V1. However, their experiment does not provide unequivocal evidence that V1 is the place where anchoring occurs. It demonstrates just one aspect of lightness perception, namely, discounting the illuminant, or input normalization, which can be initiated at the earlier model Retinal Adaptation and Contrast stages. Another factor suggesting that anchoring occurs no earlier than V2 is that long-range perceptual grouping, or boundary completion, takes place in V2 (Peterhans and von der Heydt, 1989; von der Heydt, Peterhans and Baumgartner, 1984), and influences what surfaces get filled-in: see Gilchrist *et al.* (1999) and Grossberg (1994) for further discussion. Extrastriate involvement in lightness anchoring is also suggested by the fact that global integration of information, which is needed for the BHCAW rule, needs a bigger scale of interaction than that supported by horizontal connections in V1 (Angelucci *et al.*, 2002). V2 provides a rich environment for the boundary system (interstripes) and luminance and contrast signals (thin stripes) to interact (Roe and Ts'o, 1995) to begin to form surface percepts. The data of Hung *et al.* (2001) showing a prominent Craik-O'Brien-Cornsweet effect in V2 are also compatible with this assumption.

Table 2.
Parameters used in simulations

Names	Symbols	Values
Upper bound of gain control at photoreceptor	B_z	500
Small-time scale input contribution rate for gain control	C_I	200
Large-time scale input contribution rate for gain control	$C_{\overline{I}}$	600
HC feedback constant	B_h	0.04
Photoreceptor depolarization constant	B_s	(B_z/C_I)
Saturation amplitude	a_H	6
Half-way cut-off	b_H	0.1
Shift of permeability of HC gap junction	β_p	0.08
Steepness of permeability of HC gap junction	λ_p	0.01
Size of connected neighbor for horizontal cell	ε_H	8
Activation decay	A	0.5
Depolarization constant	B	1
Hyperpolarization constant	D	1
Center spatial scale for the center-surround stage	α	0.2
Surround spatial scales, small, medium	β	3 (for small scale), 14 (for medium scale)
Amplitude of Gaussian kernel	κ	4
Vertical, horizontal widths of the ON, OFF elliptic simple cell receptive fields	γ_v, γ_h	$0.1, 5\gamma_v$,
The shift of the centers of the ON, OFF elliptic simple cell receptive fields	$Shift_{(L)}, Shift_{(R)}$	$-\gamma_v, \gamma_v$
Amplitude of activation function	a_B	0.7
Half-way cut-off of activation function	b_B	0.15
Vertical/horizontal width of the complex cell's gating field	γ_{cv}, γ_{ch}	0.3, 0.7
Small, Medium, Large Scale Weight	w_s, w_m, w_l	0.25, 0.25, 0.5
Baseline bias of multiple-scale input	b_M	0.01
Spatial constant of the cable of the filling-in unit	σ	30 (non-diffusive propagation), 11.2 (long-range diffusion)
Gating constant	ε	100 (non-diffusive propagation), 200 (long-range diffusion)
Amplitude of conductance	δ	given in equation A43 (non-diffusive propagation), 0.6 (long-range diffusion)

4.6. Area effect in natural images

The area effect is limited to simple Ganzfeld configurations. Gilchrist *et al.* (1999, p. 802) note: 'Strictly speaking, the rule applies to visual fields composed of only two regions of nonzero luminance. Application of the rule to more complex images remains to be studied'. In the model, it is assumed that when this Ganzfeld configuration was tested, the visual system of the subject adapted its multiple scales to compensate for the unusually sparse visual cues. Sections 2.2 and 4.2 noted that the model incorporates multiple spatial scales which suppress signals that are uniform with respect to each scale. Hence, given the sparse contrasts in the Ganzfeld display, the model would suppress small scales. Multiple scales were not used in the anchoring module, for simplicity. Instead, two different parameter sets were used to explain the area rule; for simple images having just two regions of non-zero luminance (Fig. 7), a bigger Gaussian kernel was used. For the other, more complex, images with smaller regions, a smaller kernel was applied (see Table 2 for parameters). The two anchoring kernel sizes were chosen that best fit the data suggested in the Anchoring theory by Gilchrist *et al.* (1999). Automatic rescaling of anchoring will be incorporated when the model fully exploits its multiple scales for purposes of 3D vision and figure-ground perception, in an extension of how multiple scales have already been used to explain related data in FACADE theory (Grossberg, 1994, 1997; Kelly and Grossberg, 2000).

Acknowledgement

Authorship in alphabetical order. SG and SH were supported in part by the Air Force Office of Scientific Research (AFOSR F49620-01-1-0397), the Defense Advanced Research Projects Agency and the Office of Naval Research (ONR N00014-95-1-0409), the National Science Foundation (NSF SBE-0354378), and the Office of Naval Research (ONR N00014-01-1-0624).

REFERENCES

Adams, M. M., Hof, P. R., Gattass, R., Webster, M. J. and Ungerleider, L. G. (2000). Visual cortical projections and chemoarchitecture of macaque monkey pulvinar, *J. Compar. Neurol.* **419**, 377–393.

Adelson, E. H. (2000). Lightness perception and lightness illusions, in: *The Cognitive Neurosciences*, Gazzaniga, M. (Ed.), pp. 339–351. MIT Press, Cambridge, MA.

Adelson, E. H. and Pentland, A. P. (1996). The perception of shading and reflectance, in: *Perception as Bayesian Inference*, Knill, D. C. and Richards, W. (Eds), pp. 409–423. Cambridge University Press, New York.

Alonso, J. M. and Martinez, L. M. (1998). Functional connectivity between simple cells and complex cells in cat striate cortex, *Nature Neurosci.* **1**, 395–403.

Alonso, J. M., Usrey, W. M. and Reid, R. C. (2001). Rules of connectivity between geniculate cells and simple cells in cat primary visual cortex, *J. Neurosci.* **21**, 4002–4015.

Anderson, B. L. (1997). A theory of illusory lightness and transparency in monocular and binocular images: the role of contour junctions, *Perception* **26**, 419–453.

Angelucci, A., Levitt, J. B., Walton, E. J., Hupé, J. M., Bullier, J. and Lund, J. S. (2002). Circuits for local and global signal integration in primary visual cortex, *J. Neurosci.* **22**, 8633–8646.
Arend, L. E. (1990). Perceived lightness, but not brightness of achromatic surfaces depends on perceived depth information, *Perception and Psychophysics* **48**, 82–90.
Arend, L. E. (1993). Mesopic lightness, brightness, and brightness contrast, *Perception and Psychophysics* **54**, 469–476.
Arend, L. E. (1994). Surface colors, illumination, and surface geometry: intrinsic-image models of human color perception, in: *Lightness, Brightness and Transparency*, Gilchrist, A. (Ed.), pp. 159–213. Lawrence Erlbaum, Hillsdale, NJ.
Arend, L. E. and Reeves, A. (1986). Simultaneous color constancy, *J. Optic. Soc. Amer.* **3**, 1743–1751.
Arend, L. E. and Spehar, B. (1993a). Lightness, brightness, and brightness contrast: 1. Illuminance variation, *Perception and Psychophysics* **54**, 446–456.
Arend, L. E. and Spehar, B. (1993b). Lightness, brightness, and brightness contrast: 2. Reflectance variation, *Perception and Psychophysics* **54**, 457–468.
Arrington, K. F. (1994). The temporal dynamics of brightness filling-in, *Vision Research* **24**, 3371–3387.
Baccus, S. A. and Meister, M. (2002). Fast and slow contrast adaptation in retinal circuitry, *Neuron* **36**, 909–919.
Barlow, H. B. (1953). Summation and inhibition in the frog's retina, *J. Physiol. (London)* **119**, 69–88.
Barlow, H. B. and Levick, W. R. (1969). Three factors limiting the reliable detection of light by retinal ganglion cells of the cat, *J. Physiol.* **200**, 1–24.
Barlow, R. B., Snodderly, D. M. and Swadlow, H. A. (1978). Intensity coding in primate visual system, *Exper. Brain Res.* **31**, 163–177.
Bartlett, J. R. and Doty, R. W. (1974). Response of units in striate cortex of squirrel monkeys to visual and electrical stimuli, *J. Neurophysiol.* **37**, 621–641.
Baylor, D. A., Hodgkin, A. L. and Lamb, T. D. (1974a). The electrical response of turtle cones to flashes and steps of light, *J. Physiol.* **242**, 685–727.
Baylor, D. A., Hodgkin, A. L. and Lamb, T. D. (1974b). Reconstruction of the electrical responses of turtle cones to flashes and steps of light, *J. Physiol.* **242**, 759–791.
Benda, J., Bock, R., Rujan, P. and Ammermuller, J. (2001). Asymmetrical dynamics of voltage spread in retinal horizontal cell networks, *Visual Neurosci.* **18**, 835–848.
Bickford, M. E., Günlük, A. E., Guido, W. and Sherman, S. M. (1993). Evidence that cholinergic axons from the parabrachial region of the brainstem are the exclusive source of nitric oxide in the lateral geniculate nucleus of the cat, *J. Compar. Neurol.* **334**, 410–430.
Bloomfield, S. A., Xin, D. and Persky, S. E. (1995). A comparison of receptive field and tracer coupling size of horizontal cells in the rabbit retina, *Visual Neurosci.* **12**, 985–999.
Bonato, F. and Cataliotti, J. (2000). The effects of figure/ground, perceived area, and target saliency on the luminosity threshold, *Perception and Psychophysics* **62**, 341–349.
Bonato, F. and Gilchrist, A. L. (1994). The perception of luminosity on different backgrounds and in different illuminations, *Perception* **23**, 991–1006.
Bonato, F. and Gilchrist, A. L. (1999). Perceived area and the luminosity threshold, *Perception and Psychophysics* **61**, 786–797.
Bressan, P. (2001). Explaining lightness illusions, *Perception* **30**, 1031–1046.
Bressan, P., Mingolla, E., Spillmann, L. and Watanabe, T. (1997). Neon color spreading: a review, *Perception* **26**, 1353–1366.
Brown, S. P. and Masland, R. H. (2001). Spatial scale and cellular substrate of contrast adaptation by retinal ganglion cells, *Nature Neuroscience* **4**, 44–51.
Budd, J. M. (1998). Extrastriate feedback to primary visual cortex in primates: a quantitative analysis of connectivity, *Proc. Roy. Soc. London Series B-Biol. Sci.* **265**, 1037–1044.
Bullier, J. and Henry, G. H. (1979). Laminar distribution of first-order neurons and afferent terminals in cat striate cortex, *J. Neurophysiol.* **42**, 1271–1281.

Bullier, J., Hupé, J. M., James, A. C. and Girard, P. (1996). Functional interactions between areas V1 and V2 in the monkey, *J. Physiol. Paris* **90**, 217–220.

Bullier, J., Hupé, J. M., James, A. C. and Girard, P. (2001). The role of feedback connections in shaping the responses of visual cortical neurons, *Prog. Brain Res.* **134**, 193–204.

Burkhardt, D. A. (1994). Light adaptation and photopigment bleaching in cone photoreceptors in situ in the retina of the turtle, *J. Neurosci.* **14**, 1091–1105.

Callaway, E. M. (1998). Local circuits in primary visual cortex of the macaque monkey, *Ann. Rev. Neurosci.* **21**, 47–74.

Calvert, P. D., Govardovskii, V. I., Arshavsky, V. Y. and Makino, C. L. (2002). Two temporal phases of light adaptation in retinal rods, *J. Gen. Physiol.* **119**, 129–145.

Caputo, G. (1998). Texture brightness filling-in, *Vision Research* **38**, 841–851.

Carpenter, G. and Grossberg, S. (1981). Adaptation and transmitter gating in vertebrate photoreceptors, *J. Theoret. Neurobiol.* **1**, 1–42.

Cataliotti, J. and Gilchrist, A. (1995). Local and global processes in surface lightness perception, *Perception and Psychophysics* **57**, 125–135.

Chapman, B. and Gödecke, I. (2002). No ON-OFF maps in supragranular layers of ferret visual cortex, *J. Neurophysiol.* **88**, 2163–2166.

Cohen, M. A. and Grossberg, S. (1984). Neural dynamics of brightness perception: features, boundaries, diffusion, and resonance, *Perception and Psychophysics* **36**, 428–456.

Cole, R. E. and Diamond, A. L. (1971). Amount of surround and test inducing separation in simultaneous brightness contrast, *Perception and Psychophysics* **9**, 125–128.

Cook, P. B. and McReynolds, J. S. (1998). Lateral inhibition in the inner retina is important for spatial tuning of ganglion cells, *Nature Neurosci.* **1**, 714–719.

Cornsweet, T. N. (1970). *Visual Perception.* Academic Press, New York.

Dacheux, R. F. and Raviola, E. (1982). Horizontal cells in the retina of the rabbit, *J. Neurosci.* **2**, 1486–1493.

Davey, M. P., Maddess, T. and Srinivasan, M. V. (1998). The spatiotemporal properties of the Craik-O'Brien-Cornsweet effect are consistent with 'filling-in', *Vision Research* **38**, 2037–2046.

Davidson, M. and Whiteside, J. A. (1971). Human brightness perception near sharp contours, *J. Optic. Soc. Amer.* **61**, 530–536.

Demb, J. B. (2002). Multiple mechanisms for contrast adaptation in the retina, *Neuron* **36**, 781–783.

De Valois, R. L. and De Valois, K. K. (1988). *Spatial Vision.* Oxford University Press, New York.

De Valois, R. L., Cottaris, N. P., Mahon, L. E., Elfar, S. D. and Wilson, J. A. (2000). Spatial and temporal receptive fields of geniculate and cortical cells and directional selectivity, *Vision Research* **40**, 3685–3702.

DeVries, S. H. and Schwartz, E. A. (1989). Modulation of an electrical synapse between solitary pairs of catfish horizontal cells by dopamine and second messengers, *J. Physiol.* **414**, 351–375.

DeVries, S. H. and Schwartz, E. A. (1992). Hemi-gap-junction channels in solitary horizontal cells of the catfish retina, *J. Physiol.* **445**, 201–230.

Dolan, R. P. and Schiller, P. H. (1994). Effects of ON channel blockade with 2-amino-4-phosphonobutyrate (APB) on brightness and contrast perception in monkeys, *Visual Neuroscience* **11**, 23–32.

Dowling, J. E. (1987). *The Retina: An Approachable Part of the Brain.* Harvard University Press, Cambridge, MA.

Dresp, B. and Grossberg, S. (1997). Contour integration across polarities and spatial gaps: from local contrast filtering to global grouping, *Vision Research* **37**, 913–924.

Dubin, M. W. and Cleland, B. G. (1977). Organization of visual inputs to interneurons of lateral geniculate nucleus of the cat, *J. Neurophysiol.* **40**, 410–427.

Edwards, D. P., Purpura, K. P. and Kaplan, E. (1995). Contrast sensitivity and spatial frequency response of primate cortical neurons in and around the cytochrome oxidase blobs, *Vision Research* **35**, 1501–1523.

Enroth-Cugell, C. and Shapley, R. M. (1973a). Adaptation and dynamics of cat retinal ganglion cells, *J. Physiol.* **233**, 271–309.

Enroth-Cugell, C. and Shapley, R. M. (1973b). Flux, not retinal illumination, is what cat retinal ganglion cells really care about, *J. Physiol.* **233**, 311–326.

Erisir, A., Van Horn, S. C., Bickford, M. E. and Sherman, S. M. (1997). Immunocytochemistry and distribution of parabrachial terminals in the lateral geniculate nucleus of the cat: a comparison with corticogeniculate terminal, *J. Compar. Neurol.* **377**, 535–549.

Erwin, E., Baker, F. H., Busen, W. F. and Malpeli, J. G. (1999). Relationship between laminar topology and retinotopy in the rhesus lateral geniculate nucleus: results from a functional atlas, *J. Compar. Neurol.* **407**, 92–102.

Eysel, U. T. and Schweigart, G. (1999). Increased receptive field size in the surround of chronic lesions in the adult cat visual cortex, *Cerebral Cortex* **9**, 101–109.

Eysel, U. T., Pape, H. C. and Van Schayck, R. (1986). Excitatory and differential disinhibitory actions of acetylcholine in the lateral geniculate nucleus of the cat, *J. Physiol.* **370**, 233–254.

Fahrenfort, I., Habets, R. L., Spekreijse, H. and Kamermans, M. (1999). Intrinsic cone adaptation modulates feedback efficiency from horizontal cells to cones, *J. General Physiol.* **114**, 511–524.

Fain, G. L. (2001). Dark adaptation, *Prog. Brain Res.* **131**, 383–394.

Felisberti, F. and Derrington, A. M. (2001). Long-range interactions in the lateral geniculate nucleus of the New-World monkey, *Callithrix jacchus, Visual Neurosci.* **18**, 209–218.

Felleman, D. J. and Van Essen, D. C. (1991). Distributed hierarchical processing in the primate cerebral cortex, *Cerebral Cortex* **1**, 1–47.

Ferster, D., Chung, S. and Wheat, H. (1996). Orientation selectivity of thalamic input to simple cells of cat visual cortex, *Nature* **380**, 249–252.

Field, D. J., Hayes, A. and Hess, R. F. (1993). Contour integration by the human visual system: evidence for a local 'association field', *Vision Research* **33**, 173–193.

Fischl, B. and Schwartz, E. L. (1997). Learning an integral equation approximation to nonlinear anisotropic diffusion in image processing, *IEEE Trans. Pattern Anal. Machine Intell.* **19**, 342–352.

Foster, K. H., Gaska, J. P., Nagler, M. and Pollen, D. A. (1985). Spatial and temporal frequency selectivity of neurones in visual cortical areas V1 and V2 of the macaque monkey, *J. Physiol.* **365**, 331–363.

Friedman, H. S., Zhou, H. and Von Der Heydt, R. (2003). The coding of uniform colour figures in monkey visual cortex, *J. Physiol.* **548**, 593–613.

Fry, G. A. (1948). Mechanisms subserving simultaneous brightness contrast, *Amer. J. Optomet. and Arch. Amer. Acad. Optomet.* **25**, 162–178.

Fry, G. A. and Alpern, M. (1953). The effect of a peripheral glare source upon the apparent brightness of an object, *J. Optic. Soc. Amer.* **43**, 189–195.

Ghose, G. M. and Ts'o, D. Y. (1997). Form processing modules in primate area V4, *J. Neurophysiol.* **77**, 2191–2196.

Gilbert, C. D. (1977). Laminar differences in receptive field properties of cells in cat primary visual cortex, *J. Physiol.* **268**, 391–421.

Gilbert, C. D. and Wiesel, T. N. (1979). Morphology and intracortical projections of functionally characterized neurones in the cat visual cortex, *Nature* **280**, 120–125.

Gilchrist, A. L. (1977). Perceived lightness depends on perceived spatial arrangement, *Science* **195**, 185–187.

Gilchrist, A. L. (1980). When does perceived lightness depend on perceived spatial arrangement? *Perception and Psychophysics* **28**, 527–538.

Gilchrist, A. L. and Bonato, F. (1995). Anchoring of lightness values in center-surround displays, *J. Exper. Psychol., Human Perception and Performance* **6**, 1427–1440.

Gilchrist, A. L. and Cataliotti, J. (1994). Anchoring of surface lightness with multiple illumination levels, *Investigat. Ophthalmol. Visual Sci.* **35**, S2165.

Gilchrist, A. L., Delman, S. and Jacobsen, A. (1983). The classification and integration of edges as critical to the perception of reflectance and illumination, *Perception and Psychophysics* **33**, 425–436.

Gilchrist, A. L., Kossyfidis, C., Bonato, F., Agostini, T., Cataliotti, J., Li, X., Spehar, B., Annan, V. and Economou, E. (1999). An anchoring theory of lightness perception, *Psycholog. Rev.* **106**, 795–834.

Gove, A., Grossberg, S. and Mingolla, E. (1995). Brightness perception, illusory contours, and corticogeniculate feedback, *Visual Neurosci.* **12**, 1027–1052.

Grossberg, S. (1980). How does a brain build a cognitive code? *Psychological Review* **87**, 1–51.

Grossberg, S. (1983). The quantized geometry of visual space: the coherent computation of depth, form and lightness, *Behav. Brain Sci.* **6**, 625–692.

Grossberg, S. (1994). 3-D vision and figure-ground separation by visual cortex, *Perception and Psychophysics* **55**, 48–120.

Grossberg, S. (1997). Cortical dynamics of 3D figure-ground perception of 2D figures, *Psycholog. Rev.* **104**, 618–658.

Grossberg, S. (1999). How does the cerebral cortex work? Learning, attention and grouping by the laminar circuits of visual cortex, *Spatial Vision* **12,** 163–186.

Grossberg, S. (2000). The complementary brain: unifying brain dynamics and modularity, *Trends Cognitive Sci.* **4**, 233–246.

Grossberg S. (2003). Filling-in the forms: surface and boundary interactions in visual cortex, in: *Filling-in: From Perceptual Completion to Skill Learning*, Pessoa, L. and De Weerd, P. (Eds). Oxford University Press, New York.

Grossberg S. and Howe, P. D. L. (2003). A laminar cortical model of stereopsis and three-dimensional surface perception, *Vision Research* **43**, 801–829.

Grossberg, S. and Kelly, F. (1999). Neural dynamics of binocular brightness perception, *Vision Research* **39**, 3796–3816.

Grossberg, S. and McLoughlin, N. (1997). Cortical dynamics of 3D surface perception: binocular and half-occluded scenic images, *Neural Network* **10**, 1583–1605.

Grossberg, S. and Mingolla, E. (1985a). Neural dynamics of perceptual grouping: textures, boundaries, and emergent segmentations, *Perception and Psychophysics* **38**, 141–71.

Grossberg, S. and Mingolla, E. (1985b). Neural dynamics of form perception: boundary completion, illusory figures, and neon color spreading, *Psycholog. Rev.* **92**, 173–211.

Grossberg, S. and Raizada, R. D. (2000). Contrast-sensitive perceptual grouping and object-based attention in the laminar circuits of primary visual cortex, *Vision Research* **40**, 1413–1432.

Grossberg, S. and Swaminathan, G. (2004). A laminar cortical model for 3D perception of slanted and curved surfaces and of 2D images: development, attention and bistability, *Vision Research* **44**, 1147–1187.

Grossberg, S. and Todorovic, D. (1988). Neural dynamics of 1D and 2D brightness perception: a unified model of classical and recent phenomena, *Perception and Psychophysics* **43**, 241–277.

Grossberg, S. and Yazdanbakhsh, A. (2003). Laminar cortical dynamics of 3D surface stratification, transparency, and neon spreading, *J. Vision* **3**, 247a.

Grossberg, S. and Yazdanbakhsh, A. (2005). Laminar cortical dynamics of 3D surface perception: stratification, transparency, and neon color spreading, *Vision Research* **45**, 1725–1743.

Grossberg, S., Mingolla, E. and Williamson, J. (1995). Synthetic aperture radar processing by a multiple scale neural system for boundary and surface representation, *Neural Network* **7/8**, 1005–1028.

Grossberg, S., Hwang, S. and Mingolla, E. (2002). Thalamocortical dynamics of the McCollough effect: boundary-surface alignment through perceptual learning, *Vision Research* **42**, 1259–1286.

Hahnloser, R. H., Douglas, R. J. and Hepp, K. (2002). Attentional recruitment of inter-areal recurrent networks for selective gain control, *Neural Computation* **14**, 1669–1689.

Helmholtz, H. von (1866). *Helmholtz's Treatise on Physiological Optics*. Optical Society of America, New York.

Helson, H. (1943). Some factors and implications of color constancy, *J. Optic. Soc. Amer.* **33**, 179–184.

Hering, E. (1920). *Outline of a Theory of the Light Sense*, trans. Hurvich, L. and Jameson, D. (1964). Harvard University Press, Cambridge, MA.

Hong, S. and Grossberg, S. (2003). Cortical dynamics of surface lightness anchoring, filling-in and perception, in: *Vision Sciences Society, Third Annual Meeting*, Sarasota, FL, p. 120.

Hong, S. and Grossberg, S. (2004). A neuromorphic model for achromatic and chromatic surface perception of natural images, *Neural Networks* **17**, 787–808.

Horn, B. K. P. (1977). Understanding image intensities, *Artificial Intelligence* **21**, 201–231.

Huang, X., MacEvoy, S. P. and Paradiso, M. A. (2002). Perception of brightness and brightness illusions in the macaque monkey, *J. Neurosci.* **22**, 9618–9625.

Hubel, D. H. and Wiesel, T. N. (1961). Integrative action in the cat's lateral geniculate body, *J. Physiol.* **155**, 385–398.

Hubel, D. H. and Wiesel, T. N. (1962). Receptive fields, binocular interaction and functional architecture in the cat's visual cortex, *J. Physiol.* **160**, 106–154.

Hung, C. P., Ramsden, B. M., Chen, L. M. and Roe, A. W. (2001). Building surfaces from borders in Areas 17 and 18 of the cat, *Vision Research* **41**, 1389–1407.

Hupé, J. M., James, A. C., Payne, B. R., Lomber, S. G., Girard, P. and Bullier, J. (1998). Cortical feedback improves discrimination between figure and background by V1, V2 and V3 neurons, *Nature* **394**, 784–787.

Hupé, J. M., James, A. C., Girard, P. and Bullier, J. (2001). Response modulations by static texture surround in area V1 of the macaque monkey do not depend on feedback connections from V2, *J. Neurophysiol.* **85**, 146–63.

Hurlbert, A. (1986). Formal connections between lightness algorithms, *J. Optic. Soc. Amer. A.* **3**, 1684–1693.

Hurlbert, A. (1989). The computation of color, Doctoral Thesis, MIT.

Hurley, J. B. (2002). Shedding light on adaptation, *J. Gen. Physiol.* **119**, 125–128.

Issa, N. P., Trepel, C. and Stryker, M. P. (2000). Spatial frequency maps in cat visual cortex, *J. Neurosci.* **20**, 8504–8514.

Jameson, D. and Hurvich, L. M. (1989). Essay concerning color constancy, *Ann. Rev. Psychol.* **40**, 1–22.

Jobson, D. J., Rahman, Z. and Woodell, G. A. (1997a). Properties and performance of a center/surround retinex, *IEEE Trans. Image Processing* (March).

Jobson, D. J., Rahman, Z. and Woodell, G. A. (1997b). A multi-scale retinex for bridging the gap between color images and the human observation of scenes, *IEEE Trans. Image Processing: Special Issue on Color Processing* (July).

Jones, E. G. (1994). GABA neurons and their role in activity-dependent plasticity of adult primate visual cortex, in: *Cerebral Cortex*, Peters, A. and Rockland, K. S. (Eds), Vol. 10, pp. 61–140. Plenum Press, New York.

Jones, H. E., Andolina, I. M., Oakely, N. M., Murphy, P. C. and Sillito, A. M. (2000). Spatial summation in lateral geniculate nucleus and visual cortex, *Exper. Brain Res.* **135**, 279–284.

Kaas, J. H. and Collins, C. E. (2001). The organization of sensory cortex, *Current Opinion in Neurobiology* **11**, 498–504.

Kahrilas, P. J., Doty, R. W. and Bartlett, J. R. (1980). Failure to find luxotonic responses for single units in visual cortex of the rabbit, *Exper. Brain Res.* **39**, 11–16.

Kamermans, M., Haak, J., Habraken, J. B. and Spekreijse, H. (1996). The size of the horizontal cell receptive fields adapts to the stimulus in the light adapted goldfish retina, *Vision Research* **36**, 4105–4119.

Kamitani, Y. and Shimojo, S. (2003). Global yet early processing of visual surfaces, in: *The Visual Neurosciencies*, Chalupa, L. M. and Werner, J. S. (Eds). MIT Press, Cambridge, MA.

Kaufman, L. (1974). *Sight and Mind.* Oxford University Press, New York, NY.

Kayama, Y., Riso, R. R., Bartlett, J. R. and Doty, R. W. (1979). Luxotonic responses of units in macaque striate cortex, *J. Neurophysiol.* **42**, 1495–1517.

Kellman, P. J. and Shipley, T. F. (1991). A theory of visual interpolation in object perception, *Cognit. Psychol.* **23**, 141–221.

Kelly, F. and Grossberg, S. (2000). Neural dynamics of 3D surface perception: figure-ground separation and lightness perception, *Perception and Psychophysics* **62**, 1596–1618.

Kimmel, R., Elad, M., Shaked, D., Keshet, R. and Sobel, I. (2003). A variational framework for retinex, *Internat. J. Computer Vision* **52**, 7–23.

Kinoshita, M. and Komatsu, H. (2001). Neural representation of the luminance and brightness of a uniform surface in the macaque primary visual cortex, *J. Neurophysiol.* **86**, 2559–2570.

Kobatake, E. and Tanaka, K. (1994). Neuronal selectivities to complex object feature, in the ventral visual pathway of the macaque cerebral cortex, *J. Neurophysiol.* **71**, 856–867.

Kobayashi, M., Imamura, K., Sugai, T., Onoda, N., Yamamoto, M., Komai, S. and Watanabe, Y. (2000). Selective suppression of horizontal propagation in rat visual cortex by norepinephrine, *Eur. J. Neurosci.* **12**, 264–272.

Komatsu, H., Murakami, I. and Kinoshita, M. (1996). Surface representation in the visual system, *Cognitive Brain Res.* **5**, 97–104.

Komatsu, H., Kinoshita, M. and Murakami, I. (2000). Neural responses in the retinotopic representation of the blind spot in the macaque V1 to stimuli for perceptual filling-in, *J. Neurosci.* **20**, 9310–9319.

Koutalos, Y. and Yau, K. W. (1996). Regulation of sensitivity in vertebrate rod photoreceptors by calcium, *Trends in Neuroscience* **19**, 73–81.

Kuffler, S. W. (1953). Discharge patterns and functional organization of mammalian retina, *J. Neurophysiol.* **16**, 37–68.

Lambert, J. H. (1760). *Photometria sive de mensura et gradibus luminis, colorum et umbrae*, Augsburg (German translation by E. Anding, Klett, Leipzig (1892)).

Lamme, V. A. (2001). Blindsight: the role of feedforward and feedback corticocortical connections, *Acta Psychologica (Amsterdam)* **107**, 209–228.

Lamme, V. A., Zipser, K. and Spekreijse, H. (1998). Figure-ground activity in primary visual cortex is suppressed by anesthesia, *Proc. Natl. Acad. Sci. USA* **95**, 3263–3268.

Land, E. H. and McCann, J. J. (1971). Lightness and retinex theory, *J. Optic. Soc. Amer.* **61**, 1–11.

Lee, B. B., Dacey, D. M., Smith, V. C. and Pokorny, J. (1999). Horizontal cells reveal cone type-specific adaptation in primate retina, *Proc. Natl. Acad. Sci. USA* **96**, 14611–14616.

Lee, A. B., Blais, B., Shouval, H. Z. and Cooper, L. N. (2000). Statistics of lateral geniculate nucleus (LGN) activity determine the segregation of ON/OFF subfields for simple cells in visual cortex, *Proc. Natl. Acad. Sci. USA* **97**, 12875–12879.

Legg, C. R. and Cowey, A. (1977a). Effects of subcortical lesions on visual intensity discriminations in rats, *Physiology and Behavior* **19**, 635–646.

Legg, C. R. and Cowey, A. (1977b). The role of the ventral lateral geniculate nucleus and posterior thalamus in intensity discrimination in rats, *Brain Research* **123**, 261–273.

Leibowitz, H., Mote, F. A. and Thurlow, W. R. (1953). Simultaneous contrast as a function of separation between test and inducing fields, *J. Exper. Psychol.* **46**, 453–456.

Livingstone, M. S. and Hubel, D. H. (1981). Effects of sleep and arousal on the processing of visual information in the cat, *Nature* **291**, 554–561

Lu, C., Zhang, D. Q. and McMahon, D. G. (1999). Electrical coupling of retinal horizontal cells mediated by distinct voltage-independent junctions, *Visual Neurosc.* **16**, 811–818.

MacEvoy, S. P., Kim, W. and Paradiso, M. A. (1998). Integration of surface information in primary visual cortex, *Nature Neurosci.* **1**, 616–620.

MacEvoy, S. P. and Paradiso, M. A. (2001). Lightness constancy in primary visual cortex, *Proc. Natl. Acad. Sci. USA* **98**, 8827–8831.

Marrocco, R. T. (1972). Maintained activity of monkey optic tract fibers and lateral geniculate nucleus cells, *Vision Research* **12**, 1175–1181.

Martin, G. R. (1983). Schematic eye models in vertebrates, *Progress in Sensory Physiology* **4**, 44–81.

Martinez, L. M. and Alonso, J. M. (2001). Construction of complex receptive fields in cat primary visual cortex, *Neuron* **32**, 515–525.

Masland, R. H. (2001). The fundamental plan of the retina, *Nature Neurosci.* **4**, 877–886.

McMahon, D. G. (1994). Modulation of electrical synaptic transmission in zebrafish retinal horizontal cells, *J. Neurosci.* **14**, 1722–1734

McMahon, D. G., Zhang, D. Q., Ponomareva, L. and Wagner, T. (2001). Synaptic mechanisms of network adaptation in horizontal cells, *Progress in Brain Research* **131**, 419–436

Mechler, F. and Ringach, D. L. (2002). On the classification of simple and complex cells, *Vision Research* **42**, 1017–1033.

Meister, M. and Berry, M. J. II (1999). The neural code of the retina, *Neuron* **22**, 435–450.

Mills, S. L. and Massey, S. C. (1995). Differential properties of two gap junctional pathways made by AII amacrine cells, *Nature* **377**, 734–737.

Mingolla, E., Ross, W. and Grossberg, S. (1999). A neural network for enhancing boundaries and surfaces in synthetic aperture radar images, *Neural Networks* **12**, 499–511.

Moore, R. Y., Weis, R. and Moga, M. M. (2000). Efferent projections of the intergeniculate leaflet and the ventral lateral geniculate nucleus in the rat, *J. Compar. Neurol.* **420**, 398–418.

Naka, K. I. and Rushton, W. A. (1967). The generation and spread of S-potentials in fish (Cyprinidae), *J. Physiol.* **192**, 437–461.

Nakamura, H., Gattass, R., Desimone, R. and Ungerleider, L. G. (1993). The modular organization of projections from areas V1 and V2 to areas V4 and TEO in macaques, *J. Neurosci.* **13**, 3681–3691.

Nakatani, K., Tamura, T. and Yau, K. W. (1991). Light adaptation in retinal rods of the rabbit and two other nonprimate mammals, *J. General Physiol.* **97**, 413–435.

Nakayama, K. and Shimojo, S. (1992). Experiencing and perceiving visual surfaces, *Science* **257**, 1357–1363.

Nawy, S. (2000). Regulation of the on bipolar cell mGluR6 pathway by Ca^{2+}, *J. Neurosci.* **20**, 4471–4479.

Neumann, H. and Sepp, W. (1999). Recurrent V1–V2 interaction in early visual boundary processing, *Biological Cybernetics* **81**, 425–444.

Neumann, H., Pessoa, L. and Mingolla, E. (1998). A neural architecture of brightness perception: non-linear contrast detection and geometry-driven diffusion, *Image and Vision Computing* **16**, 423–446.

Neumann, H., Pessoa, L. and Hansen, T. (2001). Visual filling-in for computing perceptual surface properties, *Biolog. Cybernet.* **85**, 355–369.

Newson, L. J. (1958). Some principles governing changes in the apparent lightness of test surfaces isolated from their normal backgrounds, *Quart. J. Exper. Psychol.* **10**, 82–95.

Papaioannou, J. and White, A. (1972). Maintained activity of lateral geniculate ucleus neurons as a function of background luminance, *Exper. Neurol.* **34**, 558–566.

Paradiso, M. A. and Nakayama, K. (1991). Brightness perception and filling-in, *Vision Research* **31**, 1221–1236.

Pessoa, L. and Neumann, H. (1998). Why does the brain fill-in? *Trends in Cognitive Science* **11**, 422–424.

Pessoa, L., Mingolla, E. and Neumann, H. (1995). A contrast- and luminance-driven multiscale network model of brightness perception, *Vision Research* **35**, 2201–2223.

Pessoa, L., Mingolla, E. and Arend, L. E. (1996). The perception of lightness in 3D curved objects, *Perception and Psychophysics* **58**, 1293–1305.

Peterhans, E. and von der Heydt, R. (1989). Mechanisms of contour perception in monkey visual cortex. II. Contours bridging gaps, *J. Neurosci.* **9**, 1749–1763.

Pinna, B., Brelstaff, G. and Spillmann, L. (2001). Surface color from boundaries: a new 'watercolor' illusion, *Vision Research* **41**, 2669–2676.

Pinna, B., Werner, J. S. and Spillmann, L. (2003). The watercolor effect: a new principle of grouping and figure-ground organization, *Vision Research* **43**, 43–53.

Posner, M. I., Snyder, C. R. R. and Davidson, B. J. (1980). Attention and the detection of signals, *J. Exper. Psychol.* **109**, 160–174

Przybyszewski, A. W., Gaska, J. P., Foote, W. and Pollen, D. A. (2000). Striate cortex increases contrast gain of macaque LGN neurons, *Visual Neurosci.* **17**, 485–494.

Rahman, Z., Jobson, D. J. and Woodell, G. A. (1996). Multiscale retinex for color image enhancement, in: *International Conference on Image Processing (ICIP) '96.*

Rahman, Z., Woodell, G. A. and Jobson, D. J (1997). A comparison of the multiscale retinex with other image enhancement techniques, in: *Proceedings of the IS and T 50th Anniversary Conference.*

Raizada, R. D. S. and Grossberg, S. (2001). Context-sensitive binding by the laminar circuits of V1 and V2: a unified model of perceptual grouping, attention, and orientation contrast, *Visual Cognition* **8**, 431–466.

Raizada, R. D. S. and Grossberg, S. (2003). Towards a theory of the laminar architecture of cerebral cortex: computational clues from the visual system, *Cerebral Cortex* **13**, 100–113.

Reid, R. C. and Alonso, J. M. (1995). Specificity of monosynaptic connections from thalamus to visual cortex, *Nature* **378**, 281–284.

Reynolds, J. H., Chelazzi, L. and Desimone, R. (1999). Competitive mechanisms subserve attention in macaque areas V2 and V4, *J. Neurosci.* **19**, 1736–1753.

Ribelayga, C., Wang, Y. and Mangel, S. C. (2002). Dopamine mediates circadian clock regulation of rod and cone input to fish retinal horizontal cells, *J. Physiol.* **544**, 801–816.

Ringach, D. L. (2002). Spatial structure and symmetry of simple-cell receptive fields in macaque primary visual cortex, *J. Neurophysiol.* **88**, 455–463.

Rockland, K. S. and Lund, J. S. (1982). Widespread periodic intrinsic connections in the tree shrew visual cortex, *Science* **215**, 1532–1534.

Rodieck, R. W. (1965). Quantitative analysis of cat retinal ganglion cell response to visual stimuli, *Vision Research* **5**, 583–601.

Rodieck, R. W. (1998). *The First Steps in Seeing*. Sinauer, Sunderland, MA.

Rodieck, R. W. and Stone, J. (1965). Response of cat retinal ganglion cells to moving visual patterns, *J. Neurophysiol.* **28**, 819–832.

Roe, A. W. and Ts'o, D. Y. (1995). Visual topography in primate V2: multiple representation across functional stripes, *J. Neurosci.* **15**, 3689–3715.

Roe, A. W. and Ts'o, D. Y. (1999). Specificity of color connectivity between primate V1 and V2, *J. Neurophysiol.* **82**, 2719–2730.

Roska, B. and Werblin, F. S. (2001). Vertical interactions across ten parallel, stacked representations in the mammalian retina, *Nature* **410**, 583–587.

Roska, B., Nemeth, E., Orzo, L. and Werblin, F. S. (2000). Three levels of lateral inhibition: a space-time study of the retina of the tiger salamander, *J. Neurosci.* **20**, 1941–1951.

Ross, W. D. and Pessoa, L. (2000). Lightness from contrast: a selective integration model, *Perception and Psychophysics* **62**, 1160–1181.

Rossi, A. F. and Paradiso, M. A. (1999). Neural correlates of perceived brightness in the retina, lateral geniculate nucleus, and striate cortex, *J. Neurosci.* **19**, 6145–6156.

Rossi, A. F., Rittenhouse, C. D. and Paradiso, M. A. (1996). The representation of brightness in primary visual cortex, *Science* **273**, 1104–1107.

Rozas, C., Frank, H., Heynen, A. J., Morales, B., Bear, M. F. and Kirkwood, A. (2001). Developmental inhibitory gate controls the relay of activity to the superficial layers of the visual cortex, *J. Neurosc.* **21**, 6791–6801.

Ruksenas, O., Fjeld, I. T. and Heggellund, P. (2000). Spatial summation and center-surround antagonism in the receptive field of single units in the dorsal lateral geniculate nucleus of cat: comparison with retinal input, *Visual Neurosci.* **17**, 855–870.

Sakai, K. and Tanaka, S. (2000). Spatial pooling in the second-order spatial structure of cortical complex cells, *Vision Research* **40**, 855–871.

Sakmann, B. and Creutzfeldt, O. D. (1969). Scotopic and mesopic light adaptation in the cat's retina, *Pflügers Arch.* **313**, 168–185.

Sasaki, Y., Watanabe, T., Dale, A. M. and Tootell, R. B. (2001). V1 involvement for color filling-in revealed by human fMRI, *Society for Neuroscience Abstracts* **12**, 11.

Schiller, P. H. (1992). The ON and OFF channels of the visual system, *Trends in Neuroscience* **15**, 86–92.

Schiller, P. H., Finlay, B. L. and Volman, S. F. (1976). Quantitative studies of single-cell properties in monkey striate cortex. V. Multivariate statistical analyses and models, *J. Neurophysiol.* **39**, 1362–1374.

Schiller, P. H., Sandell, J. H. and Maunsell, J. H. (1986). Functions of the ON and OFF channels of the visual system, *Nature* **322**, 824–825.

Schirillo, J., Reeves, A. and Arend, L. (1990). Perceived lightness, but not brightness, of achromatic surfaces depends on perceived depth information, *Perception and Psychophysics* **48**, 82–90.

Schwartz, E. (1980). Computational anatomy and functional architecture of striate cortex: a spatial mapping approach to perceptual coding, *Vision Research* **20**, 645–699.

Sepp, W. and Neumann, H. (1999). A multi-resolution filling-in model for brightness perception, *Ninth International Conference on Artificial Neural Networks* **1**, 461–466.

Shapley, R. and Enroth-Cugell, C. (1984). Visual adaptation and retinal gain controls, *Progress in Retinal Research* **3**, 263–346.

Sherman, S. M. (1996). Dual response modes in lateral geniculate neurons: mechanisms and functions, *Visual Neuroscience* **13**, 205–213.

Sherman, S. M. (2001). Tonic and burst firing: dual modes of thalamocortical relay, *Trends in Neuroscience* **24**, 122–126.

Shevelev, I. A. (1998). Second-order feature extraction in the cat visual cortex: selective and invariant sensitivity of neurons to the shape and orientation of crosses and corners, *Biosystems* **48**, 195–204.

Shevelev, I. A. (2000). Sensitivity of striate neurons to Y-like figures: experiment and simulation, *Biosystems* **58**, 211–217.

Shevelev, I. A., Lazareva, N. A., Sharaev, G. A., Novikova, R. V. and Tikhomirov, A. S. (1998). Selective and invariant sensitivity to crosses and corners in cat striate neurons, *Neuroscience* **84**, 713–721.

Sillito, A. M., Kemp, J. A. and Beradi, N. (1983). The cholinergic influence on the function of the cat dorsal lateral geniculate nucleus (dLGN), *Brain Research* **280**, 299–307.

Sillito, A. M., Jones, H. E. and Gerstein, G. L. (1994). West DC, Feature-linked synchronization of thalamic relay cell firing induced by feedback from the visual cortex, *Nature* **369**, 479–482.

Smirnakis, S. M., Berry, M. J., Warland, D. K., Bialek, W. and Meister, M. (1997). Adaptation of retinal processing to image contrast and spatial scale, *Nature* **386**, 69–73.

Solomon, S. G., White, A. J. and Martin, P. R. (2002). Extraclassical receptive field properties of parvocellular, magnocellular, and koniocellular cells in the primate lateral geniculate nucleus, *J. Neurosci.* **22**, 338–349.

Somers, D. C., Todorov, E. V., Siapas, A. G., Toth, L. J., Kim, D. S. and Sur, M. (1998). A local circuit approach to understanding integration of long-range inputs in primary visual cortex, *Cerebral Cortex* **8**, 204–217.

Somers, D. C., Dale, A. M., Seiffert, A. E. and Tootell, R. B. (1999). Functional MRI reveals spatially specific attentional modulation in human primary visual cortex, *Proc. Natl. Acad. Sci. USA* **96**, 1663–1668.

Spillmann, L. (1994). The Hermann Grid Illusion: a tool for studying human perceptive field organization, *Perception* **23**, 691–708.

Spray, D. C., Harris, A. L. and Bennett M. V. (1979). Voltage dependence of junctional conductance in early amphibian embryos, *Science* **204**, 432–434.

Stanley, G. B., Li, F. F. and Dan, Y. (1999). Reconstruction of natural scenes from ensemble responses in the lateral geniculate nucleus, *J. Neurosci.* **19**, 8036–8042.

Sterling, P. (1998). In: *The Synaptic Organization of the Brain*, Vol. 4, Shepherd, G. M. (Ed.), pp. 205–253. Oxford University Press, New York.

Stettler, D. D., Das, A., Bennett, J. and Gilbert, C. D. (2002). Lateral connectivity and contextual interactions in macaque primary visual cortex, *Neuron* **36**, 739–750.

Suder, K., Wörgötter, F. and Wennekers, T. (2001). Neural field model of receptive field restructuring in primary visual cortex, *Neural Computation* **13**, 139–159.

Tadmor, Y. and Tolhurst, D. J. (2000). Calculating the contrasts that retinal ganglion cells and LGN neurons encounter in natural scenes, *Vision Research* **40**, 3145–3157.

Thibos, L. N. and Werblin, F. S. (1978). The response properties of the steady antagonistic surround in the mudpuppy retina, *J. Physiol.* **278**, 79–99.

Todorović, D. (1997). Lightness and junctions, *Perception* **26**, 379–394.

Tomita, T. (1965). Electrophysiological study of the mechanisms subserving color coding in the fish retina, *Cold Spring Harbor Symposium on Quantitative Biology* **30**, 559–566.

Triesman, A. M. and Sato, S. (1990). Conjunction search revisited, *J. Exper. Psychology, Human Perception and Performance* **16**, 459–478

Troyer, T. W., Krukowski, A. E., Priebe, N. J. and Miller, K. D. (1998). Contrast-invariant orientation tuning in cat visual cortex: thalamocortical input tuning and correlation-based intracortical connectivity, *J. Neurosci.* **18**, 5908–5927.

Ts'o, D. Y., Roe, A. W. and Gilbert, C. D. (2001). A hierarchy of the functional organization for color, form and disparity in primate visual area V2, *Vision Research* **41**, 1333–1349.

Usrey, W. M., Sceniak, M. P. and Chapman, B. (2003). Receptive fields and response properties of neurons in layer 4 of ferret visual cortex, *J. Neurophysiol.* **89**, 1003–1015.

Verweij, J., Kamermans, M. and Spekreijse, H. (1996). Horizontal cells feed back to cones by shifting the cone calcium-current activation range, *Vision Research* **36**, 3943–3953.

von der Heydt, R., Peterhans, E. and Baumgartner, G. (1984). Illusory contours and cortical neuron responses, *Science* **224**, 1260–1262.

Wallach, H. (1948). Brightness constancy and the nature of achromatic colors, *J. Exper. Psychol.* **38**, 310–324.

Wallach, H. (1976). *On Perception*. Quadrangle/The New York Times Book Co., New York.

Welchman, A. E. and Harris, J. M. (2001). Filling-in the details on perceptual fading, *Vision Research* **41**, 2107–2117.

Werblin, F. S. (1971). Adaptation in a vertebrate retina: intracellular recording in Necturus, *J. Neurophysiol.* **34**, 228–241.

Werblin, F. S. (1974). Control of retinal sensitivity. II. Lateral interactions at the outer plexi form layer, *J. General Physiol.* **63**, 62–87.

Werblin, F. S. and Dowling, J. E. (1969). Organization of the retina of the mudpuppy, *Necturus maculosus*. II. Intracellular recording, *J. Neurophysiol.* **32**, 339–355.

Werblin, F. S., Roska, B. and Balya, D. (2001). Parallel processing in the mammalian retina: lateral and vertical interactions across stacked representations, *Progress in Brain Research* **131**, 229–238.

Witkovsky, P. and Dearry A. (1991). Functional roles of dopamine in the vertebrate retina, *Progress in Retinal Research* **11**, 247–292.

Wörgötter, F., Suder, K., Zhao, Y., Kerscher, N., Eysel, U. T. and Funke, K. (1998). State-dependent receptive-field restructuring in the visual cortex, *Nature* **396**, 165–168.

Wyszecki, G. and Stiles, W. S. (1982). *Color Science: Concepts and Methods, Quantitative Data and Formulae*, 2nd edn. John Wiley and Sons, New York.

Xin, D. and Bloomfield, S. A. (1999). Dark- and light-induced changes in coupling between horizontal cells in mammalian retina, *J. Compar. Neurol.* **405**, 75–87.

Xin, D. and Bloomfield, S. A. (2000). Effects of nitric oxide on horizontal cells in the rabbit retina, *Visual Neurosci.* **17**, 799–811.

Yabuta, N. H. and Callaway, E. M. (1998). Cytochrome-oxidase blobs and intrinsic horizontal connections of layer 2/3 pyramidal neurons in primate V1, *Visual Neurosci.* **15**, 1007–1027.

Yarbus, A. L. (1967). *EyeMovements and Vision.* Plenum Press.

Zhou, H., Friedman, H. S. and von der Heydt, R. (2000). Coding of border ownership in monkey visual cortex, *J. Neurosci.* **20**, 6594–6611.

Zipser, K., Lamme, V. A. and Schiller, P. H. (1996). Contextual modulation in primary visual cortex, *J. Neurosci.* **16**, 7376–7389.

APPENDIX A: MODEL EQUATIONS

The model implements 2D simulations on a 200×200 grid that represents the visual field.

A.1. Retinal adaptation

The potential s_{ij} at position (i, j) of the outer segment of the retinal photoreceptor is simulated by the equation:

$$s_{ij}(t) = I_{ij} \cdot z_{ij}(t), \tag{A1}$$

where I_{ij} is the input and $z_{ij}(t)$ is an automatic gain control term simulating negative feedback mediated by Ca^{2+} ions, among others:

$$\frac{\mathrm{d}z_{ij}}{\mathrm{d}t} = (B_z - z_{ij}) - z_{ij}\bigl(C_I I_{ij} + C_{\overline{I}}\overline{I}\bigr), \tag{A2}$$

(cf., Carpenter and Grossberg, 1981; Grossberg, 1980). In (A2), parameter B_z is the asymptote which $z_{ij}(t)$ approaches in the absence of input, and term $-z_{ij}(C_I I_{ij} + C_{\overline{I}}\overline{I}_{ij})$ describes the inactivation of z_{ij} by the present input I_{ij} and a spatial average $\overline{I}$ of all inputs that approximates the effect of recent image scanning by sequences of eye movements. The equilibrium response s_{ij} directly follows from (A1) and (A2):

$$s_{ij} = \frac{B_z I_{ij}}{1 + C_I I_{ij} + C_{\overline{I}}\overline{I}}, \tag{A3}$$

The inner segment of the photoreceptor receives the signal s_{ij} from the outer segment and also gets feedback H_{ij} from the horizontal cell (HC) at position (i, j), as in Fig. 9. HC modulation of the output of the inner segment of the photoreceptor is modeled by the equation:

$$S_{ij} = \frac{s_{ij}}{B_h \exp(H_{ij}) \cdot (B_s - s_{ij}) + 1}, \tag{A4}$$

where B_h is a small constant, and B_s is a constant close to the value (B_z/C_I). When B_s equals the value of (B_z/C_I), perfect shifts of the $\log(I_{ij}) - S_{ij}$ curve occur with varying H_{ij} (Fig. A1A). When B_s deviates from (B_z/C_I), compression occurs when $B_s > (B_z/C_I)$ (Fig. A1C). Expansion occurs when $B_s < (B_z/C_I)$ in addition to the shift. Thus to prevent expansion, which would mean excitation by the HC negative feedback, B_s needs to be bigger or equal to (B_z/C_I). Figure A2 shows the 10-Mondrian Articulation situation (see Fig. 4) with two values for B_s, one equals to (B_z/C_I), and the other to $1.2(B_z/C_I)$. This simulation demonstrates that the model is robust under this variation: compare Fig. A2 with the graph in Fig. 4G.

The equation (A4) can be generalized as follows.

$$S_{ij} = \frac{s_{ij}}{f(H_{ij}) \cdot (B_s - s_{ij}) + 1}. \tag{A4$'$}$$

Many increasing functions $f(H_{ij})$ will generate the shift property of S_{ij} as a function of $\log(I_{ij})$. Function $f(H_{ij}) = B_h \exp(H_{ij})$ was chosen because $\exp(H_{ij})$ makes the sensitivity curve shift in an accelerating manner with increasing H_{ij}, where H_{ij} is the sigmoid output of the HC at (i, j) in response to its potential h_{ij}:

$$H_{ij} = \frac{a_H h_{ij}^2}{b_H^2 + h_{ij}^2}, \tag{A5}$$

where a_H and b_H are constants. This bounded function causes the amount of shift to decrease as h_{ij} becomes large. The combination of the initial acceleration by the exponential function in the equation (A4) and the later saturation by the equation (A5) causes the S_{ij} curve to accelerate initially and later decelerate with increasing h_{ij}. Figure A1A shows an example of this shift property. The leftmost curve represents the S_{ij} curve with $h_{ij} = 0$; the other curves have h_{ij} values of $0.1, 0.2, \ldots, 0.5$, respectively. All these curves have the same average luminance $\overline{I} = 10^2$. The shift property is generated at any average luminance $\overline{I}$. Note that the leftmost curve in Fig. A1A is the same as the curve with $\overline{I} = 10^2$ in Fig. 2C. Figure A1B shows what happens when $H_{ij} = h_{ij}$ is used instead of equation (A5), with all other equations the same; it shows no deceleration. Here, h_{ij} values of 0 to 10 were used with increments of 1. Figure A1C shows a situation where the term $(B_s - s_{ij})$ in equation A4 has been replaced by 1; it shows a prominent compression. For this simulation, h_{ij} values of 0 to 0.5 with increments of 0.1 were used. Figure A1D shows a situation with $f(H_{ij}) = H_{ij}$ in equation (A4′); it does not have the smooth acceleration shown in Fig. A1A. The same h_{ij} values as for Fig. A1C were used for this simulation.

The potential of an HC connected to its neighbors through gap junctions is defined as follows.

$$\frac{dh_{ij}}{dt} = -h_{ij} + \sum_{(p,q)\in N_{ij}^H} P_{pqij}(h_{pq} - h_{ij}) + S_{ij}, \tag{A6}$$

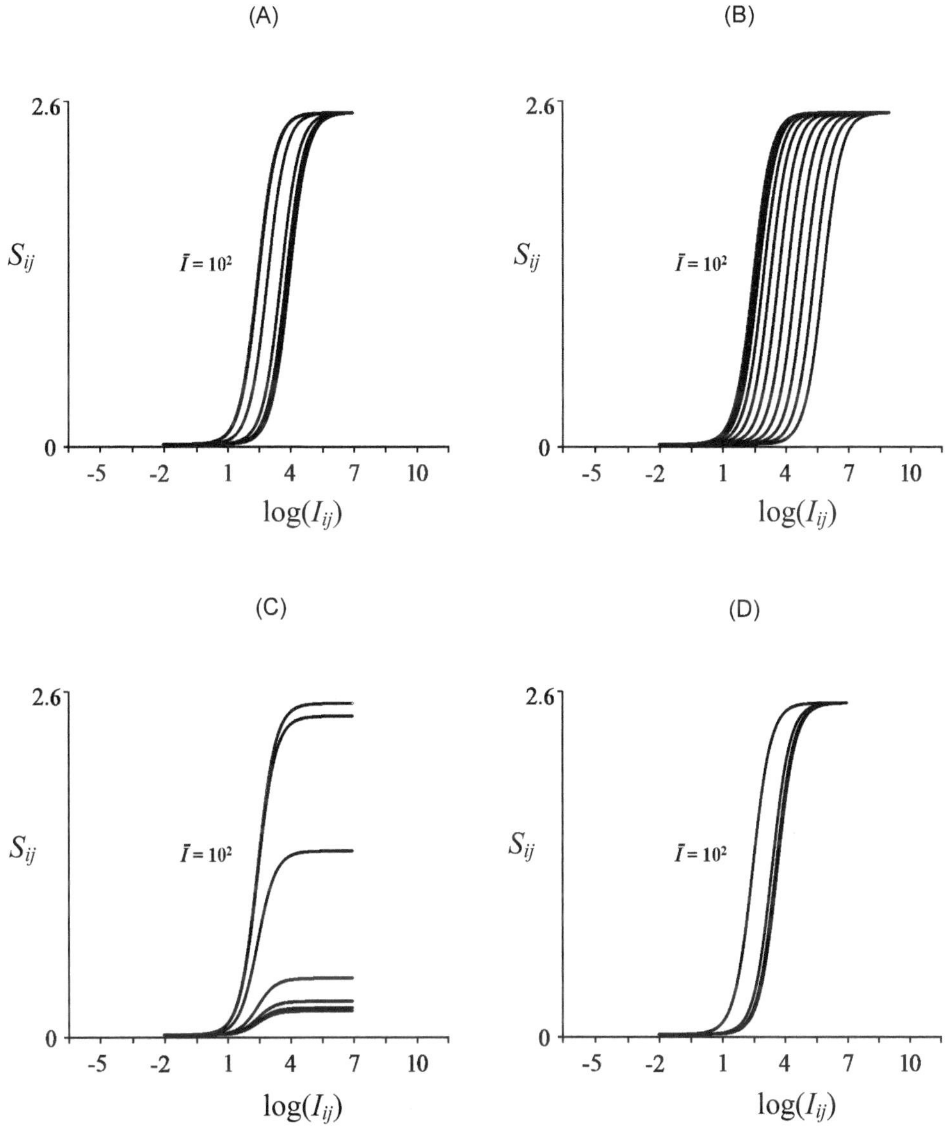

Figure A1. Shift property of spatial contrast adaptation. (A) Shift property of spatial contrast adaptation of the model. The graph shows an example where the $\log(I_{ij})$–S_{ij} curve smoothly accelerates initially and later decelerates with growing h_{ij}. These curves are generated using the equation A4. These curves and all the following curves in B–D have the same average luminance $\overline{I} = 10^2$. The curves from the left to right have h_{ij} values of 0 to 0.5 with increment 0.1. The same is true for C and D. (B) Shift property with $H_{ij} = h_{ij}$ in placed of the equation A5. The curves show no deceleration. The curves from the left to right have h_{ij} values of 0 to 10 with increment 1. (C) Shift property with no $(B_s - s_{ij})$ term in equation (A4). The curves show a prominent compression. (D) Shift property with $f(H_{ij}) = H_{ij}$ in equation (A4′). The curves do not have the smooth acceleration shown in graph A.

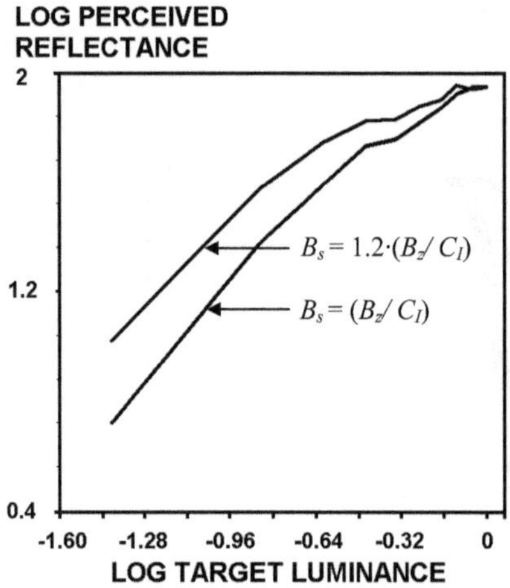

Figure A2. The curves show ten-Mondrian Articulation situation with two values for B_s, one (B_z/C_I), the other $1.2(B_z/C_I)$. While the deviation of 20% from the optimal value shows a bit of compression, the overall quality of the Articulation effect remains robust. This demonstrates that the model tolerates a fair amount of fluctuation in the value of the parameter.

where P_{pqij} is the permeability between cells at (i, j) and (p, q); namely,

$$P_{pqij} = \frac{-1}{1 + \exp[-(|S_{ij} - S_{pq}| - \beta_p)/\lambda_p]} + 1 \tag{A7}$$

Terms β_p and λ_p in (A7) are constants, and N_{ij}^H in (A6) is the neighborhood of size ε_H to which the model HC at (i, j) is connected:

$$N_{ij}^H = \left\{(p, q) : \sqrt{(i-p)^2 + (j-q)^2} \leqslant \varepsilon_H \text{ and } (p, q) \neq (i, j)\right\}. \tag{A8}$$

A.2. Center — Surround Stage

The retinally adapted signal S_{ij} is then processed by small-scale and medium-scale on-center off-surround and off-center on-surround networks. In the following, scale subscripts (e.g. x_s and x_m for small and medium scales, respectively) are omitted for simplicity. An on-center off-surround (ON) network of cell activities x_{ij}^+ that obey membrane equations is defined as follows:

$$\frac{\mathrm{d}x_{ij}^+}{\mathrm{d}t} = -Ax_{ij}^+ + (B - x_{ij}^+)C_{ij} - (x_{ij}^+ + D)E_{ij}, \tag{A9}$$

where A, B and D are constants. The on-center input obeys:

$$C_{ij} = \left(\sum_{(p,q)\in N_{ij}^C} S_{pq}C_{pqij}\right)\frac{W_C}{\sum_{(p,q)\in N_{ij}^C} C_{pqij}}, \tag{A10}$$

and the off-surround input obeys:

$$E_{ij} = \left(\sum_{(p,q)\in N_{ij}^{E}} S_{pq} E_{pqij} \right) \frac{W_E}{\sum_{(p,q)\in N_{ij}^{E}} E_{pqij}}, \tag{A11}$$

with the excitatory Gaussian on-center kernel:

$$C_{pqij} = C \exp\left\{ -\frac{(p-i)^2 + (q-j)^2}{\alpha^2} \right\}, \tag{A12}$$

and the inhibitory Gaussian off-surround kernel:

$$E_{pqij} = E \exp\left\{ -\frac{(p-i)^2 + (q-j)^2}{\beta^2} \right\}. \tag{A13}$$

Coefficients C and E in (A12) and (A13), which normalize and make the sums of the center and surround kernels the same, are defined by:

$$C = \frac{W_C}{\sum_{(p,q)\in N^C} \exp\left\{ -\frac{p^2+q^2}{\alpha^2} \right\}} \tag{A14}$$

and

$$E = \frac{W_E}{\sum_{(p,q)\in N^E} \exp\left\{ -\frac{p^2+q^2}{\beta^2} \right\}}. \tag{A15}$$

Terms α, β, W_C and W_E are constants. N_{ij}^{C} in equation (A10) is the on-center neighborhood to which the cell at (i, j) is connected:

$$N_{ij}^{C} = \left\{ (p, q) : \sqrt{(i-p)^2 + (j-q)^2} \leqslant \varepsilon_C \text{ and } 0 \leqslant p \leqslant 199 \text{ and } 0 \leqslant q \leqslant 199 \right\}, \tag{A16}$$

where ε_C is a constant defining the size of the neighbor. N^C in equation (A14) is the neighbor for the standard center kernel defined as follows.

$$N^C = \left\{ (p, q) : \sqrt{(i-p)^2 + (j-q)^2} \leqslant \varepsilon_C \right\}. \tag{A17}$$

The only difference between N_{ij}^{C} and N^C is that N_{ij}^{C} is constrained by the boundary of the image (200×200), which may cut kernels along the borders, while N^C, which defines the whole kernel, is not. For brevity, the same convention between N_{ij}^{C} and N^C is used for other equations as well. For example, N_{ij}^{E} in equation (A11) is the neighborhood for the surround kernel with a size ε_E with the same form of definition as equation (A16), and its corresponding standard neighbor is N^E with the same form of definition as equation (A17) (see Table 1 for parameters).

For each position, the normalizing factors $W_C/\Sigma C_{pqij}$ and $W_E/\Sigma E_{pqij}$ in (A10) and (A11) are constants, mostly just 1, except for the positions along the border of the image. Normalization eliminates unwanted boundary effects created by filters with a fixed kernel size. In case of a center-surround filter, for example, without normalization, halos along the border of the image can occur because of the disinhibition caused by cut kernels there.

The equilibrium activities of (A9) are:

$$x_{ij}^{+} = \frac{BC_{ij} - DE_{ij}}{A + C_{ij} + E_{ij}}. \tag{A18}$$

The corresponding equilibrium activities of the off-center on-surround (OFF) network are:

$$x_{ij}^{-} = \frac{BC_{ij}^{-} - DE_{ij}^{-}}{A + C_{ij}^{-} + E_{ij}^{-}}. \tag{A19}$$

In (A19),

$$C_{ij}^{-} = E_{ij}, \tag{A20}$$

and

$$E_{ij}^{-} = C_{ij}, \tag{A21}$$

(Grossberg *et al.*, 1995). The output signals are rectified versions of x_{ij}^{+} and x_{ij}^{-}:

$$X_{ij}^{+} = [x_{ij}^{+}]^{+}, \tag{A22}$$

and

$$X_{ij}^{-} = [x_{ij}^{-}]^{+}. \tag{A23}$$

Luminance signals L_{ij}, which constitute the large-scale of the center-surround process, are defined by:

$$L_{ij} = S_{ij}. \tag{A24}$$

Through these processes, the initial stage of the model achieves automatic gain control in all its small, medium and large scales.

A.3. Boundary System

Simple cell activities are simulated using a network of units having polarized and oriented receptive fields around a gird of pixel units. Figure A3A shows pixel units at (i, j) denoted as small filled circles, and eight surrounding numbered positions at (i', j'), where pairs of model simple cells with the same orientation but opposite contrast polarity are located. Each simple cell is represented by a half-filled and half-hollow oriented ellipse (Fig. A3B). The eight positions are as follows: $(i + 0.5, j)$, $(i + 0.5, j + 0.5)$, $(i, j + 0.5)$, $(i - 0.5, j + 0.5)$, $(i - 0.5, j)$, $(i - 0.5, j - 0.5)$, $(i, j - 0.5)$, $(i + 0.5, j - 0.5)$. A pair of simulated simple cells has one of 4 orientations: $(0, \pi/4, \pi/2, 3\pi/4)$. The even numbered positions have only two $(0, \pi/2)$ orientations; positions 3 and 7 have three orientations $(0, \pi/4, 3\pi/4)$;

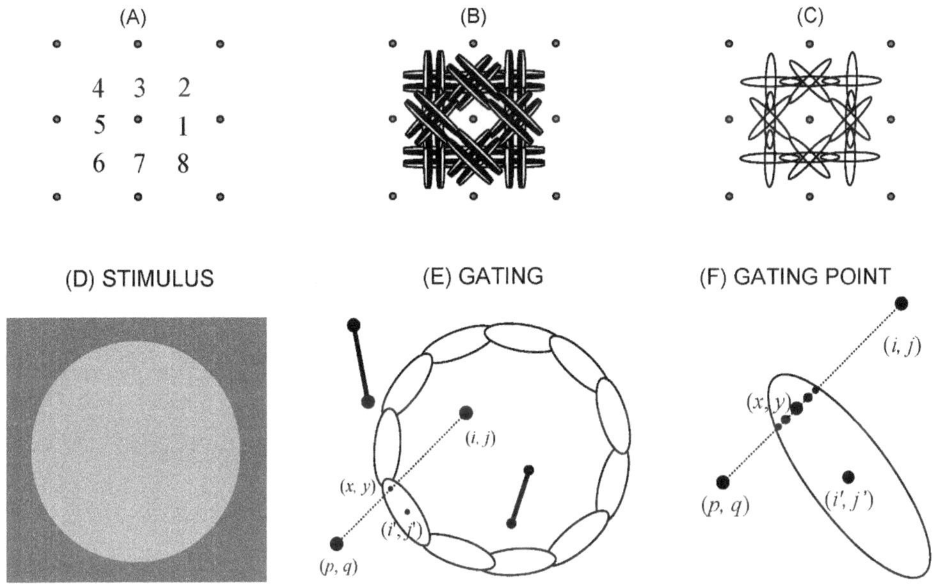

Figure A3. Model boundary system and gating mechanism. (A) Relative positions of model simple and complex cells to pixel points. The model simple and complex cells are poisoned between the pixel points. For example, for a given pixel in the middle (the small gray filled circle in the middle), there are eight surrounding positions (1 through 8) where simple and complex cells are placed. (B) Configuration of simple cell network around a pixel unit in the middle. (C) Gating field of complex cells around the pixel unit. Just one set of pixel-complex and simple cell relationship is shown for clarity. The same pixel-complex and simple cell relationship applies to other pixels. (D) Example of a stimulus. (E) Illustration of gating mechanism for stimulus D. It illustrates the resulting activations of gating components with the input in figure D. The activated complex cells that surround the disk area gate any connections crossing their gating fields represented as ellipses. The connection between (i, j) and (p, q) is gated (the doted line) by a gating signal at (x, y) in the gating field of the complex cell centered at (i', j'). The other connections are not gated, being allowed to have high conductances (solid lines). For the purpose of illustration, more orientations are shown than the four orientations used for the simulations. (F) Position of the gating point. The figure shows the blown up part of the gated part of the connection in the Fig. E. In the simulation, 10 equidistance points (5 of them are shown for clarity) along the cross-section between the ellipse and the crossing line (i, j)–(p, q) were examined to find the approximate inflection (maximum) point. The size of each dot represents the value $G^c_{xyi'j'k}$ of equation (A37) for each examined point.

and positions 1 and 5 have three orientations ($\pi/4$, $\pi/2$, $3\pi/4$). The responses of simple cells are modeled using medium-scale contrast signals. This simplification was chosen because it gives relative clean edge signals. The outputs from simple cells having light-dark and dark-light luminance polarities in their receptive fields are simulated as follows:

$$s^{LD}_{i'j'k} = \left[\left(L^+_{i'j'k} + R^-_{i'j'k}\right) - \left(R^+_{i'j'k} + L^-_{i'j'k}\right)\right]^+, \tag{A25}$$

$$s^{DL}_{i'j'k} = \left[\left(R^+_{i'j'k} + L^-_{i'j'k}\right) - \left(L^+_{i'j'k} + R^-_{i'j'k}\right)\right]^+, \tag{A26}$$

where the superscripts *LD* and *DL* indicate light-dark and dark-light luminance polarities of the model simple cell receptive fields, respectively, and k denotes the orientation. Activation of a model simple cell left and right sub-receptive fields from ON and OFF channels is modeled as follows:

$$L^{+}_{i'j'k} = \left(\sum_{(p,q)\in N^B_{i'j'}} X^{m+}_{pq} G_{pqi'j'kL} \right) \frac{W_B}{\sum_{(p,q)\in N^B_{i'j'}} G_{pqi'j'k}}, \tag{A27}$$

$$R^{+}_{i'j'k} = \left(\sum_{(p,q)\in N^B_{i'j'}} X^{m+}_{pq} G_{pqi'j'kR} \right) \frac{W_B}{\sum_{(p,q)\in N^B_{i'j'}} G_{pqi'j'k}}, \tag{A28}$$

and

$$L^{-}_{i'j'} = \left(\sum_{(p,q)\in N^B_{i'j'}} X^{m-}_{pq} G_{pqi'j'kL} \right) \frac{W_B}{\sum_{(p,q)\in N^B_{i'j'}} G_{pqi'j'k}}, \tag{A29}$$

$$R^{-}_{i'j'} = \left(\sum_{(p,q)\in N^B_{i'j'}} X^{m-}_{pq} G_{pqi'j'kR} \right) \frac{W_B}{\sum_{(p,q)\in N^B_{i'j'}} G_{pqi'j'k}}. \tag{A30}$$

Subscripts L and R indicate the two sub-receptive fields for the simple cell with L indicating the left part (to the anticlockwise) of the sub-receptive field, and the R the right part (to the clockwise) of the sub-receptive field along the axis of the orientation. Constant W_B is the sum of the standard kernel weights of the simple cell:

$$W_B = \sum_{(p,q)\in N^B} G_{pqi'j'k}. \tag{A31}$$

At each position, the normalization factor $W_B/\Sigma G_{pqi'j'k}$ is constant, mostly just 1, except for positions along the border of the image where the Gaussian kernel is incomplete. To see the size of the simple cell kernel neighbor, N^B, see ε_B in Table 1.

A pair of oriented Gaussian kernels, indicated as L and R, simulates receptive fields for the simple cell:

$$G_{pqi'j'k(L \text{ or } R)} = \kappa \exp\left\{ -\frac{[(p-i')\cos(\pi k/4) + (q-j')\sin(\pi k/4)]^2}{\gamma_h^2} - \frac{[-(p-i')\sin(\pi k/4) + (q-j')\cos(\pi k/4) + Shift_{(L \text{ or } R)}]^2}{\gamma_v^2} \right\}, \tag{A32}$$

where $Shift_{(L)}$ and $Shift_{(R)}$, which shift the sub-fields orthogonal to the axis of orientation, are constants $-\gamma_v$ and γ_v, respectively; κ is a constant; k is one of the four numbers (1, 2, 3, 4) that sets the orientation; and γ_h and γ_v are constants that define the widths of the kernel along and across the axis of orientation, respectively.

The model complex cells are also located at the eight (i', j') positions, and have oriented receptive fields, as illustrated in Fig. A3C. The model complex cell of orientation k at (i', j') pools the outputs of a pair of simple cells as follows:

$$z_{i'j'k} = s_{i'j'k}^{LD} + s_{i'j'k}^{DL}. \tag{A33}$$

This cell potential goes through an activation function:

$$Z_{i'j'k} = f(z_{i'j'k}), \tag{A34}$$

where

$$f(x) = \frac{a_B x^{1.7}}{b_B^2 + x^{1.7}}. \tag{A35}$$

The parameter 1.7 of the power of x was used in (A35) to give the optimal strength of the boundary signals across simulations. A gradual degradation of image quality occurs as the parameter deviates from the optimal value (data not shown). The complex cell gates any horizontal connections that cross its gating field. The effective gating strength at a point (x, y) along a passing horizontal connection is the product of the gating weight $(G_{xyi'j'k}^c)$ at the point and the activation of the gating complex cell at (i', j') $(Z_{i'j'k})$:

$$Z'_{xyi'j'k} = G_{xyi'j'k}^c Z_{i'j'k}, \tag{A36}$$

where x, y are continuous variables. The Gaussian kernel of the gating field, which represents the spatial spread of gating weight of complex cell axons at points (x, y) along the line (i, j)–(p, q), is defined as follows:

$$G_{xyi'j'k}^c = \exp\left\{-\frac{[(x-i')\cos(\pi k/4) + (y-j')\sin(\pi k/4)]^2}{\gamma_{ch}^2} - \frac{[-(x-i')\sin(\pi k/4) + (y-j')\cos(\pi k/4)]^2}{\gamma_{cv}^2}\right\}. \tag{A37}$$

Figures A3D and A3E show an example of the complex cell gating mechanism for a given input. For a given complex gating field, it is assumed that the gating occurs at just one point for each crossing connection. The gating point (x, y), which lies along the line (i, j)–(p, q), is chosen that gives the maximum value of equation (A37). In the simulation, 10 equidistance points along the cross-section between the ellipse and the crossing line (i, j)–(p, q) were examined to find the approximate inflection (maximum) point as shown in Fig. A3F. The size of each dot in the figure represents the value $G_{xyi'j'k}^c$ of equation (A37) for each examined point.

A.4. Filling-in

Cortical filling-in is driven by the inputs M_{ij} which are the pooled luminance and contrast signals as follows:

$$M_{ij} = \Big[w_s\big(X_{ij}^{s+} - X_{ij}^{s-}\big) + w_m\big(X_{ij}^{m+} - X_{ij}^{m-}\big) + w_l L_{ij} + b_M\Big]^+, \qquad \text{(A38)}$$

where w_s, w_m, w_l, are weighting constants, and b_M is a tonic bias term. Either of two versions of the filling-in process yield equivalent simulations of the targeted data. A long-range diffusion process, much as in the retinal HC diffusion in (A6), works well with activities F_{ij} instead of the activities h_{ij} in (A6), and inputs M_{ij} instead of the inputs S_{ij} in (A6). This long-range diffusion runs 100 times faster than previous nearest-neighbor diffusions for filling-in. In addition, the conductance values P_{pqij} are divisively gated by activated complex cells along its path. They are defined by:

$$P_{pqij} = \frac{\delta \exp[-((i-p)^2 + (j-q)^2)/\sigma^2]}{\prod_{i'j'k}(1 + \varepsilon Z'_{i'j'k})}, \qquad \text{(A39)}$$

where σ, δ and ε are constants. The numerator of (A39) describes the strengths of horizontal connections, assumed to have a Gaussian distribution, such that longer connections have smaller strengths.

Alternatively, a long-range propagation process that does not require diffusion, but is normalized in a different way, generates essentially identical simulations, which are the ones that are shown in this article. This process runs 1000 times faster than nearest-neighbor diffusion processes. The first step of the filling-in is to activate the filling-in units with the pooled multiple-scale input signals M_{ij}:

$$F_{ij} = M_{ij}. \qquad \text{(A40)}$$

Here, the filling-in activity $F_{ij}(t+1)$ equals:

$$F_{ij}(t+1) = \left(\sum_{(p,q)\in N_{ij}^F} F_{pq}(t) P_{pqij}\right) \frac{W_F}{\sum_{(p,q)\in N_{ij}^F} P_{pqij}}, \qquad \text{(A41)}$$

where the conductance P_{pqij} shares the same form of equation (A39) with different parameters (see Table 2). The constant W_F in (A41) is a sum of conductances defined as follows:

$$W_F = \sum_{(p,q)\in N^F} \delta \exp\big[-\big(i^2 + j^2\big)/\sigma_F^2\big]. \qquad \text{(A42)}$$

Since W_F is constant, the constant δ for a fixed σ is calculated as follows:

$$\delta = \frac{W_F}{\sum_{(p,q)\in N^F} \exp\{-(i^2 + j^2)/\sigma^2\}}. \qquad \text{(A43)}$$

The size of the filling-in neighborhood N^F is determined by parameter ε_F in Table 1. Equation (A41) assumes that the filling-in unit can normalize its conductances. The normalizing factor $W_F/\Sigma P_{pqij}$ affects the conductance in two ways. First, at the border of the image, the incomplete kernels get normalized to have the same size as W_F. Second, normalization compensates for the overall lost conductance caused by gating (division by the denominator in equation (A39)). By this normalization process, the sum of the effective conductances equals:

$$W_F = \sum_{pq} \left(P_{pqij} \frac{W_F}{\sum_{pq} P_{pqij}} \right). \tag{A44}$$

For example, if half of the input connections were totally blocked by gating signals, the unit would try to increase the effective input flow by doubling the efficacy of the remaining connections, keeping the sum of all the incoming conductances the same. Ten iterations of equation (A41) gives satisfactory filled-in results.

A.5. Lightness Anchoring

At the anchoring stage, the filled-in surface activity F_{ij} becomes anchored into the activity A_{ij} using the following equation:

$$\frac{\mathrm{d}A_{ij}}{\mathrm{d}t} = -B_A A_{ij} + \Psi(C_A - A_{ij})F_{ij}, \tag{A45}$$

where B_A and C_A are constants. The tonic gain control signal Ψ, which modulates all the anchoring activities A_{ij}, uses the following equation.

$$\frac{\mathrm{d}\Psi}{\mathrm{d}t} = \tau_\Psi\{-\Psi + (B_\Psi - \Psi)T_\Psi - \Psi H\}. \tag{A46}$$

The term τ_Ψ is a time constant that determines the speed of integration of equation (A46). The term $-\Psi$ is a leakage component. The next term $(B_\Psi - \Psi)T_\Psi$ is an excitatory component that drives the gain control signal Ψ toward its maximum B_Ψ until the inhibitory component ΨH kicks in due to the activation of the suppressive signal H, which is defined as follows:

$$\frac{\mathrm{d}H}{\mathrm{d}t} = \tau_H \left\{ -H + (\varphi - H) \sum_{ij} B_{ij} \right\}, \tag{A47}$$

where τ_H is a time constant. Using the equation (A47), the suppressive signal H quickly becomes activated and suppresses the gain control activity Ψ whenever there is an activated output cell at the BHCAW module, which signals the anchoring of blurred ‘highest luminance’ to white. The output of the BHCAW module B_{ij} is defined as follows:

$$B_{ij} = f^B(b_{ij}), \tag{A48}$$

Table 3.

Names	Symbols	Values
Decay rate for Anchoring	B_A	1
Depolarization constant for Anchoring	C_A	10
Time constant of modulatory unit of anchoring	τ_Ψ	0.01
Depolarization constant of modulatory unit of anchoring	B_Ψ	1.3
Recharge rate of tonic activity	T_Ψ	1
Power of BHCAW activation function	m	100
White	ϖ	0.5
Hyperpolarization constant for gain control	φ	8
Spatial scale for Anchoring	ζ_A	100 (for the area rule), 4 (for the others)
Size of connection range for the center of center-surround unit	ε_C	6 (for small scale), 28 (for medium scale)
Size of connection range for the surround of center-surround unit	ε_E	6 (for small scale), 28 (for medium scale)
Size of connection range for the half kernel of simple cell	ε_B	3
Size of connection range for the blurring kernel of Anchoring	ε_A	100 (for the area rule), 4 (for the others)
Size of connection range for the filling-in unit	ε_F	8
Sizes of various standard kernels	W_C, W_E, W_B, W_A, W_F	0.6, 0.6, 4, 1, 1

where the signal function $f^B(x)$ is a steep sigmoid:

$$f^B(x) = \frac{x^m}{\varpi^m + x^m}, \tag{A49}$$

where m and ϖ are constants (see Table 3). Function b_{ij} in (A48) is a blurred version of the anchoring signal A_{ij}:

$$b_{ij} = \sum_{(p,q)\in N_{ij}^A} G_{pqij}^A A_{pq}, \tag{A50}$$

where the blurring Gaussian anchoring kernel is defined by:

$$G_{pqij}^A = G^A \exp\left\{-\frac{(p-i)^2+(q-j)^2}{\zeta_A^2}\right\} \frac{W_A}{\sum_{(p,q)\in N_{ij}^A} G_{pqij}^A}, \tag{A51}$$

where constant

$$G^A = \frac{W_A}{\sum_{(p,q)\in N^A} \exp\left\{-\frac{p^2+q^2}{\zeta_A^2}\right\}}, \tag{A52}$$

and W_A and ζ_A are constants. The size of the blurring neighborhood N^A is determined by paramenter ε_A in Table 1. When m in equation (A49) is large, H approximates a step function

$$\begin{cases} H = \varphi & \text{whenever any } b_{ij} \geqslant \varpi \\ H = 0 & \text{otherwise,} \end{cases} \tag{A53}$$

where φ is a constant. In the simulation, equation (A53) was used in place of equations (A47) to (A49).

APPENDIX B

To generate the stimuli with different background luminance (Fig. 2C), the following formula was used:

$$I_{ij} = \rho_{ij} E_{ij}, \tag{B1}$$

where I_{ij} is the luminance at point (i, j), ρ_{ij} is the reflectance at point (i, j), and E_{ij} is the illumination on point (i, j) (Hurlbert, 1989). For a given stimulus, E_{ij} was uniform across the image. For practical purposes, ρ_{ij} in equation (B1) was replaced by the luminance at point (i, j) of the original image. This situation is roughly equivalent to a viewing situation where a picture is exposed to uniform background illumination. The range of ρ_{ij} was chosen to be –4 to 5 in log-scale for a fixed illumination level to examine the full dynamic profile of the shift property. See Fig. 2C for the values of illumination E_{ij} used for the simulations.

Color, Line, and Space, pp. 221–238 (2006)

Dissociation of color and figure–ground effects in the watercolor illusion

RÜDIGER VON DER HEYDT * and RACHEL PIERSON

Department of Neuroscience and Krieger Mind/Brain Institute, 338 Krieger Hall, Johns Hopkins University, 3400 N Charles Street, Baltimore, MD 21210, USA

Abstract—Two phenomena can be observed in the watercolor illusion: illusory color spreading and figure–ground organization. We performed experiments to determine whether the figure–ground effect is a consequence of the color illusion or due to an independent mechanism. Subjects were tested with displays consisting of six adjacent compartments — three that generated the illusion alternating with three that served for comparison. In a first set of experiments, the illusory color was measured by finding the matching physical color in the alternate compartments. Figureness (probability of 'figure' responses, 2AFC) of the watercolor compartments was then determined with and without the matching color in the alternate compartments. The color match reduced figureness, but did not abolish it. There was a range of colors in which the watercolor compartments dominated as figures over the alternate compartments although the latter appeared more saturated in color. In another experiment, the effect of tinting alternate compartments was measured in displays without watercolor illusion. Figureness increased with color contrast, but its value at the equivalent contrast fell short of the figureness value obtained for the watercolor pattern. Thus, in both experiments, figureness produced by the watercolor pattern was stronger than expected from the color effect, suggesting independent mechanisms. Considering the neurophysiology, we propose that the color illusion follows from the principles of representation of surface color in the visual cortex, while the figure–ground effect results from two mechanisms of border ownership assignment, one that is sensitive to asymmetric shape of edge profile, the other to consistency of color borders.

Keywords: Watercolor illusion; figure–ground segregation; border ownership; surface color; neural coding of contour.

INTRODUCTION

The watercolor illusion (Pinna *et al.*, 2001) is a phenomenon that can be observed when a figure is defined by a contour consisting of a pair of parallel lines on white background — a dark line, and a lighter colored line on the inside. The interior

*To whom correspondence should be addressed. E-mail: von.der.heydt@jhu.edu

of the figure then appears slightly tinted with the hue of the lighter colored line. Pinna and coworkers further showed that configurations of lines that produce the watercolor illusion also influence the perception of figure and ground. In displays like the one shown in Fig. 1A, the watercolor regions tend to be perceived as 'figures' and the surrounding regions as ground (Pinna *et al.*, 2003).

These findings might suggest that the 'figureness' effect is a consequence of the color illusion. The appearance of the illusory color tint suggests a filling-in process: color signals generated by the light-colored lines spread in visual cortex until they reach a contour boundary (Gerrits and Vendrik, 1970; Cohen and Grossberg, 1984); the dark lines of the watercolor displays form boundaries that limit the spread of color signals which thus fill the enclosed space evenly. The resulting color signals might then act exactly like real color signals, and the filled-in space would thus stand out as a figure against the surrounding white space, just like a colored region stands out against a white background.

An alternative possibility is that illusory color and figureness are produced in parallel, by independent mechanisms that are both activated by the double lines. The color might be the result of one process (e.g. filling-in, as sketched above), and figure–ground segregation might be produced by another process that is also triggered by the double lines. For example, the double lines of different contrast might activate cortical neurons that respond selectively to edges with asymmetric luminance (or color) profile and such neurons might contribute to the definition of border-ownership (Zhou *et al.*, 2000).

The aim of the present study was to distinguish between these alternative hypotheses. If the figure–ground effect of the watercolor pattern is a consequence of a color filling-in process it should covary with the strength of the color illusion. If the illusory color behaves just like a veil of physical color, it should be possible to null the color contrast subjectively, for example by adding the appropriate amount of physical color to the white of the surrounding space, and the figure–ground effect should then disappear. However, if a figure–ground effect persists after nulling of the subjective color contrast, this would indicate that at least some of the figure–ground process is independent of the process that generates the color illusion.

GENERAL METHODS

The stimuli were designed and presented using commercial software (Corel Draw, MS Powerpoint), and displayed on a 19″ computer monitor (Gateway VX900), with a resolution of 1280 × 1024. Subjects were seated in a dimly lit room and viewed the monitor at a distance of 210 cm.

The stimuli consisted of 6 adjacent compartments defined by wavy double lines, as shown in Fig. 1A–B. The colors of the lines were dark purple and orange. In three of the compartments, the orange line was on the inside, and in the alternate compartments it was on the outside. The pattern subtended 7.5 by 5 deg of visual angle. The width ratio of alternating compartments was varied between 1 : 1, 1 : 2, and 1 : 4 (1 : 2 illustrated). Patterns as shown in Fig. 1A were used in Experiments 1 and 2, and patterns like the one shown in Fig. 1B were used in Experiments 1 and 3. In Experiment 4 displays similar to that of Fig. 1B were used, but without the orange lines. The colors in CIE (1931) coordinates were (4.4, 0.26, 0.20) for the purple lines, (33, 0.42, 0.43) for the orange lines, and (64, 0.30, 0.32) for the background, where the first number of each triplet is the luminance in cd/m^2, and the following numbers are the chromaticities x and y. In the following we use the CIE 1976 ($L^*a^*b^*$) color space which was designed to reflect better the perceptual differences between colors (Wyszecki and Stiles, 1982). The ($L^*a^*b^*$) coordinates of our colors were (31.5, 25.3, −25.7) for the purple lines, and (76.8, 5.5, 53.7) for the orange lines, with (100, 0, 0) for the background. The widths of the purple and orange lines subtended approximately 3 and 5 arc min of visual angle, respectively.

RESULTS

We performed four experiments to test the relationship between color illusion and figure–ground effect in the watercolor display. In Experiment 1 we measured the illusory color change using a matching technique. In Experiments 2 and 3 we determined, for two different configurations, whether figure–ground segregation persists after subjectively nulling the color contrast. In Experiment 4 we reversed the strategy. If the figure–ground segregation is the result of the illusory tint, then a comparable figure–ground effect should result also from a physical tint.

For Experiments 1–3 we used displays of 6 adjacent compartments, each defined by wavy double lines consisting of a dark purple and an orange color component (Fig. 1A–B). In three of the compartments, the orange line was on the inside (producing the watercolor illusion), and in the alternate compartments it was on the outside. This pattern was modeled after Pinna *et al.* (2003) (who used a row of 8 compartments). For Experiment 4 we used displays without color illusion and measured the strength of figure–ground segregation produced by coloring alternate compartments with a slight tint comparable to the illusory watercolor.

Figure 1. (See color plate X) Examples of the stimuli used to study figure–ground organization in the watercolor illusion. The stimuli consisted of six adjacent compartments outlined by wavy double lines. The compartments that were lined orange on the inside are called 'watercolor compartments', the others 'alternate compartments'. The relative width of the two compartments was varied; a ratio of 2 : 1 is illustrated. Arrows and numbers at the top served to identify the two types of compartments and were displayed 1s after stimulus onset. Stimuli of type A were used in Experiment 2, stimuli of type B in Experiments 3 and 4. Color matching (Exp. 1) was performed for both types of stimuli.

MEASURING THE ILLUSORY COLOR

Experiment 1

For the color matching, the alternate compartments were tinted with a comparison color, and subjects compared this color with the illusory tint in the watercolor compartments. We used a pattern with a 2 : 1 width ratio between watercolor and alternate compartments. Based on pilot experiments we assumed that the illusory color could be matched by a color located in color space on the straight line connecting the background white with the color of the orange line of the watercolor pattern. A recent study published after the present results were submitted confirmed this assumption (Devinck *et al.*, 2005). A set of 20 comparison colors distributed along that line was used. Because pilot experiments had shown that some observers perceive very little watercolor tint, if any, we chose a background color that was slightly off-white, so that the range of comparison colors could extend beyond the background color. This assured that the matching color was well within the range of comparison colors for each subject.

Participants. Seventeen subjects participated in this experiment, 16 who were naive about the goals of the study, ranging in age from 13 to 28 yr (median: 16 yr), and one of the authors (RvdH) whose age was 59. All had 20/20 corrected or uncorrected visual acuity.

Procedure. The subjects were instructed as follows: 'You will see a computer screen on which slides will be presented. Each slide will show a pattern of six fields separated by wavy double lines. The interiors of the fields have whitish colors, but some have a slight orange tint compared to the others. Two of the fields are numbered at the top. Tell the experimenter the number of the field that has the orange tint. If you cannot tell for sure, guess.'

A simple staircase method was used. After showing a blank field, the watercolor display in which the alternate stripes were tinted with a comparison color was presented. After one second, the numbers 1 and 2 appeared at the top of the two compartments in the center and the subject (S) indicated which of them looked more colored. Then, after display of a blank field for about one second, the watercolor display was shown again with a different comparison color, and so on. E began the sequence with a comparison color from one end of the scale, at which the colors clearly differed for all subjects, as S gave the expected response, say '1'. E then incremented (or decremented) the comparison color until S responded '2' for 3 consecutive presentations. E then recorded that color and reversed the direction of changes until 3 consecutive '1' responses were given, and so on. This was continued for a total of 10 reversals (5 in each direction).

Results. Figure 2 shows the color matching results for the 17 subjects. The ordinate shows the relative saturation of the comparison color (length of vector from background to comparison color divided by length of vector from background to color of orange lines; negative values represent colors in the direction opposite to orange). Dots represent the means of the reversal points which we took as the matching colors. Vertical lines connect the means of the upper and lower reversal points. These colors were perceived as clearly different from the watercolor tint, either more saturated, or less saturated. The connected pairs of dots represent data from repeated tests of the same subjects taken 5 months apart.

It can be seen that for most of the subjects, the watercolor illusion was equivalent to a tint of about 3% of the inducing color. However, the matching color varied considerably between subjects, with two observers showing weak or no illusory color shift (data points on the left), and others showing an illusion of 6% (data points on the right). The close agreement between the data from the repeated tests indicates that the color matches of the individual observers were highly reproducible.

Ten of the data sets were obtained with the original watercolor pattern (to be used in Experiment 2), and the other ten were obtained with displays in which the top and bottom of the pattern were occluded (to be used in Experiment 3). The mean estimated strengths of illusion for the two display types were 3.2% and 3.5%, which was not significantly different (ANOVA, $F(1, 18) = 0.32$, $p = 0.58$).

The horizontal dotted lines in Fig. 2 represent the test colors that were used in Experiments 2 and 3. Two different sets of test colors were used for the subjects with weak color illusion (Numbers 1–3 on the abscissa) and for the rest of the subjects.

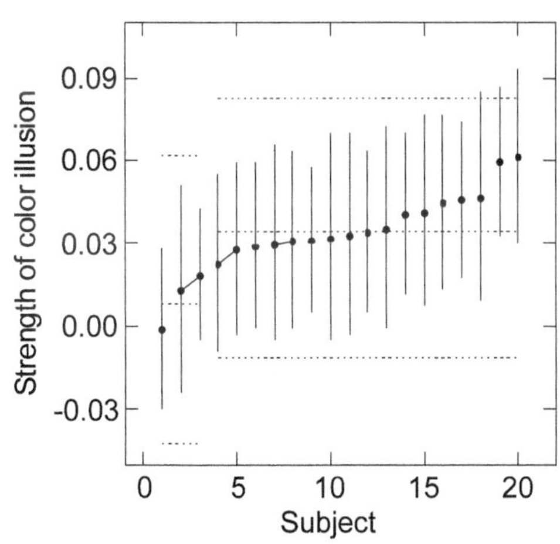

Figure 2. The results of the color matching experiment. Dots indicate the contrast of the physical color in the alternate compartments that subjectively matched the illusory watercolor, for each of 17 subjects, including repeated tests in three subjects (dots connected by lines). Vertical lines connect the means of the reversal points in the matching procedure, that is, the two colors that looked clearly lighter and darker (more saturated), respectively, than the watercolor. Dotted horizontal lines indicate the three colors used in the figure–ground experiments (two sets of colors for two groups of subjects).

FIGURE–GROUND SEGREGATION WITH MATCHING COLORS

Experiment 2

In the following experiments we measured the direction and strength of figure–ground organization. The stimuli were derived from the watercolor display of Fig. 1A which was similar to that used by Pinna *et al.* (2003). Note that we connected the orange lines between compartments, rather than the purple lines, so as to avoid T-junctions of the purples lines which could have induced a figure bias in favor of the watercolor compartments.

The alternate compartments were filled with one of three colors: one that approximately matched the watercolor tint, one that was clearly more saturated, and one that was clearly less saturated, for every subject, as determined in Experiment 1. Patterns without watercolor illusion (no orange lines) and without added tint were included to provide a baseline.

To compare the watercolor illusion with a known factor in figure–ground segregation we also varied the relative width of watercolor and alternate compartments. Width ratios of 1 : 1, 2 : 1, and 4 : 1 were applied, while the sum of the widths was kept constant. With the 2 : 1 and 4 : 1 ratios, the expected figure–ground effect of the watercolor illusion was pitted against the Gestalt effect of proximity (*cf.* Pinna *et al.*, 2003). To control for a possible influence of spatial order, a left-right mirror image version was also presented for each pattern and color, so that in half of the conditions the pattern began with a watercolor compartment on the left and ended with an alternate compartment on the right, and in the other half the order was reversed. Thus, a total of 24 patterns ((3 colors + 1 baseline) × 3 width ratios × 2 spatial orders) were used.

A between-subjects design was used. This design yields the reliability of the conclusions in the case that the experiments would be repeated with other subjects. A 4-way analysis of variance (ANOVA) was performed on the data from the watercolor conditions (factors: color, width ratio, spatial order, and subject), and another 4-way ANOVA to compare the watercolor condition in which the illusory color has been matched with the baseline condition (factors: test condition, width ratio, spatial order, and subject). A significance level of 0.05 was adopted.

Participants. Eight subjects participated — seven who were naive about the purpose of the study and one of the authors, RvdH. All had previously performed the color matching task (Experiment 1).

Procedure. The phenomenon of figure–ground organization was first explained to each subject with the help of five figures from Kanizsa (1979): Figs 5.19, 5.20a–b, 5.21a–b. These figures are also composed of rows of compartments of alternating figureness, similar to the watercolor patterns, but based on light and dark shades of gray. They were designed to demonstrate the influences of symmetry and convexity of contour on figure–ground perception and produce obvious figure–ground segregation in most observers. The subjects were then instructed as follows:

'You will see a computer screen on which slides will be presented. Each slide will show a pattern of six fields separated by wavy lines. In some of the slides, three of the fields may appear as 'figure', alternating with fields that appear as 'ground'. Your task is to decide for each slide which of the fields you perceive as 'figure'. For each presentation, after one second, two arrows numbered 1 and 2 will appear at the top. Tell the experimenter which arrow points to a figure. Try to decide before the arrows appear. Ignore the varying tints or color fringes that may be present.'

We preferred a simple binary response over the rating method used in previous studies of the watercolor illusion (Pinna *et al.*, 2001, 2003) because we thought these responses would be quick and reflect the perception more directly than the rating procedure which requires more effort and might invoke reasoning, for example, about the role of the distances between contours, or the color of the compartments.

Each of the 24 stimulus conditions was presented eight times in random order (except for three subjects who received only four presentations/condition).

Results. Figure 3 summarizes the results of this experiment. The filled symbols represent the degree of 'figureness' of the watercolor compartments, as measured by the probability of this compartment being perceived as figure. Figureness is plotted as a function of the width ratio between watercolor and alternate compartments, with the color tint of the alternate compartments as a parameter. Open circles represent the baseline condition, in which all compartments were defined by a single dark line and filled with the background white.

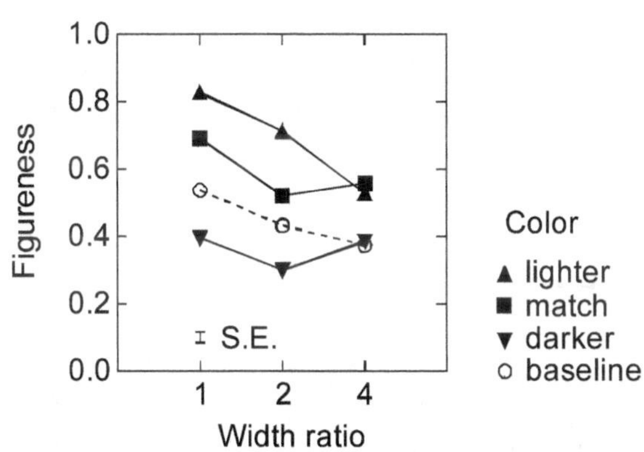

Figure 3. 'Figureness' (probability of being perceived as figure) of the watercolor compartments plotted as a function of the width ratio between watercolor and alternate compartments. Filled symbols and solid lines represent the results for the three color conditions, one in which the color of the alternate compartments matched the illusory color (squares), one in which it was whiter (upward pointing triangles), and one in which it was darker, or more saturated (downward pointing triangles). Open symbols and dashed lines represent the baseline condition in which the compartments were defined by the dark lines only (without the orange lining). It can be seen that figureness of the watercolor compartments tended to decrease as the width of the alternate compartments was reduced (effect of proximity). Figureness decreased also as the color of the alternate compartments was made more saturated.

It can be seen that the highest value of figureness was obtained for the lightest color (upward pointing triangles), which was virtually equal to the background white, and a width ratio of 1. This is the figure–ground effect of the watercolor illusion that was described earlier (Pinna *et al.*, 2003). As the width ratio is increased, making the alternate compartments narrower than the watercolor compartments, figureness of the watercolor compartments decreases. This is because the watercolor effect is pitted against the effect of proximity (Pinna *et al.*, 2003).

When the alternate compartments are tinted with the color that matched the illusory tint, figureness of the watercolor compartments drops (squares), and when the color of the alternate compartments is made more saturated, figureness of the watercolor compartments is further decreased (downward pointing triangles). However, it can be seen that figureness of the watercolor compartments in the color matching condition (filled squares) is consistently higher than that of the corresponding compartments in the baseline condition (open circles) ($F(1, 7) = 10.8$, $p < 0.025$).

The color dependence of figure–ground segregation can be seen more clearly in Fig. 4 where figureness is plotted as a function of the compensatory color for the three width ratios. The dashed lines indicate the corresponding probabilities of figure responses in the baseline condition. The graphs show that the figure–ground effect of the watercolor illusion would be neutralized only with colors that are more saturated than the matching color.

Analysis of the data of the watercolor conditions showed a significant effect of the compensatory color ($F(2, 14) = 4.2$, $p < 0.05$), but not the width ratio ($F(1, 14) = 2.19$, $p > 0.1$). Figureness was slightly higher for compartment #1 (the third from the left) than for compartment #2 (the fourth from the left), suggesting a left-to-right segmentation strategy, but the effect of spatial order did not reach significance ($F(1, 7) = 5.48$, $0.05 < p < 0.1$).

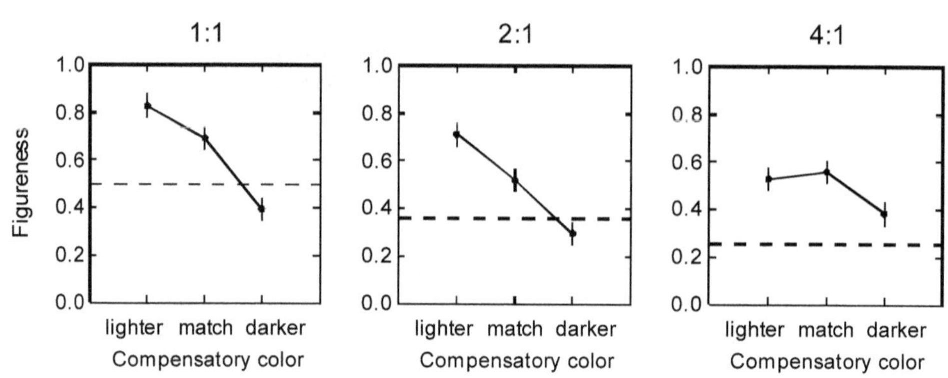

Figure 4. Figureness of the watercolor compartments as a function of the color condition, for three width ratios. Dashed line represents baseline condition. Data replotted from Fig. 3. Note that figureness of the watercolor compartments in the color-matching condition was higher than figureness of the corresponding compartments in the baseline condition.

Experiment 3

We were surprised that the width ratio did not produce a more reliable effect. The Gestalt principle of proximity would predict the alternate compartments to become figures for higher width ratios, thus decreasing the figureness of the watercolor compartments. Such a decrease can be seen in the curves of Fig. 3, but it is not very consistent. We thought that the reason for the modest variation of figureness in Experiment 2 might be the continuity of the contours at the top and bottom which might hinder perceptual segregation of the compartments. We therefore performed another experiment in which the top and the bottom of the compartments were occluded, thus eliminating the horizontal borders and the contour junctions (Fig. 1B).

Participants and procedure. Ten subjects (nine naive, and RvdH) participated in this experiment. Seven were new subjects, while three had already served in Experiment 2. All had previously performed the color matching (Experiment 1). The instructions and the procedure were the same as in Experiment 2. Each of the 24 stimulus conditions was presented eight times in random order.

Results. Figures 5 and 6 summarize the results of this experiment. The figureness values for the watercolor conditions are slightly higher in this experiment than in Experiment 2, producing an even larger separation from the baseline data, but otherwise the results are similar. Figureness of the watercolor compartments in the color matching condition (filled squares) is considerably higher than that of the corresponding compartments in the baseline condition (open circles) ($F(1, 9) = 11.4$, $p < 0.01$). As in Experiment 2, the compensatory color had a significant effect ($F(2, 18) = 12.3$, $p < 0.01$), but the effect of width ratio was not reliable ($F(2, 18) = 1.74$, $p > 0.1$). The factor of spatial order was not significant ($F(1, 9) = 2.9$, $p > 0.1$). Thus, the experiment confirmed the findings of

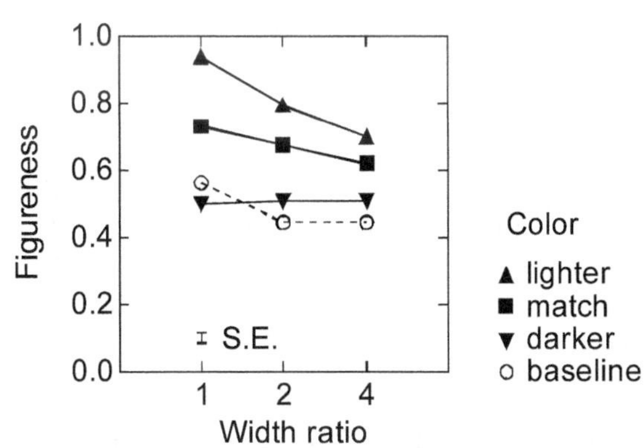

Figure 5. Results of a similar experiment as described in Fig. 3, but for displays in which the top and bottom of the compartments were occluded by gray bars (Fig. 1B). The figureness values were slightly higher than without the occluding bars, but the results were otherwise similar.

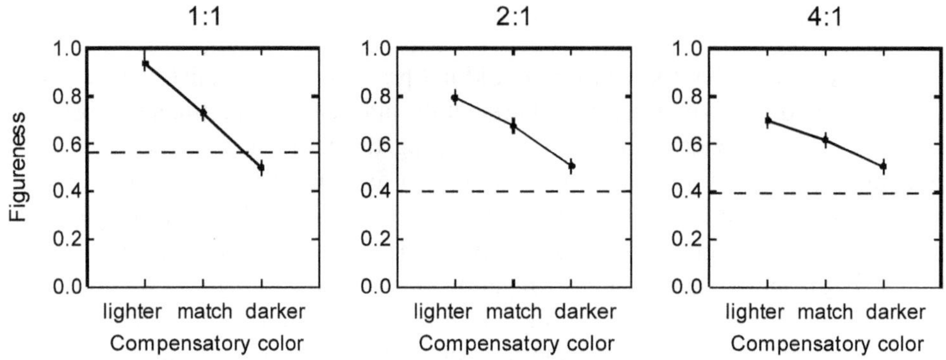

Figure 6. Figureness of the watercolor compartments as a function of the color condition for displays in which the top and bottom of the compartments were occluded by gray bars (Fig. 1B). Data replotted from Fig. 5. Note that figureness of the watercolor compartments in the color-matching condition was higher than figureness of the corresponding compartments in the baseline condition.

Experiment 2. Occluding the contour junctions in the watercolor pattern did not change the pattern of results in any significant way. The figure–ground effect of the watercolor illusion survives compensation of the color illusion.

COLOR-INDUCED FIGURE–GROUND SEGREGATION

Experiment 4

The hypothesis formulated at the beginning, that figure–ground segregation in the watercolor illusion is a consequence of the (illusory) color spreading, implies that filling a region with a tint of color makes it stand out as figure. Although we have seen in the previous experiments that compensating the illusory tint does not completely cancel the figure–ground effect, it did reduce figureness of the watercolor compartments. Thus, adding tint in the alternate compartments increased their figureness.

In Experiment 4 we studied the effect of color on figure–ground segregation in itself. We used a similar display as in Experiment 3, but without the orange lines that induce the watercolor illusion, and measured the strength of figure–ground segregation produced by various color tints in alternate compartments. This enabled us to compare the strength of figure–ground segregation produced by the tint that matched the watercolor illusion with the strength of figure–ground segregation in the actual watercolor display.

Participants and procedure. Eight subjects participated in this experiment – seven who were naive about the goals of the experiment, and one of the authors (RvdH). Each of them had also participated in Experiment 2 or 3. Instructions and procedure were the same as in Experiment 2.

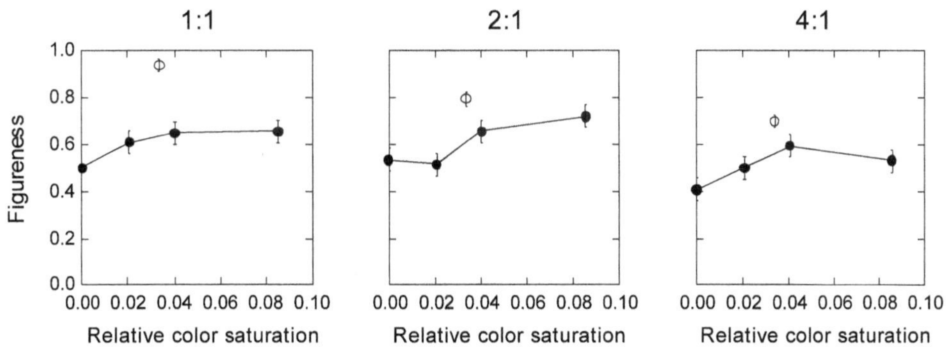

Figure 7. Figureness produced by color tint. Compartments were defined by single dark lines and alternate compartments were tinted with diluted orange. Filled symbols show the probability of the tinted compartments being perceived as figures as a function of the color contrast. Open symbols show the figureness values for the watercolor displays at the equivalent contrast. It can be seen that the watercolor illusion produced higher figureness values than the corresponding color tint.

Results. Figure 7 shows the probability of 'figure' responses for the tinted compartment as a function of the color contrast, for each of the three width ratios. For the 2 : 1 and 4 : 1 ratios only the data for the case that the wider compartments were tinted are shown. It can be seen that figureness increases with contrast, but only weakly. The width ratio has the opposite effect, decreasing figureness, as expected from the law of proximity (figureness = 1 means wider compartments are figure), but this effect is also quite weak. Neither effect reached significance. (However, if all the data from this experiment are taken together, and contrast is coded positive if the wide compartments were tinted, and negative if the narrow compartments were tinted, the contrast effect was significant: $F(6, 42) = 2.97$, $p < 0.025$).

The open circles in Fig. 7 indicate the figureness values obtained for the similar displays with the watercolor illusion (data from Experiment 3). It is plotted at the contrast that was found to be equivalent to the illusory tint (Experiment 1). It is apparent that tinting alternate compartments with the color equivalent to the color of the illusion produces far less of a figure effect than the watercolor display itself.

DISCUSSION

We tested the hypothesis that the watercolor display is a consequence of the illusory color spreading. We determined the color tint that is perceptually equivalent to the illusory watercolor (Experiment 1) and then applied this tint to test the figure–ground effect with two different strategies. First, we compensated the subjective color of the watercolor compartments by adding physical color to the alternate compartments, thus nulling the subjective contrast between the compartments, and then measured figure–ground segregation (Experiments 2 and 3). Second, we applied the equivalent tint to alternate compartments of a pattern that does not produce a watercolor illusion to see how much of a figure–ground effect would result from the coloring itself (Experiment 4).

The results of both types of test showed that the figure–ground effect of the watercolor illusion was stronger than predicted from the hypothesis. Nulling the subjective contrast did not neutralize the figureness of the watercolor compartments, and the figureness produced by just the color tint fell short of the figureness measured for the watercolor display. Thus, the results of these experiments argue against the hypothesis, and we conclude that at least part of the figure–ground effect of the watercolor illusion is due to a mechanism that is independent of the one that generates the color illusion.

This is not to say that the illusory color tint is neutral with respect to figure–ground perception. Experiment 4 showed that even slight coloring of a region relative to the surrounding areas does increase figureness of that region, and Experiments 2 and 3 showed that figureness of the watercolor compartments is significantly reduced when the color contrast is compensated. Thus, color tint, even of low contrast, enhances figureness, but in all three experiments this effect was weak compared to that of the watercolor pattern.

By varying the width of the stripes we tested the Gestalt factor of proximity. The effect of this factor, although clearly visible in the averaged results, was not statistical reliable across subjects. The figure–ground effect of the watercolor illusion was stronger than the effect of proximity, as shown previously (Pinna *et al.*, 2003).

While we achieved a dissociation of the figure–ground effect from the color illusion by compensating the illusory color, Pinna (2005) has recently used watercolor patterns in which the color of the lighter lines varied within a compartment so as to weaken the color illusion. He also concludes that the figure–ground effect is relatively independent of the color illusion.

A possible explanation for the watercolor illusion

The watercolor illusion seems to reveal important characteristics of the neural representation of visual surfaces. Our results, showing that the figure–ground effect and color illusion can be dissociated, indicate that the two phenomena are due to at least partially different mechanisms. In the following we will discuss the possible origin of the two aspects of the illusion.

The color illusion. The observation that a thin color seam along the contour of a figure can change the color over the whole area of the figure demonstrates the importance of border signals in color surface perception. Phenomenal spreading of color over large distances in visual space can be observed also under a variety of other conditions (Spillmann and Werner, 1996). We have previously argued (von der Heydt *et al.*, 2003a) that these phenomena can in principle be explained by two alternative theories: one that assumes that color and brightness signals actually spread within a retinotopic neural representation (isomorphic coding theory), and another that postulates that color and brightness are represented at a higher level as attributes of object surfaces, and assumes that the phenomenal color spreading corresponds to a change in this abstract representation (symbolic coding theory). A review of the literature on color filling-in shows that it is not easy to distinguish between these alternatives by psychophysical experiments (Spillmann and Werner, 1996; Pessoa *et al.*, 1998; von der Heydt *et al.*, 2003a).

Neurophysiological experiments have shown that uniform image regions are represented in the visual cortex by activity corresponding to the borders of the regions rather than the interior (Hubel and Wiesel, 1968). This holds not only for light and dark, but also for colored stimuli (Michael, 1978a, b; Friedman *et al.*, 2003). Also, and perhaps more importantly, the vast majority of color coded neurons in the top layers of area V1 and in area V2 are orientation selective (Gegenfurtner *et al.*, 1996; Friedman *et al.*, 2003). This is contrary to earlier claims that color and form tend to be separated in the visual cortex (Livingstone and Hubel, 1984; Hubel and Livingstone, 1987). The recent studies show that the same neurons that represent the form of a stimulus to a large extend code also for its color. This does not fit well with the isomorphic theory in which the highly non-uniform spatial distribution of brightness and color signals is considered only as a transitory representation which is subsequently transformed into a uniform distribution. A comparison of the color coding in V1 and V2 shows that the contour-based color coding scheme is preserved, and nothing suggests such a transformation (Friedman *et al.*, 2003).

It could be argued that the (relatively sparse) neural activity corresponding to the interior of uniform regions might be the result of neural signal spreading, and this 'color surface activity' might thus correlate with the perception of surface color. Von der Heydt *et al.* (2003a) have examined this possibility by recording the neural activity under conditions in which the perceived color of a stimulus patch changes spontaneously (for example, from red to green) due to perceptual filling-in (Troxler's effect). Behavioral experiments indicated that macaques perceive filling-in under these conditions just like humans (Friedman *et al.*, 1999). The recordings showed that the color surface signals in V1 and V2 did not change in parallel with the perceptual color change, but stayed rather constant, just like the color signals at the entrance to the cortex. In contrast, the signals of oriented color neurons representing the color step at the boundary of the test patch did show a decline that correlated with the time course of the perceived filling-in.

While the isomorphic filling-in theory is clearly at variance with the behavior of neuronal signals during Troxler fading, it has received support from neurophysiological studies using brightness induction and brightness illusion paradigms with temporally modulated stimuli (Rossi and Paradiso, 1999; Kinoshita and Komatsu, 2001; Roe *et al.*, 2005). These studies found that the activity of neurons of V1 and V2 could often be modulated by distant surround stimuli even when the receptive field was located within a region of constant luminance (although in many neurons the response modulation was in anti-phase with the brightness illusion). We are not aware of neurophysiological studies of the watercolor illusion.

Under the symbolic coding theory, the diverse findings of phenomenal color spreading are easy to reconcile with the basic neurophysiological data. In the case of Troxler's effect, filling-in is explained by the decay of the color border signals of the test patch. When these signals decay, the representation of the test patch vanishes altogether, and what remains is the representation of the 'underlying' surface which is sustained by its proper border signals (von der Heydt *et al.*, 2003a). The color effect of the watercolor illusion can be explained by the responses of the various color edge selective neurons to the double-line contours: cells with narrow, even-symmetrical receptive fields will signal the colored purple and orange lines, whereas cells with relatively broad, odd-symmetrical receptive fields (Friedman *et al.*, 2003) will signal a small color contrast (in the direction purple-orange) across the contour, and their signals give rise to the illusory color veil.

How exactly the color border signals have to be integrated to explain the watercolor illusion needs to be worked out by a quantitative model. Devinck *et al.* (2005) have measured how the illusory color depends on the color and luminance of the inducing lines. Further details on the influence of the widths of the lines, the size of the compartments, the undulations of the contour etc. would be most helpful. Sometimes, complex perceptual phenomena can be explained by rather simple and physiologically plausible mechanisms (Blakeslee and McCourt, 2004). Montag (1997) proposed a model of color signal integration that might explain the watercolor illusion, but it is not clear how it can be related to the physiology.

The figure–ground effect. What could be the mechanisms underlying the figure–ground segregation in the watercolor illusion? Psychophysical studies have emphasized the importance of border assignment in vision. Images of 3D scenes consist of regions bounded by contours, and the regions often correspond to surfaces at different depths. As a consequence, a region can be bounded by 'intrinsic' and 'extrinsic' contours (Nakayama *et al.*, 1989). A contour is intrinsic to a region if it is the contour of the corresponding physical surface; it is extrinsic if it is the contour of an occluding object. It is clear that the understanding of an image depends on the ability to distinguish these contours. Thus, the assignment of 'border ownership' (a term that goes back to the early Gestalt psychologists) may be the primary process underlying figure–ground segregation. The way the borders are assigned profoundly influences how an image is interpreted, which in turn determines the recognition of objects in the image. It was recently discovered that border ownership is encoded in the firing rate of edge selective neurons in visual cortex (Zhou *et al.*, 2000; Qiu and von der Heydt, 2005). In essence, each contrast edge is represented by two groups of cells with opposite border-ownership preference (Fig. 8). The differential activity between the two reflects the border-ownership assignment at this stage of processing. This simple coding scheme might be the basis of perceptual figure–ground organization (von der Heydt, 2003; von der Heydt *et al.*, 2003b).

Local and global mechanisms. To understand the watercolor illusion it is important to understand that cortical neurons generally combine different mechanisms for the exploitation of different figure–ground cues. The interpretation of images is generally ambiguous and the visual system has to infer border ownership from the given information which can be quite variable. Sometimes, the depth order of surfaces is given by stereoscopic cues, or, in the case of motion, by the pattern of dynamic occlusion. This information can in principle be derived by processing a small neighborhood of a border (i.e. by small receptive fields). It specifies border ownership directly (the nearer surface owns the border). In other situations local depth cues may be absent, and the system then seems to rely on the image context, using rules that specify what shapes are likely to be object shapes, and which of the contours are likely to belong to the same object. In general, the system combines local and global cues for the assignment of border ownership. For example, neurons of V2 use stereoscopic depth and global form criteria (Qiu and von der Heydt, 2005). We argue that the figure–ground effect in the watercolor illusion results from two such mechanisms — one that analyzes the local edge profile, and another that evaluates consistency of color along contours. The first is a local mechanism; the second requires global processing.

The special form of the watercolor contours, consisting of a pair of lines of high and low contrast, creates an asymmetric edge profile similar to edge profiles that are characteristic of 3D objects (Fig. 8A–B). In the case of the 3D objects, the gradual luminance transition is on the object side, and the abrupt transition on the background side. Asymmetric edge profiles can be detected by local processing, and it is likely that neurons with appropriate selectivity exist in the visual cortex. We assume that these neurons provide part of the input to the border ownership cells of V2.

This mechanism could explain some of the figure–ground effect in the watercolor illusion. However, the results of Experiment 2, showing that figureness of the watercolor compartments is reduced when the illusory color is compensated by color in the adjacent compartments, indicate that another mechanism must be involved that is based on color. This was confirmed by Experiment 4, showing that figureness of a region can be enhanced by a tint of color (Fig. 7). Surface color *per se* may not be a strong predictor of objectness (objects of color A seen against a background of color B may be as common as objects of color B seen against a background of color A). However, consistent edge color along a stretch of contour indicates objectness. This rule is based on the assumption that color tends to vary less within objects than between objects, and hence, since backgrounds are usually composed of multiple objects, color variation tends to be greater on the background side of contours than on the foreground side. Accordingly, we

Figure 8. (See color plate X) Hypothetical explanation of the figure–ground effect in the watercolor illusion as a result of the neural processing of border ownership. See Discussion for further explanation.

assume that border ownership cells receive contributions from mechanisms that sense the global consistency of color edges (Fig. 8C). The existence of such mechanisms is suggested by studies of the formation of illusory contours, showing that consistent edge polarity enhances amodal completion (He and Ooi, 1998). The neurophysiology shows that a large proportion of cells that integrate image context for border ownership assignment are color coded and selective for the polarity of edge contrast (Zhou *et al.*, 2000). This indicates that edge color and polarity are important ingredients in the neural process of border assignment. Interestingly, Pinna's (2005) results, principally showing that a strong figure–ground effect can be obtained without consistent edge color, also indicate that figureness of the watercolor compartments was reduced when the light-colored line varied in hue along the contour, compared to the regular watercolor pattern in which the color was the same all around. This is further evidence that consistent edge color contributes to figureness.

Recent theoretical studies demonstrate that the neurophysiological findings and a range of observations on perceptual figure–ground organization can be simulated by simple, neurophysiologically plausible models (Craft *et al.*, 2005; Zhaoping, 2005). In essence, these models implement Gestalt rules, such as good continuation of contour, compactness of form, and proximity. The model of Craft *et al.* (2005) assumes a network of 'grouping cells' which integrate local edge signals according to concentric templates and in turn regulate the amplitude of the local edge signals, thereby enhancing or reducing the local edge responses when a figure is detected, as observed in the neurons of V2. The principle of consistent edge color could be easily implemented in this model by making some of the input to the grouping cells selective for edge color and polarity.

As can be seen in Fig. 8D, the pattern of the watercolor illusion simultaneously activates these two hypothetical figure–ground mechanisms: the first is stimulated by the asymmetrical profile of the contour, and the other is activated by the consistent purple-orange gradient all around the contour. The convergence of the two mechanisms explains the robust figure–ground bias towards the compartments that are lined with a bright color on the inside. It also explains why compensating the illusory color reduces the dominance of the watercolor compartments, but does not eliminate it completely, and why a color tint by itself enhances figureness.

The response properties of cortical neurons also explain the effect of the width ratio of the compartments on figure–ground perception (Experiments 2–4). In the case of a simple figure such as a square, the strength of the border-ownership signals tends to decrease when the size of the figure increases (Qiu and von der Heydt, 2005). Thus, in our experiment, the narrow compartments probably generated stronger border-ownership signals than the wide compartments (Gestalt principle of proximity).

In conclusion, our results suggest that the color illusion and the figure–ground effect of the watercolor display are due to partially separate mechanisms. Based on recent neurophysiological data we propose that the color illusion can be explained by the general principle that the system computes surface color from color border signals, the same principle that explains phenomenal color filling-in under various other conditions. We further propose that the figure–ground effect results from two different mechanisms of border ownership assignment; one is sensitive to asymmetric shape of edge profiles, the other to consistency of edge color inside contours. This theory predicts that both kinds of mechanisms should be detectable in border ownership selective neurons and that their assignment of side of ownership should agree.

Acknowledgements

This research was supported by NIH RO1 grant EY02966.

REFERENCES

Blakeslee, B. and McCourt, M. E. (2004). A unified theory of brightness contrast and assimilation incorporating oriented multiscale spatial filtering and contrast normalization, *Vision Research* **44**, 2483–2503.

Cohen, M. A. and Grossberg, S. (1984). Neural dynamics of brightness perception: features, boundaries, diffusion, and resonance, *Perception and Psychophysics* **36**, 428–456.

Craft, E., Schuetze, H., Niebur, E. and von der Heydt, R. (2006). A neural model of figure-ground organization, *J. Neurophysiol.*

Devinck, F., Delahunt, P. B., Hardy, J. L., Spillmann, L. and Werner, J. S. (2005). The watercolor effect: quantitative evidence for luminance-dependent mechanisms of long-range color assimilation, *Vision Research* **45**, 1413–1424.

Friedman, H. S., Zhou, H. and von der Heydt, R. (1999). Color filling-in under steady fixation: Behavioral demonstration in monkeys and humans, *Perception* **28**, 1383–1395.

Friedman, H. S., Zhou, H. and von der Heydt, R. (2003). The coding of uniform color figures in monkey visual cortex, *J. Physiol.* **548**, 593–613.

Gegenfurtner, K. R., Kiper, D. C. and Fenstemaker, S. B. (1996). Processing of color, form, and motion in macaque area V2, *Visual Neuroscience* **13**, 161–172.

Gerrits, H. J. M. and Vendrik, A. J. H. (1970). Simultaneous contrast, filling-in process and information processing in man's visual system, *Exper. Brain Res.* **11**, 411–430.

He, Z. J. and Ooi, T. L. (1998). Illusory-contour formation affected by luminance contrast polarity, *Perception* **27**, 313–335.

Hubel, D. H. and Livingstone, M. S. (1987). Segregation of form, color, and stereopsis in primate area 18, *J. Neurosci.* **7**, 3378–3415.

Hubel, D. H. and Wiesel, T. N. (1968). Receptive fields and functional architecture of monkey striate cortex, *J. Physiol.* **195**, 215–243.

Kanizsa, G. (1979). *Organization in Vision. Essays on Gestalt Perception.* Praeger, New York.

Kinoshita, M. and Komatsu, H. (2001). Neural representation of the luminance and brightness of a uniform surface in the macaque primary visual cortex, *J. Neurophysiol.* **86**, 2559–2570.

Livingstone, M. S. and Hubel, D. H. (1984). Anatomy and physiology of a color system in the primate visual cortex, *J. Neurosci.* **4**, 309–356.

Michael, C. R. (1978a). Color vision mechanisms in monkey striate cortex: Simple cells with dual opponent-color receptive fields, *J. Neurophysiol.* **41**, 1233–1249.

Michael, C. R. (1978b). Color-sensitive complex cells in monkey striate cortex, *J. Neurophysiol.* **41**, 1250–1266.

Montag, E. D. (1997). Influence of boundary information on the perception of color, *J. Opt. Soc. Am. A, Opt. Image Sci. Vis.* **14**, 997–1006.

Nakayama, K., Shimojo, S. and Silverman, G. H. (1989). Stereoscopic depth: its relation to image segmentation, grouping, and the recognition of occluded objects, *Perception* **18**, 55–68.

Pessoa, L., Thompson, E. and Noe, A. (1998). Finding out about filling-in: a guide to perceptual completion for visual science and the philosophy of perception, *Behav. Brain Sci.* **21**, 723–748.

Pinna, B. (2005). The role of the Gestalt principle of similarity in the watercolour illusion, *Spatial Vision* **21**, 1–8.

Pinna, B., Brelstaff, G. and Spillmann, L. (2001). Surface color from boundaries: a new 'watercolor' illusion, *Vision Research* **41**, 2669–2676.

Pinna, B., Werner, J. S. and Spillmann, L. (2003). The watercolor effect: a new principle of grouping and figure–ground organization, *Vision Research* **43**, 43–52.

Qiu, F. T. and von der Heydt, R. (2005). Figure and ground in the visual cortex: V2 combines stereoscopic cues with Gestalt rules, *Neuron* **47**, 155–156.

Roe, A. W., Lu, H. D. and Hung, C. P. (2005). Cortical processing of a brightness illusion, *Proc. Natl. Acad. Sci. USA* **102**, 3869–3874.

Rossi, A. F. and Paradiso, M. A. (1999). Neural correlates of perceived brightness in the retina, lateral geniculate nucleus, and striate cortex, *J. Neurosci.* **19**, 6145–6156.

Spillmann, L. and Werner, J. S. (1996). Long-range interactions in visual perception, *Trends in Neuroscience* **19**, 428–434.

von der Heydt, R. (2003). Image parsing mechanisms of the visual cortex, in: *The Visual Neurosciences*, Werner, J. S. and Chalupa, L. M. (Eds), pp. 1139–1150. MIT Press, Cambridge, Mass.

von der Heydt, R., Friedman, H. S. and Zhou, H. (2003a). Searching for the neural mechanisms of color filling-in, in: *Filling-in: From Perceptual Completion to Cortical Reorganization*, Pessoa, L. and De Weerd, P. (Eds), pp. 106–127. Oxford University Press, Oxford.

von der Heydt, R., Zhou, H. and Friedman, H. S. (2003b). Neural coding of border ownership: Implications for the theory of figure-ground perception, in: *Perceptual Organization in Vision: Behavioral and Neural Perspectives*, Behrmann, M., Kimchi, R. and Olson, C. (Eds), pp. 281–304. Lawrence Erlbaum Associates, Mahwah.

Wyszecki, G. and Stiles, W. S. (1982). *Color Science: Concepts and Methods, Quantitative Data and Formulae*. John Wiley, New York.

Zhaoping, L. (2005). Border ownership from intracortical interactions in visual area V2, *Neuron* **47** (in press).

Zhou, H., Friedman, H. S. and von der Heydt, R. (2000). Coding of border ownership in monkey visual cortex, *J. Neurosci.* **20**, 6594–6611.

Color, Line, and Space, pp. 239–271 (2006)

Lighting, backlighting and watercolor illusions and the laws of figurality

BAINGIO PINNA [1,*] and ADAM REEVES [2]

[1] *Department of Sciences of Languages, University of Sassari, via Roma 151, I-07100, Sassari, Italy*
[2] *Department of Psychology, Northeastern University, Boston MA 02115, USA*

Abstract—We report some novel 'lighting' and 'backlighting' effects in plane figures similar to those which induce the 'watercolor illusion', that is, figures made with outlines composed of juxtaposed parallel lines varying in brightness and chromatic color. These new effects show 'illumination' as an emergent percept, and show how arrangements of 'dark and light' along the boundaries of various plane figures model the volume and strengthen the illusion of depth. To account for these various effects we propose several phenomenological 'laws of figurality' to add to the Gestalt laws of organization and figure–ground segregation. We offer a set of meta-laws which are speculative but which serve to integrate and organize the phenomenological laws. These laws indicate how luminance gradient profiles across boundary contours define both the 3D appearance of figures and the properties of the light reflected from their volumetric shapes.

Keywords: Watercolor illusion; Gestalt principle of grouping; figure–ground segregation; color spreading; light.

1. INTRODUCTION

The aim of this work is twofold. First, we report two new phenomena we call 'lighting' and 'backlighting' effects in plane figures similar to those which induce the watercolor illusion (Pinna, 1987). Such figures are outlines composed of the juxtaposition of parallel lines varying in brightness and chromatic color. The novel effects were obtained by changing the number of lines (from 2 to 6) and their luminance profiles. These effects illustrate 'illumination' as an emergent percept, which depends on the way the illusory illumination appears to fall upon an object and on how arrangements of 'dark and light' along the object's boundaries model the perceived volume and depth. (Earlier work has shown that perceived lightness of achromatic surfaces depends on perceived depth, e.g. Schirillo *et al.*,1990). Second,

*To whom correspondence should be addressed. E-mail: baingio@uniss.it

we attempt to extract from these boundary profiles some 'laws of figurality'. These are similar in spirit to the well-known laws of organization of Wertheimer (1923) and of figure–ground segregation of Rubin (1915, 1921), but they add to these figural processes rules to determine the phenomenal appearance of the figure — its color and spatial volume, seen under an apparent illumination — and how this depends on luminance gradient profiles across boundary contours. We next remind the reader of the watercolor illusion, which serves as a useful starting point.

2. THE WATERCOLOR ILLUSION

The watercolor illusion is a long-range assimilative spread of color emanating from a thin colored line running contiguous to a darker chromatic contour and imparting a figure–ground effect across a large area (Pinna, 1987; Pinna *et al.*, 2001; Pinna *et al.*, 2003; Spillmann *et al.*, 2004; Pinna, 2005).

In Fig. 1, purple undulated contours flanked by orange edges are perceived as two undefined shapes evenly colored by an opaque light veil of orange tint, with a clear surface color property (*Erscheinungweise*, Katz, 1911, 1930) spreading from the orange edges. The two undefined shapes manifest a strong figure–ground organization and a solid figural appearance comparable to a bas-relief illuminated from the top. Thus the two shapes appear similar to rounded surfaces segregated in depth that extend out from the flat surface. The complementary regions appear as holes or empty spaces.

In Fig. 2 the Gulf of Mexico (a) and the Mediterranean Sea (b) immediately and clearly emerge as figures and appear illusorily colored with the same orange tint as in Fig. 1. These effects are obtained by reversing purple and orange lines, i.e. the purple/orange wiggly lines in Fig. 1 are replaced by orange/purple lines in Fig. 2. The Gulf and Sea which are easily visible in Fig. 2 are initially imperceptible in Fig. 1, and remain difficult to discern even with continued presentation.

By replacing the orange lines with purple ones in Fig. 3, the Gulf of Mexico (a) and the Mediterranean Sea (b) still emerge, but less strongly. Under these conditions the figure–ground organization can be easily reversed. It is worth noting

Figure 1. (See color plate XI) *The watercolor illusion*: purple undulated contours flanked by orange edges appear as two undefined shapes similar to rounded surfaces segregated in depth and extending out from the flat surface and evenly colored by an opaque light veil of orange tint.

Figure 2. (See color plate XI) By reversing purple and orange lines of Fig. 1, the Gulf of Mexico (a) and the Mediterranean Sea (b) immediately and clearly emerge as figures and appear colored with the same orange tint as in Fig. 1.

Figure 3. (See color plate XI) By replacing the orange lines of Fig. 2 with purple ones, the Gulf of Mexico (a) and the Mediterranean Sea (b) still emerge but with a much weaker figural and volumetric appearance. The figure–ground organization can be reversed much more easily than in Fig. 2.

that the two geographical regions now lose the coloration effect, the strong figural properties, and the bas-relief appearance which emerged in Fig. 2. Passing from Fig. 2 to Fig. 3, not only is the figure–ground segregation weakened, but also the set of peculiar phenomenal properties of the watercolor illusion that we define as 'figurality' effects, which refer to the 3D appearance, to the strong coloration with surface and object color attributes, and to the apparent illumination which depicts the three-dimensional shape of the object. Comparing Figs 2 and 3, two different kinds of figure–ground segregations are perceived with different figural effects that, in order to clarify the distinction between them, can be respectively called 'figure–ground' and 'object–hole' segregations. The object–hole, i.e. the emergence of empty space comparable to a hole, is much stronger in Fig. 2 (and in Fig. 1) than in Fig. 3 (see Grossberg *et al.*, 2005).

These figures illustrate the four main phenomenal effects that the watercolor illusion manifests: coloration, figure–ground, object–hole and illumination effects. We now summarize what has been discovered about the first two effects in detail, as a preparation for the method of the current study. Discussion of the object–hole and illumination effects is deferred.

3. THE COLORATION EFFECT

The main properties of color spreading in the water color illusion (Pinna *et al.*, 2001) are the following: (i) the coloration is approximately uniform; (ii) it extends up to about 45 deg; (iii) it is complete by 100 ms; (iv) it can be generated by all the basic colors; (v) it occurs on both colored and black backgrounds; (vi) the optimal line thickness is approx. 6 arc min; (vii) it is stronger with wiggly lines, but it also occurs with straight lines and with chains of dots (see Pinna *et al.*, 2001); (viii) high luminance contrast between inducing lines increases the coloration effect, but color spreading is clearly visible at near equiluminance (Pinna *et al.*, 2001; Devinck *et al.*, 2005); (ix) the line with a less luminance contrast relative to the background spreads proportionally more than the line with a higher luminance contrast; (x) the color spreads in directions other than the line orientation; (xi) by reversing the colors of the two adjacent lines, the coloration reverses accordingly; (xii) phenomenally, the coloration appears solid, impenetrable and epiphanous as a surface color (Katz, 1930); (xiii) similarly to neon color spreading (Varin, 1971; van Tuijl, 1975; Bressan *et al.*, 1997), the watercolor illusion induces a complementary color when one of the two juxtaposed lines is achromatic and the other chromatic (Pinna and Grossberg, 2005). The coloration effect is also scale-invariant, in that it appears to be approximately independent of viewing distance up to the point that the line colors are no longer resolvable.

4. THE FIGURE–GROUND EFFECT

The main figure–ground qualities of the watercolor illusion are the following: (i) it is hard to reverse — the figure–ground segregation is unequivocal; (ii) it determines grouping and figure–ground segregation more strongly (Pinna *et al.*, 2001) than the Gestalt principles of proximity, good continuation, prägnanz, closure, symmetry, convexity, past experience, similarity, surroundedness and parallelism; (iii) by reversing the colors of the two adjacent lines, figure–ground segregation reverses accordingly (see Figs 1 and 2); (iv) the figure–ground effect occurs on colored and black backgrounds (Pinna *et al.*, 2001); (v) by reversing the background from white to black, or black to white, while the luminance contrast of the lines is kept constant, the figure–ground segregation reverses (see Fig. 7, below) — this is in contrast to Rubin's claim that the currently figural region in the face-vase illusion is maintained even in the face of a black/white reversal; and (vi), the watercolor illusion strongly enhances the 'unilateral belongingness of the boundaries' (Pinna, 1987), the phenomenal property that the boundaries belong to the figure but not to the background (Rubin, 1915).

The phenomenal properties of coloration and figure–ground effects raise some questions. Can the two effects be considered relatively independent? Under which conditions does a possible dissociation occur? How does the dissociation of one effect, say the coloration, influence the figure–ground effect and vice versa? When dissociated, which are the phenomenal properties of each effect taken individually? Do new properties emerge when the integration of the two effects occur? The answer to these questions may help to understand if and how the two effects belong to different mechanisms, if and how they can integrate and at which processing stage. To answer these questions, spontaneous descriptions of subjects were recorded under the controlled conditions described next.

5. GENERAL METHOD

5.1. Stimuli

Stimuli were composed by the figures illustrated within the paper. The mean overall size of the stimuli was about 21 by 15 deg of visual angle. The luminance of the white (background) paper was 80.1 cd/m^2. The CIE x, y chromaticity coordinates for the main chromatic colors used were: (purple) 0.30, 0.23; (orange) 0.57, 0.42; (red) 0.62, 0.34; blue (0.201, 0.277), green (0.3, 0.5), yellow (0.46, 0.42). Black components had a luminance contrast of 0.03. The stroke width was approx. 6 arc min. Stimuli were presented on a computer screen illuminated by Osram Daylight fluorescent light (250 lux, 5600 K), and were observed binocularly from a distance of 50 cm with free viewing.

5.2. Observers

Unless otherwise stated, different groups of fourteen naive observers each described only one stimulus. (This was done in case one stimulus might influence the perception of another.) All the observers had normal color vision on the Ishihara test. Both male and female undergraduates at the University of Sassari participated.

5.3. Task

The subject's task was to report freely what they perceived by giving an exhaustive description of the main visual properties. Preceding the experiment there was a training period, where subjects viewed some known coloration effect like the figures reported by Pinna *et al.* (2001), classical examples by Gestalt psychologist of figure–ground segregation, and some examples of different color spreading effects, like neon color spreading, chromatic assimilation, simultaneous color contrast and complementary color induction. Observation time was unlimited. All the reports were quite spontaneous. The descriptions reported in quotation marks below are translations from Italian edited for brevity and typicality. There were some individual differences, especially for the weaker effects, but we report only those descriptions generated by at least 11 of the 14 subjects in each group. Our edited descriptions were judged by three other students, who were naive as to the purposes of the present research, to provide a fair representation of those provided by the observers. All the edited descriptions were approved afterwards by all subjects and evaluated as equivalent to those provided during the experimental session.

Studies of color constancy have shown the task given to an observer has a large effect on the report of color appearance; when asked to act as a colorimeter by making 'direct' matches between two surfaces seen under different illuminants, observers do so nearly perfectly; and when asked to make 'paper' matches between the surfaces, in effect discounting the illuminant and reporting the surface color alone, some observers (though not all) are capable of this as well (Arend and Reeves, 1986), even when luminance variations are introduced between the surfaces (Arend *et al.*,1991). In the current work we did not impose any restrictions on the observer's judgements. However, the descriptions that they produced are compatible with a 'paper match' or surface–color mode of judgement rather than with the alternative colorimetric mode.

RESULTS

6. DISSOCIATION BETWEEN COLORATION AND FIGURE–GROUND EFFECTS

The purpose of this section of the Results is to demonstrate, by introducing new cases of the watercolor illusion, that coloration and figure–ground are relatively independent effects. A strict dissociation implies that (i) coloration is perceived without the figure–ground effect and, conversely, (ii) the figure–ground effect is perceived without any color spreading in the inside edges. If these conditions hold, the coloration effect is not a necessary condition for the figure–ground one, and vice versa. Besides strict dissociation, other, weaker, but not less interesting, cases can be considered, with the aim of understanding if the two effects interact and how their properties might change by modifying their binding.

6.1. Coloration without figure–ground effect

Coloration without a figure–ground effect can be induced through a dichoptic composite stimulus, where the orange component is presented to one eye while the purple one is presented to the other eye (see Figs 4a and 4b). By combining dichoptically the two differently-colored borders on the left with those on the right, the 'coloration' (11 out of 14) is clearly perceived within the inner edges. However, it manifests different properties from those described in Section 3: it loses the surface color properties and appears diaphanous like a 'colored light floating (Fig. 4a) or more stable (Fig. 4b) and diffusing homogeneously within the wiggly figure' (11 out of 14). Moreover, the figure–ground effect, peculiar to the watercolor illusion under the previous conditions, is totally absent. These results suggest that coloration and figure–ground effects can be dissociated, and that by dissociating the coloration from the figure–ground effect, the phenomenal properties of coloration can change. They also suggest, perhaps not surprisingly, a cortical origin of the coloration effect.

The binding between the coloration and figure–ground effects can be weakened and modified by reversing the figure–ground organization where the coloration occurs. This can be done by making the regions where the color spreads appear as background rather than figure. Under these conditions, the coloration should not manifest the surface color quality and the solid, impenetrable and opaque appearance previously described. Figure–ground reversal can be achieved by strengthening the Gestalt grouping and figure–ground principles pitted against the watercolor illusion, so that the former wins against the latter. Hence, what is perceived as a figure should be induced by Gestalt principles, while the complementary watercolored region should be perceived as a background. Given these conditions, one may ask: Does the coloration persist? If it persists, what is its phenomenal appearance? If it changes in appearance, can coloration really be considered independent from the figure–ground effect?

Figure 4. (See color plate XI) Coloration without figure–ground effect can be induced through a dichoptic composite stimulus, where the orange component is presented to one eye while the purple one to the other eye: colored light floating (Fig. 4a) or more stable (Fig. 4b) and diffusing homogeneously within the wiggly figure is perceived, while the figure–ground effect, peculiar to the watercolor illusion, is totally absent.

In Fig. 5a, the inset square contains circles with outer and inner boundaries respectively of red and purple, while the surrounding squared frame contains circles with outer and inner boundaries of blue and purple. Under these conditions, several Gestalt factors (closure, symmetry, contrast, surroundedness, prägnanz) cooperate and win against the watercolor illusion, inducing a clear figure–ground effect in the circles. Phenomenally, 'the circles appear' not as holes or annuli, but 'as bright circular figures with purple boundaries emerging from a background, where a dense smoggy light fills the interspaces and creates an inset reddish and a surrounding blue backlighting effect' (11 out of 14). It is worthwhile noting that only 'the purple annuli' (i.e. the contours with the highest luminance contrast) 'become the boundaries of each circle', thus 'defining their shape' (11 out of 14). Conversely, 'the red and/or blue lighter-colored boundaries belong to the background and appear like haloes due to a colored light coming from behind and surrounding the purple boundaries of the circle' (11 out of 14). In Fig. 5b, the colors of the outer/inner boundaries of Fig. 5a are reversed so that all the figure–ground principles operate synergistically with the watercolor illusion. As a result 'the circles both with inner red and blue boundaries appear as solid volumetric colored circles clearly segregated in depth like red and blue spheres' (11 out of 14).

By filling the empty circles of Fig. 5a with red and blue colors, their figural properties are enhanced, becoming surfaces (see Fig. 6a). However, while in Fig. 5a there are two boundaries for each circle that split as previously described, in Fig. 6a only one boundary surrounds each circular surface. Phenomenally, 'the colored fringes around the filled circles appear as the boundaries of the circular shapes and not as components of the background' (11 out of 14). Nevertheless, 'the coloration persists but it assumes neither the surface color properties nor the appearance of colored light coming from behind. It is perceived as a diaphanous colored background' (11 out of 14). A similar result can be perceived in Fig. 6b.

A further case of coloration without a figural effect is illustrated in Figs 7 and 8. By viewing Fig. 7 from the top to the bottom, the perceived objects switch respectively from rows of eight-pointed stars (top) to rows of crosses (bottom).

This result is obtained by filling the background with a linear luminance gradient going from black to white. What is interesting for our purpose is that along the luminance gradient there is a gray range in the middle of the gradient. In the

Figure 5. (See color plate XII) In Fig. 5a, the circles appear as bright circular figures with purple boundaries emerging from a background, where a dense smoggy light fills the interspaces and creates an inset reddish and a surrounding blue backlighting effect. In Fig. 5b, the circles both with inner red and green boundaries appear as solid volumetric colored circles clearly segregated in depth like red and blue spheres.

Figure 6. (See color plate XII) In Fig. 6a, the colored fringes around the filled circles appear as the boundaries of the circular shapes and not as components of the background. The coloration persists but it assumes neither surface color properties nor the appearance of colored light coming from behind. It is perceived as a diaphanous colored background. A similar result can be perceived in Fig. 6b.

gray range 'neither the stars nor the crosses prevail as a figure against the other, thus creating a reversible condition (where) the coloration effect occurs' (12 out of 14). Coloration occurs, but not the figural effect, at least not in terms of the strong and univocal figure–ground segregation peculiar to the watercolor illusion: 'the figural appearances of the perceived objects (stars and/or crosses) are flat and not bulging as in the top or bottom regions' (11 out of 14) of Fig. 7. In other words, in about the middle region of Fig. 7 'if the crosses are perceived as figures, the stars appear as a background but, compared to what is perceived in the top and bottom regions of the figure, the crosses do not show the clear volumetric effect and the stars are not perceived as deep holes' (11 out of 14). Besides the figural effect, the coloration manifests different surface color properties: in the top and bottom regions the spreading colors appear as 'surface colors, impenetrable and solid' (11 out of 14), while in the critical central region the spreading colors (yellow for stars, and blue for crosses) appear 'empty and transparent' (11 out of 14).

Another kind of dissociation between coloration and figure–ground effects can be perceived in Fig. 8a, where quasi-equiluminant adjacent lines (gray and red) spread their colors but show 'a flat and reversible figure–ground organization or a tessellation of crosses and stars' (13 out of 14), which can be considered as a limiting case of figure–ground segregation. These results are better appreciated by comparing these regions with the complementary ones, where adjacent lines with strong luminance contrast (black and red) induce volumetric objects and holes, not comparable in terms of figurality with those induced by quasi-equiluminant adjacent lines. By comparing Figs 8a and 8b, the coloration appears 'more strongly as a surface color in Fig. 8b than in Fig. 8a' (14 out of 14).

Altogether, these results suggest that coloration can be dissociated from, or altered in the way it is connected and integrated with, the figure–ground effect. However, by splitting or changing their binding, the phenomenal properties of the coloration effect change accordingly, testifying that the dissociation from the figure–ground effect does not leave the coloration unchanged, but is strongly affected by it. Therefore, the two effects, even if they belong to different processes, may interact at a subsequent processing stage.

The aim of the next section is to ask, can the figure–ground effect be isolated or dissociated from the coloration one, and if so, do its properties change?

Figure 7. (See color plate XII) Along the luminance gradient there is a gray range, about halfway between the two colored lines and in the middle of the gradient, where neither the stars nor the crosses prevail as figure, thus creating a reversible condition where the coloration effect occurs but not the figural one. The figural appearances of the perceived objects (stars and/or crosses) are flat, not bulging as in the top or bottom regions.

Figure 8. (See color plate XIII) In Fig. 8a, quasi-equiluminant adjacent lines (gray and red) spread their colors but show a flat and reversible figure–ground organization or a tessellation of crosses and stars. In Fig. 8b, adjacent lines with strong luminance contrast (black and red) induce volumetric objects and holes. In Fig. 8b, the coloration appears more strongly as a surface color than in Fig. 8a.

6.2. Figure–ground without coloration effect

In Fig. 9, an example of figure–ground effect without coloration is illustrated. Under these conditions, 'the crosses emerge as figures' (11 out of 14) while the stars lose their boundaries and amodally complete them as a background. It is worthwhile noting that the physically homogenous gray appears not exactly as it really is, but 'the gray within the crosses is part (the inner region) of them' (11 out of 14), i.e. 'it is captured' (surface capture) and 'segregated in front of the complementary gray that is perceived as a dark empty gray space and as a background' (11 out of 14). The captured 'grays do not manifest any coloration' effect coming from the lines, so 'they do not look chromatically different from the complementary grays, but appear bright and solid, assuming the surface and opaque color properties' (12 out of 14).

The absence of the coloration effect is likely due to the admixture of the colors (see also Fig. 11) belonging to the four pairs of wiggly lines that phenomenally become 'the inner boundaries of the crosses' (13 out of 14). Under these conditions, the figure–ground effect without coloration does not present 'the bulging volumetric effect proper to Fig. 8b, where the crosses emerge from the background creating a rounded convexity', but 'the crosses bulge as upland, creating a plateau, i.e. rounded along the boundaries but on the gray elevated surfaces, whose volumetric effect is flat, without further rounded convexities' (11 out of 14). This plateau property is due to the kind of luminance gradient operating on the boundaries, where the luminance value of the gray surfaces is about intermediate between the two colored lines.

Returning to the coloration effect and to the luminance profile of Fig. 8b, the rounded volumetric effect is restored even if the gray rectangle is replaced by an oblique striped pattern (see Fig. 10). Under these conditions, 'the oblique lines within the crosses, strongly perceived as figures, appear bulging, curved in depth and segregated from those within the stars that are perceived as placed in the background or perceived through holes' (11 out of 14). This is so even if one set of oblique lines is the continuation of the other set, as in a texture. Another possible perceptual result is 'a matrix of transparent watercolored crosses above a background made of oblique parallel lines' (13 out of 14). What is interesting

Figure 9. (See color plate XII) Figure–ground effect without coloration: The crosses emerge as figures. The gray within the crosses does not manifest any coloration, and is captured (surface capture) and segregated in front of the complementary gray that is perceived as a dark empty gray space and as a background.

Figure 10. (See color plate XIII) The oblique lines within the crosses, strongly perceived as figures, appear bulging, curved in depth and segregated from those within the stars that are perceived as placed in the background or perceived through holes.

Figure 11. (See color plate XIII) The crosses clearly emerge as rounded convex figures while the coloration is absent, i.e. the inner edge of each cross appears as a dirty white, denser and brighter than the one within each star but not completely opaque. It does not appear as a white surface but fluffy and soft. The white within the stars is on the contrary completely empty and transparent.

is that even if the crosses are perceived as transparent, 'the oblique lines appear bulging and belonging to the crosses' (11 out of 14).

Considering Fig. 11, in which the gray of Fig. 9 is replaced with white to restore the luminance contrast gradient of Fig. 8 (see Pinna, 2005), 'the crosses clearly emerge as rounded convex figures' (12 out of 14), while the coloration is absent, i.e. the inner edge of each cross appears as 'a dirty white, denser and brighter than the one within each star but not completely opaque. It does not appear as a white surface but as fluffy and soft' (11 out of 14). In contrast, 'the white within the stars is on the contrary completely empty and transparent' (11 out of 14). These results also suggest that the absence of the coloration affects figure–ground segregation.

Given the previous findings, the luminance gradient along the boundaries can indeed be considered the crucial factor for the figure–ground effect. In the watercolor illusion, different conditions of the gradient strongly affect the set of properties that define a figure to be segregated in depth from a background. Figure–ground segregation happens independently of the coloration effect that is induced with quasi-equiluminant lines (see Fig. 8a), i.e. without figure–ground segregation. Note that the coloration effect is not induced when differently-colored adjacent lines defining the boundaries of a figure are used (see Figs 9 and 11), i.e. with a clear figure–ground segregation.

6.3. The lighting illusion

By increasing the length of the gradient by adding new adjacent lines, another case of the figure–ground effect without coloration is obtained. In Fig. 12a, six adjacent blue lines, placed in succession on the ascending steps of a luminance gradient, define the boundaries of an undulated irregular shape that is perceived as 'a blue object similar to an insect, to an amoeba, or to a fish bone with very strong bulging and three-dimensional appearance illuminated from the top by a bright light' (12 out of 14). The coloration effect is very weak, almost invisible, and 'the inner edges appear as light and bright fuzzy white, whiter than the white of the background and like a shine or a lighting: a light reflected by a three-dimensional blue shape' (11 out of 14). The new conditions not only define univocally what is figure and what is background, but also model the volume and strengthen the illusion of depth peculiar to the watercolor illusion by creating a 3D object–hole effect. The object–hole effect is sculptured by the inhomogeneous reflected light that depicts light and shade (the illumination effect) and leaves the color of the object unaltered and constant, i.e. a homogeneous blue. We asked the subjects 'after discounting the illumination effect, what is the color of the fish bone?' Six (of 14) picked the fourth lightest blue, and

Figure 12. (See color plate XIV) *The lighting illusion.* In Fig. 12a, a blue object similar to a fish bone with very strong bulging and three dimensional appearance illuminated by a bright light. The coloration effect is very weak, almost invisible, and the inner edges appear as light and bright fuzzy white, whiter than the white of the background and like a shine or lighting: a light reflected by a three dimensional blue shape. In Fig. 12b holes within a 3D green object are clearly perceived.

eight, the fifth lightest blue. (The 6 lines were clearly distinguishable in the original drawing, although they may blur together in the reproduction here). This result emerged after a comparison between Fig. 12a and six figures where the shape of Fig. 12, without adjacent lines, was homogeneously filled with the color of each of the lines. The task was to choose the color of the fish bone shape among the six filled figures after having discounted the illumination creating the lighting.

In short, by increasing the number of adjacent lines arranged on a luminance staircase, (i) the coloration effect, considered as an assimilative color spreading similar to the watercolor illusion, is absent; (ii) new kinds of long-range brightness and lightness inductions emerge in place of the coloration (see also Fig. 21); (iii) these boundary brightness and lightness effects are perceived as an illumination effect or as a glare, a shine and, more generally, as a 'lighting illusion' (see also Fig. 21); and (iv) the figure–ground segregation and the volume proper to the watercolor illusion are increased and the 3D appearances bulge much more strongly, creating a clear object–hole effect. (Note that blue is not required; green will also do, as in Fig. 12b.)

Altogether, the phenomenal results of this section indicate that the figure–ground effect can be dissociated from the coloration effect and that, by separating the two effects, new properties (illumination and lighting illusion, clear volumetric effect, brightness and lightness inductions, etc.) emerge, testifying that their independence is not absolute as both of them strongly influence each other. We infer that the two underlying processes interact and converge at a subsequent processing stage where the coded color and figure–ground properties are integrated and processed in a new way, which will be the main topic of Section 10.

The previous points suggest that it may be more important from a theoretical point of view to demonstrate how all these effects interact to create the integrated properties of our visual world, and new emergent properties, than to demonstrate their independence. The aim of the next section is to show new and emergent properties referred to the object–hole and to the illumination effects and, in a more integrated way, to figurality. The questions are: by changing the hue and the luminance across the boundaries, how does the figurality change? And, how does the phenomenal appearance of the object–hole and of the illumination effects change as a result?

7. THE OBJECT–HOLE AND THE ILLUMINATION EFFECTS

The phenomenal figurality of the surfaces, induced by the watercolor illusion and, more compellingly, by the new cases shown in Figs 12a and 12b, manifest unique emergent properties, never shown before in previous studies of figure–ground segregation. (i) The physical conditions eliciting these effects, i.e. the juxtaposition of parallel lines, not only induce a coloration effect and clearly define what is 'figure' and what 'background', but also model the volume and strengthen the illusion of depth in a flat surface by inducing a 3D appearance. (ii) Increasing the number of adjacent lines increases volumetric figurality. Going from the two-line watercolor illusion (Fig. 1), to the six-line lighting illusion (Fig. 12), the object–hole effect becomes much stronger: the resulting surface appears thick, solid, opaque and dense, and rounded similar to a high-relief. The object–hole effect is one of the new properties emerging under these conditions. (iii) With enough juxtaposed lines (e.g. six), the resulting object clearly shows new kinds of brightness and lightness inductions, in directions other than the line orientations. The continuation of the luminance gradient does not appear to stop when it reaches white, but continues increasing in brightness and lightness. (iv) The brightness and lightness inductions are perceived as an illumination effect through the figural organization of shaded and bright regions along the long gradient and within the perceived object. The illumination effect is a new and emergent property, which is not clearly perceived when only two lines are placed parallel and contiguous to each other. The fact that six juxtaposed lines weakens the coloration effect — as seen in the watercolor illusion with only two lines — may result from the illusory illumination enhancing the inner brightness and lightness of the object. It is worth noting that (v) the resulting figurality, i.e. the combination of object–hole and illumination effects, looks like that obtained through the Renaissance painting technique of *chiaroscuro*: the modeling of volume by depicting light and shade (see Pinna, 2005). A highlight marks the point where the light is more directly (orthogonally) reflected, moving away from this highlight, light hits the object less directly and therefore has a darker value of color (lower luminance).

7.1. The backlighting illusion

A further demonstration of the illumination effect and of the previous emergent properties is shown in Fig. 13, where the abrupt luminance jump of the external boundary side of Fig. 12 is reduced by replacing the black outer line with an orange one and the darkest blue line with a black one. Under these conditions 'the black line appears as the boundary of the fish bone that maintains a strong volumetric and 3D appearance. The inner blue luminance gradient is perceived like the shade of the bone and the orange line like its backlighting. The color of the light coming from behind is orange. The object appears illuminated also from the front with a white light' (11 out of 14). Backlighting mainly occurs when the source of light is coming from behind the subject, namely, when the observer is looking into the light source: the perceived object, intercepting the light source, should appear at least partially in shadow. This is not the case in Fig. 13, where two light sources are perceived. It is worth observing that the 'orange backlight manifests the coloration effect by coloring the background that manifests a certain density, like a fog' (11 out of 14). This is another example of coloration dissociated from the figure–ground effect, and, as previously shown, both effects assume new perceptual roles and new emerging 'meanings'. The color of the backlight varies by changing the hue of the external line. If the hue is blue like the internal lines, or gray, the backlighting effect is absent.

The color of the front light can also be changed by changing the chromatic color of the innermost line. In Figs 14a and 14b 'the reflected light appears green and orange' (11 out of 14). The color of back and front lighting can be combined at will, for example one orange and the other green (not illustrated). Pilot data suggest that the hue of the innermost (outermost) lines most strongly influences the color of the front (back) lighting, whereas the luminance contrast determines the perceived intensity of the light, but not the color.

In Fig. 15, the luminance gradients in Fig. 13 increase and decrease more gradually and are more similar on both sides of the black line, but still differ slightly. Under these conditions, the line with the highest luminance contrast is again perceived as the 'boundaries of the object' (14 out of 14). However, 'the volumetric object appears fuzzy and blurred' (13 out of 14) (fuzziness effect) compared to the one perceived in Fig. 13. When the black line is placed in the middle of the other lines, so that it is the line of symmetry of the luminance gradient, the fuzziness effect reaches its maximum value (not illustrated).

Another way to create a fuzziness effect is illustrated in Fig. 16a, where the background of Fig. 12a is black. Under these conditions going from one side to the other side of the luminance gradient and vice versa, the luminance contrast gradually decreases. Phenomenally, 'a white bright fuzzy object bounded by blue neon light on a black background' (12 out of 14) is perceived. 'The boundary of the object is none of the blue lines but the entire bunch of lines appears like a beam of blue neon light surrounding and bounding a white fish bone or amoeboid object' (11 out of 14). A combination of Fig. 12a and Fig. 16a is illustrated in Fig. 16b, which

Figure 13. (See color plate XIV) *The backlighting illusion.* The black line appears as the boundary of a fish bone that maintains a strong volumetric and 3D appearance. The inner blue luminance gradient is perceived like the shade of the bone and the orange line like its backlighting. The color of the light coming from behind is orange. The object appears illuminated also from the front with a white light. The orange backlight manifests the coloration effect by coloring the background that manifests a certain density, like a fog.

Figure 14. (See color plate XIV) The color of the front light can be changed by changing the chromatic color of the innermost line. In Figs 14a and 14b the reflected light appears green and orange.

Figure 15. (See color plate XV) The line with the highest luminance contrast is perceived as the boundaries of the fish bone that appears fuzzy and blurred (fuzziness effect) compared to the one perceived in Fig. 13.

Figure 16. (See color plate XV) In Fig. 16a, a white bright fuzzy object bounded by blue neon light on a black background is perceived. The boundary of the object is not one of the blue lines but the entire bunch of lines, which appears like a beam of blue neon light surrounding and bounding a white fish bone or amoeboid object. Fig. 16b allows a direct comparison of all the figurality properties perceived on the left and right sides.

allows a direct comparison of all the figurality properties perceived on the left and right sides.

By reversing the inner and outer white and black and the blue luminance gradient of Fig. 16, 'a black fuzzy fish bone against a blue light emerges' (11 out of 14) (see Fig. 17). The backlighting effect perceived under these conditions is not the complement of the results obtained in Fig. 16a, where no backlight is reported. It is likely to be a direct consequence of the backlighting phenomena where the object illuminated strongly from behind appears in deep shadow. Usually, there must be enough light and details to recognize what the object is, and prevent it from looking like a flat silhouette against a strong light, but not so strong that the dramatic effect of backlighting is lost. Backlighting is especially effective when combined with reflected light, causing interesting shadows as illustrated in Fig. 13. (This is not the case in Fig. 17, where the object maintains a clear 3D appearance, perhaps due to the luminance gradient along the boundaries that casts lights of different intensities.)

Note that a phenomenal demonstration of the belongingness of the lines to the fish bone (whether or not they belong to the object) comes from the comparison of the size or width of the resulting object: for example, the width of the bone is greater in Figs 14a and 14b than in Figs 15 and 17. By increasing the number of lines belonging to the object, the width of the resulting object increases accordingly. A similar demonstration can be applied to the size of the circles perceived in Figs 5a and 5b.

In Figs 16a and 17 the juxtaposed lines appear as neon lights, not as the boundaries of the object: even if they surround the object, they are not 'the object'. This is due to the fact that the luminance gradient does not present any abrupt luminance variation on either side of the black or white regions, so there is no abrupt luminance termination to assume the role of boundary of the perceived object. The inverse case of Fig. 16a (and 17) is illustrated in Fig. 18, where the extremes of the gradient end with very high luminance contrast variations.

Phenomenally, in Fig. 18, 'both the inner and outer edges appear as empty spaces: The outer space as a background and the inner one as a black hole. Only the bunch of lines is perceived as a figure' (12 out of 14). It is worthwhile noting that two different things are perceived, dark holes and white background, even if the conditions of the gradient terminations are about the same. This result is likely due to the closure and surroundedness Gestalt principles, which make the inside edge of the gradient appear as a figure. However, since the abrupt luminance termination strongly determines the perception of background in this region, the combination of the two groups of principles makes the inner edges appear as holes — 'figures', that

Figure 17. (See color plate XV) A black fuzzy fish bone appears against a blue light.

Figure 18. (See color plate XV) Phenomenally, both the inner and outer edges appear as empty spaces: The outer space as a background and the inner one as a black hole. Only the bunch of lines is perceived as a figure.

is, whose boundaries belong not to the figure, as expected from Gestalt principles, but rather to the complementary region. (This point will be discussed further in Section 9.3.) This interpretation applies also to Fig. 12b, where the blobs within the objects are not perceived as background, even with very deep complementary boundary differentiation, but as holes. Another factor that may play a role in Fig. 18 is that black is more easily perceived as empty space than is white; when the black and white regions of Fig. 18 are reversed (not illustrated), the perception of the inner edges as a hole is weakened.

The results presented in this section may be summarized as follows. (i) Besides the figure–ground segregation effect suggested by Rubin (1921), the so-called 'object–hole and illumination effects' emerge: when juxtaposed lines create a luminance gradient and define the boundaries of a figure, the figure becomes an illuminated 3D object and the background appears deeper — similar to a hole. (ii) Not only is the luminance of the juxtaposed lines involved in the figurality effect, but so is the hue that determines the color of the illumination. (iii) Object–hole and illumination effects cannot be easily dissociated because, as the previous examples showed, they change dramatically their 'meaning' when tiny variations within the luminance gradient are introduced. This strong link between the two effects and the fact that both of them strongly contribute to determine the figurality of an object makes it reasonable to introduce an overall effect, called 'figurality effect', that includes all the properties belonging to the object–hole and to the illumination effects. (iv) Most of the necessary information needed to define the figurality properties is in the boundaries and, more specifically, in the gradient surrounding and defining a perceptual object. Thus, varying the luminance profile of the gradient permits discovery of the laws of figurality that determine not only what is figure and what is background, as the figure–ground segregation principles do (Rubin, 1921), but also the phenomenal appearance of what is perceived as a figure within the three-dimensional space and under a defined illumination. The aim of the next section is to clarify further the necessity of introducing the laws of figurality in addition to the Gestalt grouping and to figure–ground segregation laws.

8. THE PLACE OF FIGURALITY IN A WORLD OF GROUPING AND FIGURE–GROUND SEGREGATION

On the basis of these phenomenal results several general principles can be introduced, following the example traced by Gestalt psychologists like Rubin and Wertheimer. Before defining and naming the laws of figurality, it is necessary to clarify the role of figurality in a perceptual world ruled by the well-known Gestalt laws of grouping and figure–ground segregation.

8.1. The problem of perceptual grouping

Wertheimer (1923) in his pioneering study of the factors underlying visual perception approached this problem in terms of grouping. The question he answers is:

how do the elements in the visual field 'go together' to form an integrated, holistic (*Gestalt*) percept? How do elements group in parts that in their turn group in overall objects? How do individual elements create larger wholes separated from others? In a set of classical and elegant experiments, he discovered the ten basic principles, called 'grouping factors', mentioned in the Introduction: proximity, similarity, good continuation, closure, symmetry, convexity, prägnanz, past experience, common fate, and parallelism. These principles (like Rubin's, below) represent '*ceteris paribus*' laws, namely principles defining what determines the emerging whole when all else is equal. In addition to these classical grouping factors, a number of new principles have been identified (Palmer, 1992, 1999; Palmer and Rock, 1994), but despite their novel appearance, these principles can be traced back to the classical ones, as they represent a special or limiting case of at least one known factor.

8.2. The problem of figure–ground segregation

One cannot talk about grouping without also mentioning figure–ground segregation. Rubin (1921) was the first to name the principles ruling what appears as a figure and what as a background, one being the 'surroundedness principle' stating that a surrounded region tends to become a figure while the surrounding region becomes background. (The other principles were size, orientation, contrast, symmetry, convexity, and parallelism (Note 1)). Rubin went beyond the laws of figure–ground segregation by identifying the main properties which belong to the figure but not to the background. (i) The figure appears closer to the observer than the background, implying a role for depth perception; (ii) its color appears denser than the same physical color on the background; and (iii) it assumes the shape traced by the contour, implying that the contour belongs unilaterally to the figure, not to the background. The unilateral belongingness of the boundaries is a basic property that is common to all the principles. We therefore regard it as a meta-principle, one of the rules governing all the laws of figure–ground segregation, rather than as just another property. Often ignored, it is linked to the problem of figurality and constitutes its prelude. It is considered here by studying (i) different gradient profiles across the boundary contours, (ii) the belongingness of the juxtaposed lines to the figure or to the background, and (iii) how they group or split to create emergent properties like the volumetric and illumination effects.

Unfortunately, the problem of the boundaries was not further developed by Rubin and, until now, no new principles have been added to Rubin's original ones. This is where the watercolor illusion and related cases come in. Like Wertheimer and Rubin, we used the phenomenological approach by first describing what we see (previous Results sections), then defining the boundary conditions under which all the effects and, more specifically, the figurality properties occur (next sections). Finally, we put forward a hypothesis to explain the problem of figurality (last section).

8.3. The problem of figurality

The laws of figurality complement the grouping and figure–ground principles. As shown in this work, the main figural properties defined by Rubin, and particularly the unilateral belongingness of the boundaries, can be considered as the starting point to understand what is next, i.e. the figurality properties. We do not aim to define the logical or the temporal processing order of grouping, figure–ground, and figurality, but to highlight the problem of figurality and to suggest principles that govern it. The main question the laws of figurality intend to answer is how variations in boundaries and luminance gradient profiles define both the 3D appearance of the object (its specific figural properties) and the lighting — the perceived illumination and light reflected by the object. Thus, figurality proceeds beyond the properties of the figure as defined by Rubin's laws.

9. THE LAWS OF FIGURALITY

The following laws are incomplete and require further study and integration. Still, they are not merely summary statements. Rather, they are Janus-faced, both pointing back towards the purely phenomenal (but reliable) descriptions of the figures discussed earlier, and pointing forward towards statements of regular, analytic relations (those with predictable consequences) that will eventually need explanation at a mechanistic level. These laws are consistent with the Gestalt usage of the term 'law' that does not necessarily concern causes and effects. Nevertheless, like other laws governing different scientific fields, the figurality laws represent regularities for all practical purposes that apply to certain classes of phenomena (e.g. figurality) or generalizations based on recurring facts or events. Their specific formulations, as we shall see below, describe relationships that are observed to be invariable between or among phenomena for all cases in which the specified conditions are met.

The figurality laws currently refer to just four gradient conditions, which are the ones previously considered. (i) A luminance gradient profile where the line having the highest luminance contrast is placed at one extreme of the gradient and the lowest at the other. The watercolor and the lighting effect exemplify this, the 'high–low' gradient. (ii) A luminance gradient profile with two lower luminance contrast lines placed at the extremes and the line with the highest luminance contrast is between them. The backlighting and the fuzziness effects seen in Figs 13 and 15 illustrate this, the 'low–high–low' gradient. (iii) A luminance gradient profile like the high–low one but with the extremities ending on a black or white surrounding region such that luminance contrast is low at both extremities. This is illustrated in Figs 16a and 17, the 'low–low' gradient. (iv) By making the black surrounding region of the low–low gradient white and the white region black, a high luminance contrast transition is created on both extremes of the gradient, as illustrated in Fig. 18. We call this the 'high–high' gradient. The names used to indicate the four kinds of gradients are just labels necessary to distinguish them and simplify the following descriptions.

A limiting case of these conditions is the one with only one boundary contour like those studied by Rubin. Rubin's unilateral belongingness of the boundaries, also called 'border ownership' (Nakayama and Shimojo, 1990) can be considered as the first law of figurality.

Law 1: The boundary contours belong unilaterally to the figure, but not to the background.

This law can be applied easily to the figural conditions where only one boundary contour is present, and it states that this contour belongs only to the region determined as a figure by Rubin's laws. The unilateral belongingness of the boundaries is very effective under reversible conditions, or when different laws determine complementary regions as a figure thus creating equilibrium between them. This law does not determine what appears as a figure and what as a background; thus it is not a figure–ground segregation law, but it defines the belongingness rule within figurality. Therefore it is the first of the figurality laws. These can be considered as belongingness laws, in the sense that, depending on their belongingness (to the figure or to the background), the juxtaposed lines assume different perceptual roles (lighting, backlighting, shadows, etc.). In this approach, unilateral belongingness is valid not just with only one boundary contour, as studied by Rubin, but also with many juxtaposed lines.

The second law is directly related to the first one. Given many juxtaposed lines, this law specifies which line becomes the most external unilateral boundary of the perceived object.

9.1. The role of the line with the highest luminance contrast

9.1.1. Boundary line.

Law 2: The juxtaposed line with the highest luminance contrast in relation to the homogenous surrounding regions tends to appear as the outermost boundary of the figure. We call this line the 'boundary line'.

This law clearly applies to the watercolor illusion (e.g. Figs 1 and 2), where the purple line appears as the boundary of the watercolored object, and to the equiluminant limiting case where a tessellation or a reversible figure–ground organization is perceived (Fig. 8a). Figurality also refers to the coloration and, particularly, to the surface color properties. In fact, if one line assumes the role of 'boundary', the other line specifies something else, for example, the color or shade of the object (the topic of the third law). Under high luminance contrast conditions between the two lines, not only is the role of the boundary affected, but there is an asymmetry even in the amount of color spreading from the two lines. The line with less luminance contrast relative to the background spreads proportionally more than the line with higher luminance contrast. The color spreads in directions other than the line orientation. In quasi-equiluminant conditions, both lines spread similarly, in opposite directions (Fig. 8a). Coloration and boundary belongingness proceed in agreement.

As a consequence of Law 2, the weaker surface color property under quasi-equiluminant conditions is linked to the weaker figure–ground and object–hole effects. This law applies also when coloration is absent, as illustrated in Figs 9 and 11. It also applies to the other figures, except Figs 16 and 17, where there is no line with the highest luminance contrast (see Law 8). The lighting, backlighting and fuzziness effects fall within this law in relation to the line with most contrast. Figure 18 is a special case because it has two lines with the highest luminance contrast. However, due to Law 2, both lines become the boundaries delimiting an object; therefore the object is everything in between these boundaries, i.e. the bunch of lines. Following this observation, a corollary of Law 2 can be introduced to extend this law to the cases where more than one or two lines are present within the gradient.

COROLLARY 1. *Given more than one line with the highest luminance contrast, all of them appear as boundaries of the figure.*

This corollary applies even to special cases as those illustrated in Fig. 19, where 'a flat figure with three boundaries' (13 out of 14) is perceived. The presence of three boundaries makes the figure lose its volumetric and 3D properties. Nevertheless, the coloration effect is much stronger than the condition where only the inner purple and orange lines are present. This is an interesting coloration case on which we are working.

If the line with the highest luminance contrast becomes the boundary of the figure, so that the perceptual role of the boundary is taken by this kind of line, the next question is: what is the perceptual role of the other lines? (We will assume that the same role cannot be played by different 'things'; therefore, if one 'thing' plays a specified role, an adjacent different 'thing' should play a different role. The meta-laws of section 10 clarify the need for this basic assumption.) On the basis of the previous phenomenal results, Law 3 can answer this question.

9.2. *The role of the lines with lower luminance contrast*

Law 3: Given the high–low gradient, the juxtaposed lines with lower luminance contrast in relation to the homogenous surrounding regions affect the perception of the bounded area. They do this in at least three ways: (i) *they determine the color,* (ii) *they model the volume depicting lights and shades, and* (iii) *they define the properties of the lighting (both apparent illumination and reflected light) through highlights and dark values.*

When the line with most contrast assumes the role of the boundary (Law 2), the lines with lower contrast lose their individual boundary contour status and become a single component of the bounded object, whose percept they qualify according to Law 3.

9.2.1. The role of color in Law 3. The lines with lower contrast define the color of the resulting object, as illustrated in the watercolor illusion. (An exception occurs in Figs 9 and 11, where the two adjacent lines of the crosses are divided in four different chromatic colors. None of these pairs of lines alone can define the color of the crosses due to the presence of the other three, and an admixture or even absence of coloration is obtained.) As illustrated in Fig. 12, the lower-contrast lines define the color in the lighting effect, even when the coloration is absent.

Figure 14, where the color of the light is perceived, and Fig. 13, where the backlighting emerges with a proper color, were not covered by Law 2, because the lines in these figures do not define the main object directly. Moreover, only particular lines are involved, like the line terminating the gradient in Fig. 14, or the line placed next to the line with the highest luminance contrast (see Laws 4 and 5). Nevertheless, in general, Law 3 can be applied to these figures as well.

It is interesting that it is sufficient to change the color of only one line to define the color of the object. For example, by replacing the third brighter blue line in Fig. 12a with a green one (see Fig. 20) the entire object now appears 'aquamarine' (12 out of 14) colored. The object color changes by varying the color of any of the other lines (not illustrated), except for the brightest one that defines the color of the light. The resulting object color is the resultant of the admixture of the colors of all the single components.

9.2.2. Modeling volume in Law 3. In Figs 9 and 11, where no coloration occurs, and in Fig. 18, where a high–high gradient is illustrated and where the bunch of lines appears as 'a shaded cylinder depicting the perimeter of a fish bone shape' (12 out of 14), the lines clearly model the volume. Note in particular the surface capture effects in Figs 9 and 10 that clearly show this role. By varying the luminance profile of the lines, and thus the amplitude of the bulge and of the relief, the volume of the resulting object appears to vary. This is the topic of Law 10 and Corollary 10.1, which expand on Law 3.

9.2.3. Defining the properties of the lighting in Law 3. Our phenomenal results showed the emergence of the illumination and of the lighting effect, mostly in the 'lighting illusion'. The number of lines composing the gradient is an important

Figure 19. (See color plate XVI) The modified 'watercolor' figure lose its volumetric and 3D properties. Nevertheless, the coloration effect is much stronger than the condition where only the inner purple and orange lines are present.

Figure 20. (See color plate XVI) By replacing the third brighter blue line in Fig. 12a with a green one the entire object appears now aquamarine.

factor in determining the strength of the lighting effect (see Law 10). The watercolor conditions with only two juxtaposed lines show a weaker or absent lighting effect. By increasing the number of the lines, the illumination emerges more and more. Figure 21a illustrates an indirect demonstration of the role played by the lines in producing the illumination effect. If the inner region of Fig. 12a is physically filled with a light chromatic tint (e.g. light red), 'the inner region appears white' (11 out of 14), and 'the light red totally discolors' (12 out of 14). The discoloration illusion (Pinna, in press) — meaning the loss of the coloration — disappears when the gradient is reversed (not illustrated), or when the length is reduced (see Fig. 21b).

The previous observations reveal that it is not necessary for all three roles (color, volume and light) to be present at the same time. For this reason a corollary has been introduced.

COROLLARY 1. *The juxtaposed lines with lower luminance contrast define at least one of the three phenomenal properties of color, volume and light.*

9.2.4. Lighting line. Not all the juxtaposed lines play the same role along the high–low gradient with six lines, as illustrated in Fig. 12. The extremes play a particular role. The line with the highest luminance contrast defines the object and its boundaries (Law 2), while the role of the line with the lowest value is defined in Law 4.

Law 4: Given the high–low gradient, the line at the extreme of the gradient with the lowest luminance contrast in relation to the homogenous surrounding region defines not only the object properties, described in Law 3, but also the properties of the lighting (its color and intensity). We call this extreme line the 'lighting line'.

The extremes of the gradient are placed in the borderland between different regions, namely, the juxtaposed lines and a large white homogeneous area, so they delimit and form the boundaries of these two regions. These extremes separate the object from 'something else': that is, the line with the highest luminance contrast separates the object from the background, and the line with the lowest contrast separates the object from the most lighted area. Thus the former defines the object, the latter defines the property of the light. The other lines less strongly define the boundary and the illumination, presumably because they are placed in between the extremes.

However, they determine the color more powerfully, and also the shade and the volumetric shape of the object — these being properties not only of the object but also of the lighting, as shown by *chiaroscuro*. The different roles of the lines on the basis of their spatial positions within the gradient are demonstrated by the fact that variations in the color of the lowest line can change the color of the light, while

Figure 21. (See color plate XVI) *The discoloration illusion.* In Fig. 21a, when the inner region of Fig. 12a is physically filled with a light red tint, the inner region appears white: the light red totally discolors (Fig. 21b is a control).

color variations of the other lines change mostly the color of the object. The closer the intermediate lines are to one of the two extremes, the more their role is similar to them.

9.2.5. Backlighting line. The backlighting effect occurs in the other condition in which one or more lines define the role of the light. This is defined in Law 5; Law 7 will cover the related backlighting gradient condition.

Law 5: Given the low–high–low gradient, the line at the extreme of the gradient with a luminance contrast lower than the boundary line (and beyond) is the 'backlighting line'. This line does not define object properties, but only the properties (color and intensity) of the illumination coming from behind.

In Fig. 13, the backlighting line splits from the others both structurally, because it is placed beyond the boundary line, and phenomenally, because it is stronger when it has a chromatic color different from all the other lines. If the boundary line behaves as a watershed, the line placed beyond it does not 'talk' about the object, but rather about the apparent lighting, and more precisely the backlighting, because (once again) it is beyond the boundary line.

These laws defining the various roles of the lines are so far incomplete, as the gradient properties that determine the object–hole effect have not yet been defined. This is the topic of the next section.

9.3. The role of the gradient profiles

The following principles define (i) which regions appear as objects and which as holes depending on the properties of the luminance gradient, (ii) how the object–hole and volumetric effects vary by changing these properties, and (iii) under which gradient conditions new emergent effects (lighting, fuzziness, etc.) are created.

Law 6: Given the high–low gradient (e.g. Figs 2 and 12), *the region whose luminance transition is less abrupt is perceived as a volumetric object, relative to the complementary region with a more abrupt transition which is perceived as an empty 3D space.*

The phenomenal and physical asymmetric luminance transition across the juxtaposed lines makes the figural effect due to the watercolor and to the lighting illusions stronger than in the classical figure–ground condition, and prevents reversibility of figure–ground segregation by strengthening the unilateral belongingness of the boundaries (Pinna, 2005). In fact, so that a boundary contour can belong only to the figure (and not to the background) under the conditions studied by Rubin or as illustrated in Fig. 3, it is necessary that the visual system induces some kind of figurality asymmetry on both sides of the contour, so this contour can become the boundary of only one of the two complementary regions separated by it.

Based on this phenomenological reasoning, it follows that the luminance asymmetry between boundary contours, physically present in the high–low gradient, strengthens the belongingness of the line of highest contrast to the region where the luminance value of the lines decreases more gradually. If this is true, each line of the gradient reinforces this unidirectional figural process by creating a cascade connection due to the same asymmetrical conditions present in each pair of lines. The interruption of this structure, for example by reversing the gradient profile in some of the pairs of lines or at the extremes of the gradient, may cause phenomenal variations and the emerging of new properties. One of these cases is defined in the next law:

Law 7: Given the low–high–low gradient (see Figs 13 and 15), *the region whose luminance transition is longer is perceived as the volumetric object relative to the complementary region, with the shorter transition perceived as an empty 3D space illuminated by backlighting on the volumetric object whose color and intensity are defined by this shorter transition.*

This kind of gradient reveals the different perceptual roles played by the two sides of the line (the one with the highest luminance contrast) dividing the two regions. On the side where the gradient created by juxtaposed lines is longer, object properties emerge, whereas on the other side, background properties emerge. Therefore, the lines placed on the background regions appear as light rather than manifesting object properties, but since they are placed all around the boundary line, they appear as backlighting. As a consequence, they modulate the object, as described in the next corollary.

COROLLARY 1. *Given the low–high–low gradient, all else being equal, when the luminance transitions on both sides of the boundary line have the same length and the color of the lines placed on opposite sides of the boundary line are different, an equilibrium appears: each side can be perceived as an object or as a background, alternatively and reversibly. Under these conditions, the backlighting effect is absent. When the color of the lines on both sides is the same, what emerges is not a reversible object-background segregation but a fuzziness effect: an object with blurred boundaries (not illustrated).*

9.3.1. Neon Lighting. Law 8 describes a special case of Law 2, in which there is no boundary line.

Law 8: Given the low-low gradient (see Figs 16 and 17), *due to the absence of a line with a highest luminance contrast, none of the lines assume the role of the boundary line. However, all of the lines surround an object, which is segregated on the basis of Rubin's figure–ground laws. The lines now become not boundaries but rather a neon light, either surrounding* (Fig. 16) *or coming from behind* (Fig. 17).

The absence of a boundary line creates a reversible figure–ground segregation. This is not in fact perceived in Figs 16 and 17, where closure and surroundedness segregate the inner fish-bone region as a figure. However, by reproducing the low-low gradient under different geometrical conditions, where these factors are counterbalanced, a reversible figure–ground segregation can be perceived (not illustrated).

This law is a special case of Law 2, depending on the absence of a line with the highest luminance contrast. For this reason it could become corollary of 2. However, to underline the properties of the gradients, we preferred to keep this law in this section. The same reason applies to the next law, which could be read as another corollary of Law 2.

Law 9: Given the high–high gradient (see Fig. 18), *the presence of two extremes with the highest luminance contrast forces both of them to assume the role of boundary lines, inducing the perception of a bunch of lines placed within outer and inner regions perceived as empty space.*

Again, Rubin's principles play some role in Fig. 18 by inducing within the inner region a more figural appearance, making it appear as a hole that, paradoxically, can be considered as a figure whose boundaries belong not to it but to the complementary region that is, in its turn, a figure (see also Palmer, 1999; Bertamini, 2001; Bertamini and Croucher, 2003; Grossberg *et al.*, 2005; Palmer and Nelson, 2005).

9.4. The role of the properties of the gradient profile

The role of the color of the lines with lower luminance contrast placed in different positions of the gradient has been described by introducing Laws 3, 4 and 5. Not only the color, but also the length of the gradient plays a basic role in defining figurality. This role is described in the Law 10 specifically for the high–low gradient, but it applies also to low–high–low and low–low gradients (not illustrated).

Law 10: Given the high–low gradient, by increasing the length of the gradient, i.e. by enhancing the number of lines (e.g. from two lines, as in the watercolor illusion, up to six lines, as in the lighting effect), (i) *the coloration effect decreases,* (ii) *the object–hole and volumetric effects increase, and* (iii) *the lighting effect increases.*

These results accord with intuition, because increasing the gradient length increases the salience and the length of the shade that phenomenally involve all the other properties, as shown through the *chiaroscuro* technique. By increasing the length, the luminance jump between each line changes accordingly. By reducing the number of lines in Fig. 12, the bulging effect is reduced and the object becomes flatter. This observation suggests the next corollary.

COROLLARY 1. *Given the high–low gradient, by changing the luminance value of each line so that the spatial profile is speeded up or slowed down (while keeping the same extremes), the volume of the perceived object changes accordingly.*

What is predicted by this corollary is what the Renaissance painters used to do with the *chiaroscuro* technique to create and modify the volume of their subjects. Note that even though variations in the luminance gradient specified in Corollary

10.1 can change the volumetric effect, the other figurality properties do not change qualitatively. For this reason the precise luminance values of the lines within the figures shown here are not critical (and indeed have not been specified.)

9.5. Grouping and figure–ground laws involved in the figurality laws

The ten figurality laws complement and follow the laws of grouping and figure–ground segregation in other senses, different from those already described in Section 8. In fact, these laws operate within the juxtaposed lines, segregating some of them from others. For example, the salience of the lines of highest contrast that appears as a border line can be considered as a special case of the contrast law of figure–ground segregation suggested by Rubin. Again, the separation between the two kinds of gradients proper to the low–high–low profile can be due to the 'similarity in the direction of the luminance', which is a limiting case of Wertheimer's similarity law. The similarity of color operates in the backlighting effect and in the segregation of the lighting line. In general, the laws of Wertheimer and Rubin seem to operate at the low level of the color and luminance gradient profile by creating the conditions necessary to assign different roles to segregated or ungrouped lines.

Nevertheless, the figurality laws cannot be considered as special cases of grouping and figure–ground segregation laws, because even if the Gestalt laws group, ungroup or segregate several lines within the gradient, they do not say anything about the figurality properties of the resulting object. As a consequence, the figurality laws cannot be reduced to the Gestalt ones without leaving out many phenomenal figurality properties. Indeed, the figurality laws operate at a meta-level which defines how the 'things' (lines, gradient profiles, etc.) grouped and segregated by the Gestalt laws represent emergent properties and objects like 'shade', 'boundary line', 'illumination', 'lighting', and 'backlighting'. Thus they implicitly contain interesting general statements, placed at a meta-level of explanation, of how the visual system works and how numerous stimulus variations are organized into a small set of constant objects with emergent properties.

A problem with these laws is related to the fairly large number of them; like the Gestalt laws, which are also numerous, they may come into conflict. Further theoretical work and research is needed to develop meta-level laws to regulate such potential conflicts, in part by reducing the number of laws to fewer, but more general, principles, and to elucidate how these laws interact. What is important in this reduction process will be to retain the detailed, phenomenal level of the original laws at the 'parent' level, not to blur them into more general but less precise statements. Each subsequent set of laws should then be explained by (or reduced to a meta-level of) meta-laws that are derived from the parent and permit rebuilding it. The following section is an attempt to formulate a set of such meta-laws, but it is far from definitive. Because they represent new super-ordinate principles, the phenomenological language needed to describe them is also new, at least in places.

10. THE FIGURALITY SPLIT–UNION META-LAW OF EMERGENT THINGS

In general, a law allows the prediction with a certain precision of what happens under specified conditions. This was the aim of the ten figurality laws. However, in section 6, it was shown how difficult it is to predict exactly how the coloration should appear when the figure–ground effect is dissociated from it and, vice versa, how the figure–ground effect should appear after dissociation from the coloration. It is even more difficult to predict exactly the phenomenal result when the four effects, including the object–hole and the illumination effects, are reciprocally dissociated, as they cannot be completely disjoined without changing or losing their properties and, more importantly, without losing the new properties emerging through their union. As a consequence it can be much more interesting to understand, and define with meta-laws, how these effects create integrated properties, so that the resulting effects are 'explained' through the emergent qualities and objects.

If it is difficult to predict the exact appearance created by a specific effect, especially if multiple percepts are possible, still, the phenomenal structure common to a set of specific effects or stimuli can be determined. Such a structure has to describe the variability of appearances as well as the central tendency. We attempt next to specify the existence and nature of such variability by appealing to some 'meta-laws'. These will indicate which of the ten figurality laws is dominant, and which modulatory, and in which conditions. The specific meta-laws are somewhat speculative, but they illustrate the potential of laws at this level. Several meta-laws governing the phenomenal variability within the laws of figurality are now suggested.

Meta-law 1: Stimulus variations split in two independent and opposite phenomenal components: one component ('invariant') is perceived not directly but amodally, while the other ('variant') is perceived directly and modally. The one component acts to discount or annul the variations, while the other subsumes all the variations and attributes them to a different 'thing'.

We call this the 'split meta-law'. The first component is represented, for example, by the amodally perceived homogeneous blue color of the fish bone in Fig. 12, which is obtained by discounting the dark and bright variations among the lines of the gradient and the inner white region. The second component is represented by the directly perceived bright light, which in some sense 'explains' the variations by attributing them to the effects of lighting, in particular becoming responsible for the illumination and discoloration of the blue homogeneous surface color of the fish bone.

These examples show that the split both creates two independent emerging things', i.e. the color of the object and the lighting, and creates two 'things' that are strongly united and mutually dependent. The term 'thing' represents a perceived entity or quality that is not fixed or cannot be named specifically for all cases but can indicate a large number of properties or emergent objects. The following meta-law defines the union between the split components.

Meta-law 2: The invariant and variable components emerging from the split meta-law are mutually dependent. They reinforce their role reciprocally and join together like the two sides of the same phenomenal reality at a higher level.

We call this the 'union meta-law'. Within the previous examples, the perceived bright light determines the perception of a specific color which in its turn reinforces the perception of the bright light. They are like two sides of the same illuminated 3D perceptual reality. So, they are inseparably united, in that: they are integrated and segregated at the same time; one exists only by virtue of the other; they cannot be dissociated or disconnected without changing; one cannot be phenomenally changed without changing the other; and, finally, they contribute together to create the 3D perceptual reality. In these terms, the paradox suggested by the words 'split' and 'union' is only apparent; in fact they refer to different things belonging to different perceptual levels: modal/amodal; direct/indirect (the gradient and the lighting); and level/meta-level (the light 'talks about' the object, that in its turn 'talks about' the light). This basic relationship is described by the next meta-law.

Meta-law 3: The union between the split components requires that a certain variation of one component is counterbalanced by the contrary variation of the other component.

We call this the 'constant ratio meta-law'. By increasing the lighting intensity within the fish bone the coloration decreases. Variations from this constant ratio have to be reconsidered through the previous meta-laws, thus creating a new set of emerging constant and variable sides. The emergence of new 'things' is the topic of the next meta-law.

Meta-law 4: The two sides of the split–union meta-laws, that is, the amodal (invariant) and indirect (variant) 'things' emerging from the split–union process, are able to 'explain' the variations within the stimulus. In other words, the emergent 'things' can discount, simplify, regulate, and predict the variations within the stimulus.

We call this the 'meta-law of emergent things'. To illustrate this meta-law, the perceived lighting of Fig. 12 'explains', in the sense that it 'defines' or 'rules over' the variations within the luminance gradient and the inner white region, thus determining the surface color of the fish bone. Similarly, the amodal color of the object 'explains' the perceived gradient and the properties of the lighting. Both lighting and color are emergent things, not directly present within the stimulus pattern but the result of a meta-process.

These emergent things are only two among many others that in their turn emerge through the same rules and from the same stimulus pattern. Others things are: the boundary line that separates the object from the empty background, the shadow, the lighting line, the backlighting, the object–hole segregation, etc. The coexistence of so many emergent things suggests the next recursive meta-law.

Meta-law 5: Each stimulus variation is subjected to Law 4, the split–union meta-law of emergent things, thus creating, for each variation, pairs of emergent things that are nested one inside the other and are globally and recursively ruled by the same Law 4.

This meta-law implies that two emergent 'things', like the hole (the constant side) and the object (the variable side), are nested inside another emergent thing, the boundary line, which is nested inside the shadow, and so on. The variations are respectively the highest luminance contrast with respect to the region that appears as an empty background, the highest luminance contrast with respect to its juxtaposed line, the juxtaposed lines among them, and so on. What is interesting in this law is that, due to the recursive and nested process, the emergent things usually do not contradict one another and do not create any perceptual paradox. Studies, aimed at understanding under which conditions paradoxical things do emerge, are in progress.

These meta-laws are reminiscent of the phenomenal scission (*Spaltung*) suggested by Gestalt psychologists (Koffka, 1935; Metzger, 1954, 1963, 1975). However, like the laws of figurality, they are aimed at continuing and enlarging the Gestalt work within the context of figurality and the phenomenal and neural processes of the formation of emergent 'things'.

11. CONCLUSIONS

In this work we have reported new cases of the watercolour illusion and introduced new effects, like the lighting and the backlighting effects, suggesting what we called laws and meta-laws of figurality. By defining the conditions of the luminance gradient profiles across boundary contours, these laws define the 'figurality': the color and the volume of the object with lighted and shaded regions, and the direction and the color of the light emerging from the object. Furthermore, the definition of the figurality laws and meta-laws in these terms may provide key insights into the neural mechanisms underlying the problem of figurality.

The relative separation between the coloration and figural effects studied in Section 6 suggests the existence of parallel but strongly related mechanisms. The FACADE model (Grossberg, 1987a, b, 1994, 1997) posits that boundary grouping and surface filling-in processes can explain the two effects in the watercolor illusion. They are substantiated by the cortical interblob and blob streams, respectively, within cortical areas V1 through V4. These boundary and surface processes show complementary properties (Grossberg, 2000) and their interaction generates a consistent perceptual representation that overcomes the complementary deficiencies of each stream, acting on its own. Boundary and surface processes are modeled by the Boundary Contour System (BCS) and by the Feature Contour System (FCS), (Cohen and Grossberg, 1984; Grossberg and Mingolla, 1985a, b; Grossberg and Todorovic, 1988).

The figures illustrated and the laws described suggest that most of the basic information about the figurality resides in the boundaries and, more specifically, in the luminance gradient peculiar to these illusions. Recent findings (Zhou *et al.*, 2000; Friedman *et al.*, 2003; von der Heydt *et al.*, 2003) showed that neurons in V2 respond with different strength to the same contrast border, depending on the side of the figure to which the border belongs, implying a neural correlate process linked to the unilateral belongingness of the boundaries. Several findings showed that figure–ground segregation may be processed in areas V1 and V2 (Zhou *et al.*, 2000; von der Heydt *et al.*, 2003; Friedman *et al.*, 2003), in inferotemporal cortex (Baylis and Driver, 2001) and the human lateral occipital complex (Kourtzi and Kanwisher, 2001). Zhou *et al.* (2000) reported that approximately half of the neurons in the early cortical areas are selective in coding the polarity of color contrast, e.g. a neuron may respond to a red–gray border, but not to a gray–red border. Von der Heydt and Pierson (2006) suggest that not only the figure–ground segregation, but also the color tint of the watercolor illusion might have its explanation in the cortical representation of borders.

The laws of figurality and, more particularly, the meta-laws go further by suggesting a possible neural scenario where multiple juxtaposed lines may stimulate neurons, selective for different asymmetric luminance profiles and signaling not only the unilateral belongingness of the boundaries and the coloration effect but also the volumetric and the illumination effects. Due to the gradient variations at different scales, it can be suggested that these neurons, at the beginning undifferentiated, become more and more specialized by assuming different roles (for example, the role of boundary or lighting lines) but at the same time becoming part of a more global integrated process described in Meta-law 2. This specialization and integration process may be possible in terms of neural processing stages, where invariant and variant components of the stimulus at different scale are organized through mechanisms that trace the phenomenal structure reported within the meta-laws. More particularly, on the basis of the split–union meta-law, it can be conjectured that variations within the stimulus pattern stimulate the same set of neurons at different scale, among which some annul or discount the variations by minimizing them and some signal those variations. These opposite processes may depend on the fact that some cells signal stimulus variations at a specified scale that are not necessarily variations at a different scale (more local or global) but invariants. All these different signals (local *vs.* global), signaling or annulling stimulus variations, are integrated at a global processing stage making them to perceptually coexist within a whole 'thing' and to assume specialized properties and roles. All of them define synergistically the emerging thing at different levels and through different specific roles (boundary line, etc.) assumed by each component. Thus, even if these neurons signal pairs of local vs. global emergent components, constant and varying at different scales, they are integrated through a neural mechanism that is the correlate of the union meta-law and that make them becoming inseparably united and reinforcing one another.

Beyond the last conjectures just sketched out, the nature of the neural correlates of the laws and meta-laws of figurality needs to be reconsidered and deepened especially with reference to the emergent effects: what they really are and how they emerge from lower processing levels. Nevertheless, the advantage of the laws and meta-laws of figurality is that they suggest precise questions and experiments in phenomenology, psychophysics and neurophysiology that prelude to find out precise processes and mechanisms. This is due to the fact that our laws are not '*ceteris paribus*' laws like the Gestalt ones, i.e. they do not concern a list of properties (like proximity, similarity, surroundedness, etc.) useful to predict the perceptual results only when everything else is equal (see Palmer, 1999), but they provide phenomenal integrated dynamics and processes underlying whole and emergent properties that can be translated in terms of neural mechanisms. Furthermore the phenomena and the laws here shown can be interesting to: understand processing strategies centered around the problem of a 3D interpretation of 2D images (see also Zhou *et al.*, 2000; von der Heydt *et al.*, 2003; Friedman *et al.*, 2003); suggest combinations of mechanisms that also explain other phenomena (e.g. brightness and lightness induction, neon color spreading, see Pinna and Grossberg, 2005), but with a unique contextually-driven balance of the same mechanisms; determine the role of the new principles relative to the classical gestalt principles; study how coloration, figure–ground, object–hole and illumination effects are related depending on boundary conditions; clarify how the watercolor, lighting and backlighting illusions increase the figural strength of a region by creating an unambiguous and unilateral direction of the margins much stronger than in the gestalt figure–ground principles, and, finally, stimulate the study of neurons selective for some gradient border ownership cues, but not others (see also Zhou *et al.*, 2000; von der Heydt *et al.*, 2003; Friedman *et al.*, 2003).

Acknowledgements

We thank Massimo Dasara, Maria Tanca and Mario Pintus for assistance in testing the subjects.

This work was supported by Fondazione Banco di Sardegna, Alexander von Humboldt Foundation, PRIN ex 40% Cofin. es. 2005 (prot. 2005112805_002) and Fondo d'Ateneo (ex 60%) to Pinna; AFOSR grant FA9550-04-1-0244 to Reeves.

NOTES

1. It is reasonable to think that figure–ground segregation must operate before grouping (Hoffman and Richards, 1984; Palmer, 1999). For example, the line elements on which grouping acts must be already segregated as a figure from the ground, otherwise the visual system would not know which regions to group. However, lines do not possess the figural properties of holistic organized and segregated figures; rather, they appear as elementary components necessary to

create boundaries. Thus the lines creating the luminance gradient of (say) Fig. 12 should first be grouped to become the boundary of the fish bone, but they do not manifest figural properties: they are not surfaces, but something like perceptual 'bricks' necessary to create something more holistic. We therefore do not commit ourselves to the view that grouping must always precede segregation at the processing level. Instead we consider this as a phenomenological order between the two dynamics, grouping and figure–ground segregation, when defining the role and the place of figurality in a perceptual world built up by grouping and figure–ground segregation. Thus, we are concentrating on what the Gestalt laws state (which is phenomenological) rather than on their causal status. What the reader needs most is concrete examples to set the Gestalt laws against the 'laws of figurality' to see how they really differ.

REFERENCES

Arend, L. and Reeves, A. (1986). Simultaneous color constancy, *J. Opt. Soc. Amer. (A)* **3**, 1743–1751.

Arend, L. E., Reeves, A., Schirillo, J. and Goldstein R. (1991). Simultaneous color constancy: papers with diverse Munsell values, *J. Opt. Soc. Amer. (A)* **8**, 661–672.

Baylis, G.C. and Driver, J. (2001). Shape-coding in IT cells generalizes over contrast and mirror reversal, but not figure–ground reversal, *Nature Neuroscience* **4**, 937–942.

Bertamini, M. (2001). The importance of being convex: an advantage for convexity when judging position, *Perception* **30**, 1295–1310.

Bertamini, M. and Croucher, C. E. (2003). The shape of holes, *Cognition* **87**, 33–54.

Bressan, P., Mingolla, E., Spillmann, L. and Watanabe, T. (1997). Neon color spreading: a review, *Perception* **26**, 1353–1366.

Cohen, M. and Grossberg, S. (1984). Neural dynamics of brightness perception: features, boundaries, diffusion, and resonance, *Perception and Psychophysics* **36**, 428–456.

Devinck, F., Delahunt, P. D., Hardy, J. L., Spillmann, L. and Werner, J. S. (2005). The watercolor effect: quantitative evidence for luminance-dependent mechanisms of long-range color assimilation, *Vision Research* **45**, 1413–1424.

Friedman, H. S., Zhou, H. and von der Heydt, R. (2003). The coding of uniform color figures in monkey visual cortex, *J. Physiol. (London)* **54**, 593–613.

Grossberg, S. (1987a). Cortical dynamics of three-dimensional form, color, and brightness perception I: monocular theory, *Perception and Psychophysics* **41**, 97–116.

Grossberg, S. (1987b). Cortical dynamics of three-dimensional form, color, and brightness perception II: binocular theory, *Perception and Psychophysics* **41**, 117–158.

Grossberg, S. (1994). 3-D vision and figure–ground separation by visual cortex, *Perception and Psychophysics* **55**, 48–120.

Grossberg, S. (1997). Cortical dynamics of three-dimensional figure–ground perception of two-dimensional pictures, *Psycholog. Rev.* **104**, 618–658.

Grossberg, S. (2000). The complementary brain: unifying brain dynamics and modularity, *Trends in Cognitive Science* **4**, 233–245.

Grossberg, S. and Mingolla, E. (1985a). Neural dynamics of form perception. Boundary completion, illusory figures and neon color spreading, *Psycholog. Rev.* **92**, 173–211.

Grossberg, S. and Mingolla, E. (1985b). Neural dynamics of perceptual grouping: textures, boundaries, and emergent segmentations, *Perception and Psychophysics* **38**, 141–171.

Grossberg, S. and Todorovic, D. (1988). Neural dynamics of 1-D and 2-D brightness perception: a unified model of classical and recent phenomena, *Perception and Psychophysics* **43**, 241–277.

Grossberg, S., Dasara, M. and Pinna, B. (2005). The problem of the perception of holes and figure–ground segregation in the watercolor illusion (Abstract), *J. Vision* **5**, 54a; http//journalofvision.org/5/8/54/, doi: 10.1167/5.8.54.

Hoffman, D. D. and Richards, W. A. (1984). Parts of recognition. Special Issue: Visual Cognition, *Cognition* **18**, 65–96.

Katz, D. (1911). Die Erscheinungsweisen der Farben und ihre Beeinflussung durch die individuelle Erfahrung, in: *Zeitschrift für Psychologie*, Ergänzungsband **7**, pp. 6–31. Barth, Leipzig.

Katz, D. (1930). *Die Erscheinungsweisen der Farben*, 2nd edn. (Translation into English: R. B. MacLeod and C. W. Fox, 1935, *The World of Color*, Kegan Paul, London.)

Koffka, K. (1935). *Principles of Gestalt Psychology*. Harcourt Brace, New York.

Kourtzi, Z. and Kanwisher, N. (2001). Representation of perceived object shape by the human lateral occipital complex, *Science* **293**, 1506–1509.

Metzger, W. (1954). *Psychologie: Die Entwicklung ihrer Grundannahmen seit der Einfuhrung des Experimentes*, 2nd edn. Steinkopff, Darmstadt.

Metzger, W. (1963). *Psychologie*. Steinkopff Verlag, Darmstadt.

Metzger, W. (1975). *Gesetze des Sehens*. Kramer, Krankfurt-am-Main.

Nakayama, K. and Shimojo, S. (1990). Towards a neural understanding of visual surface representation, *Cold Spring Harbor Symposia on Quantitative Biology LV*, pp. 911–924.

Palmer, S. E. (1992). Common region: a new principle of perceptual grouping, *Cognit. Psychol.* **24**, 436–447.

Palmer, S. E. (1999). *Vision Science: Photons to Phenomenology*. The MIT Press, Cambridge, Massachusetts.

Palmer, S. E. and Nelson, R. (2005). The hole paradox: perceiving and remembering the shapes of intrinsic *vs.* accidental holes, in: *VSS Conference 2005*, Sarasota.

Palmer, S. E. and Rock I. (1994). Rethinking perceptual organization: the role of uniform connectedness, *Psychonomic Bull. Rev.* **1**, 29–55.

Pinna, B. (1987). Un effetto di colorazione, in: *Il laboratorio e la città. XXI Congresso degli Psicologi Italiani*, Majer, V., Maeran, M. and Santinello, M. (Eds), p. 158.

Pinna, B. (2005). The role of Gestalt principle of similarity in the watercolour illusion, *Spatial Vision* **21**, 1–8.

Pinna, B. (in press). The discoloration illusion, *Visual Neurosci.*

Pinna, B. and Grossberg, S. (2005). The watercolor illusion and neon color spreading: a unified analysis of new cases and neural mechanisms, *J. Opt. Soc. Amer. (A)* **22**, 1–15.

Pinna, B., Brelstaff, G. and Spillmann, L. (2001). Surface color from boundaries: a new 'watercolor' illusion, *Vision Research* **41**, 2669–2676.

Pinna, B., Werner J. S. and Spillmann, L. (2003). The watercolor effect: a new principle of grouping and figure–ground organization, *Vision Research* **43**, 43–52.

Rubin, E. (1915). *Synsoplevede Figurer*. Glydendalske, Kobenhavn.

Rubin, E. (1921). *Visuell wahrgenommene Figuren*. Gyldendalske Boghandel, Kobenhavn.

Schirillo, J., Reeves, A. and Arend, L. E. (1990). Perceived lightness, but not brightness, of achromatic surfaces depends on perceived depth information, *Perception and Psychophysics* **48**, 82–90.

Spillmann, L., Pinna, B. and Werner J. S. (2004). Form-from-watercolour in perception and old maps, in: *Seeing Spatial Form*, Jenkin, M. R. M. and Harris, L. R. (Eds). Oxford University Press.

van Tuijl, H. F. J. M. (1975). A new visual illusion: neon-like color spreading and complementary color induction between subjective contours, *Acta Psychologica* **39**, 441–445.

Varin, D. (1971). Fenomeni di contrasto e diffusione cromatica nell'organizzazione spaziale del campo percettivo, *Rivista di Psicologia* **65**, 101–128.

von der Heydt, R. and Pierson, R. (2006). Dissociation of color and figure–ground effects in the watercolor illusion, *Spatial Vision* **19**, 323–340.

von der Heydt, R., Zhou, H. and Friedman, H. S. (2003). Neural coding of border ownership: implications for the theory of figure–ground perception, in: *Perceptual Organization in Vision:*

Behavioral and Neural Perspectives, Behrmann, M., Kimchi, R. and Olson, C. R. (Eds), pp. 281–304. Lawrence Erlbaum Associates, Mahwah, NJ.

Wertheimer, M. (1923). Untersuchungen zur Lehre von der Gestalt II, *Psychologische Forschung* **4**, 301–350.

Zhou, H., Friedman, H. S. and von der Heydt, R. (2000). Coding of border ownership in monkey visual cortex, *J. Neurosci.* **20**, 6594–6611.

Color, Line, and Space, p. 273 (2006)

Subject Index

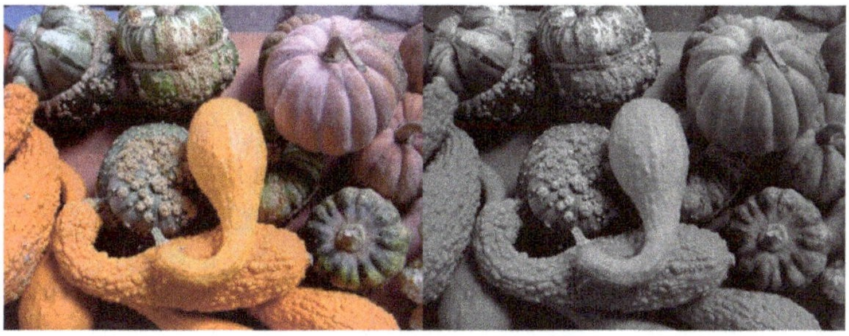

Pinna, Figure 1.

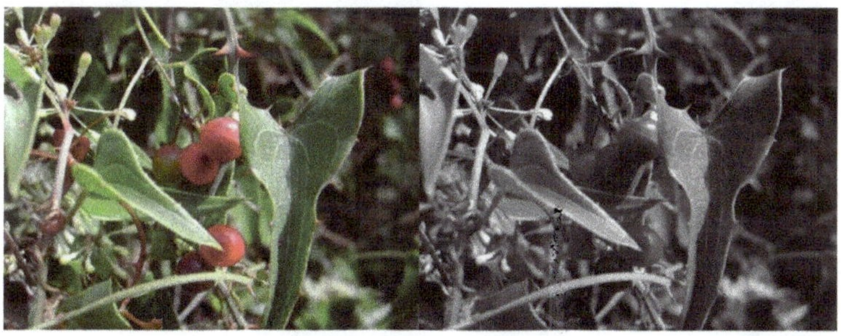

Pinna, Figure 2.

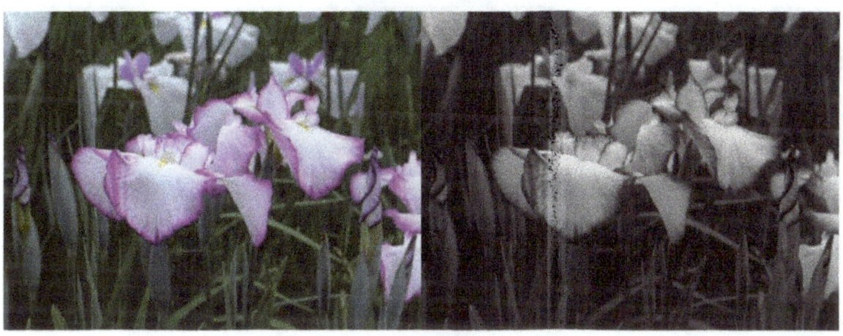

Pinna, Figure 3.

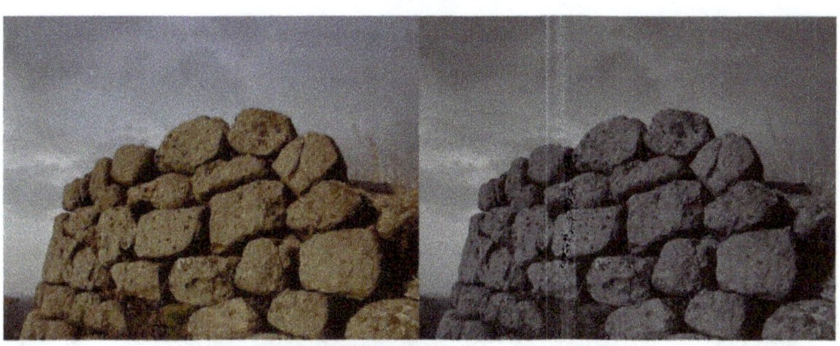

Pinna, Figure 4.

Plate II

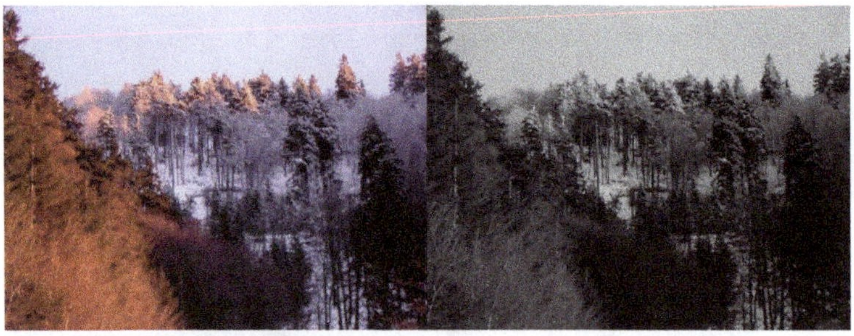

Pinna, Figure 5.

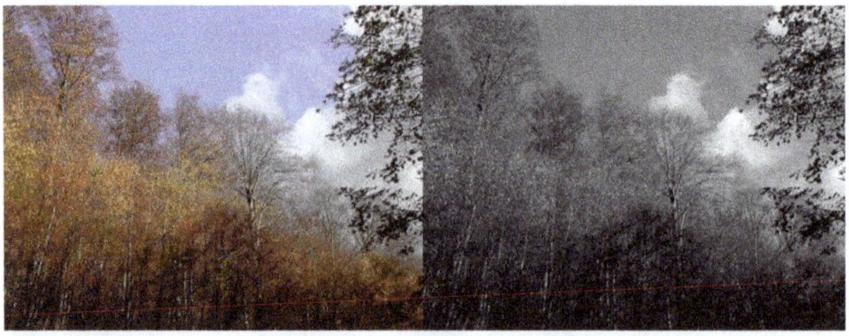

Pinna, Figure 6.

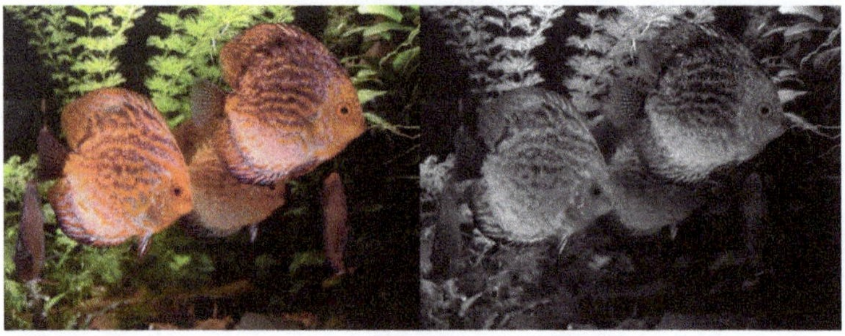

Pinna, Figure 7.

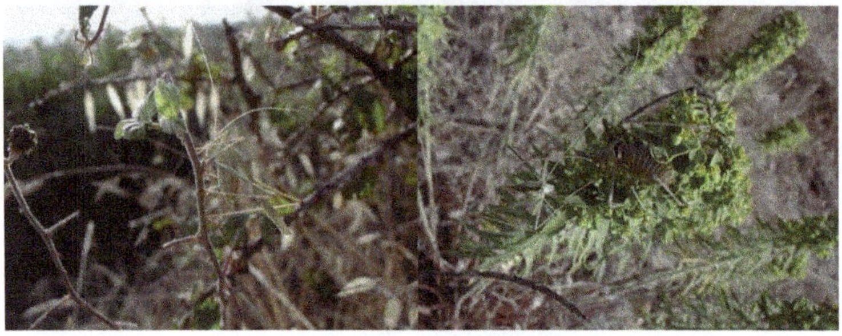

Pinna, Figure 8.

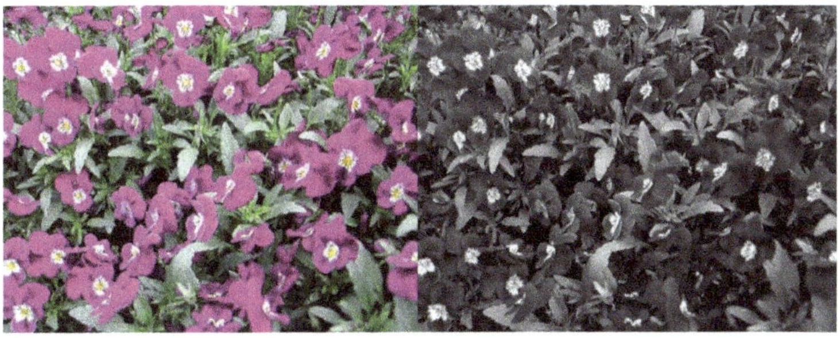

Pinna, Figure 9.

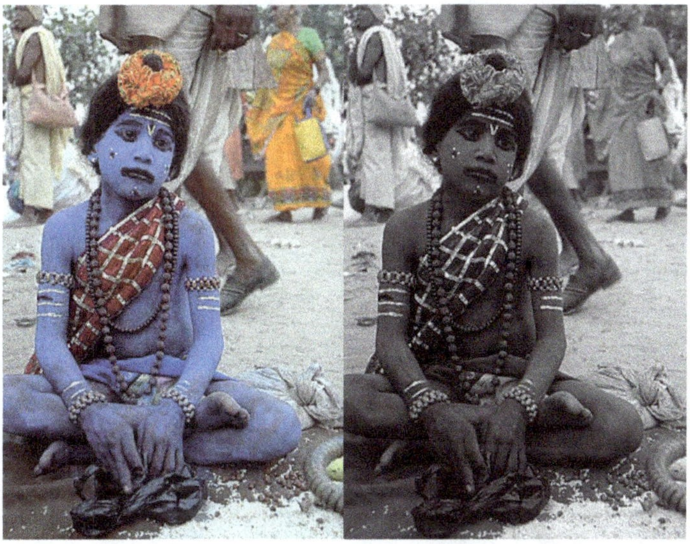

Pinna, Figure 10.

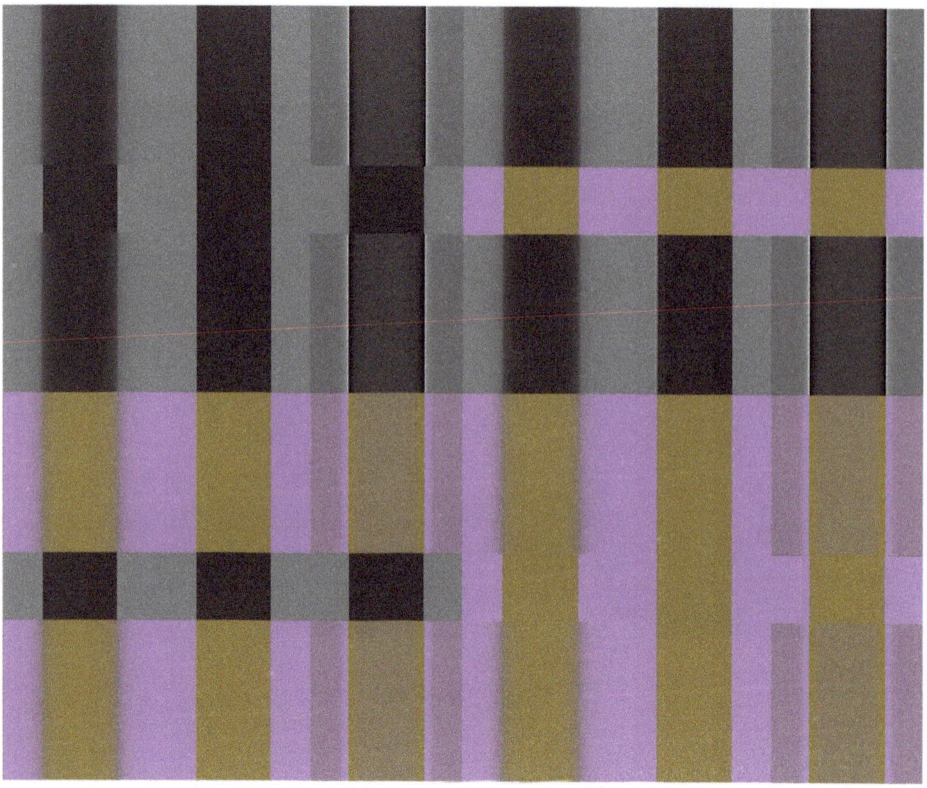

Webster *et al.*, Figure 6.

Plate V

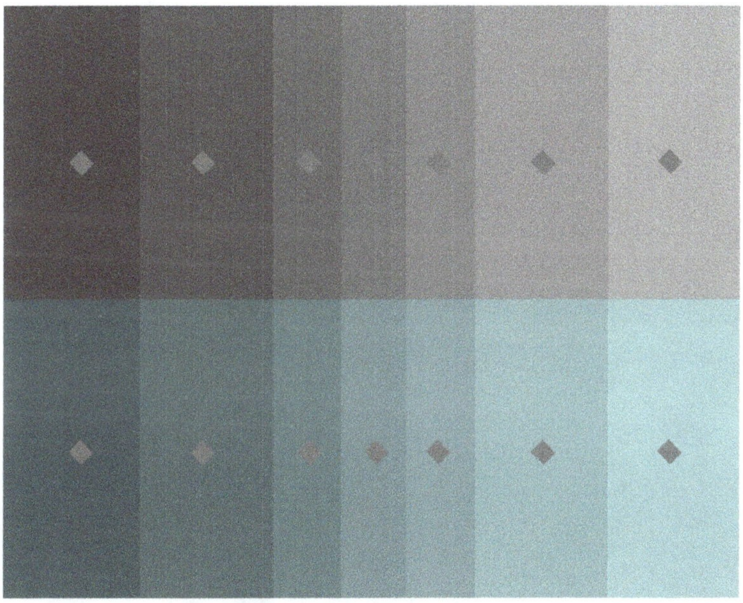

Gordon and Shapley, Figure 6.

Gordon and Shapley, Figure 1.

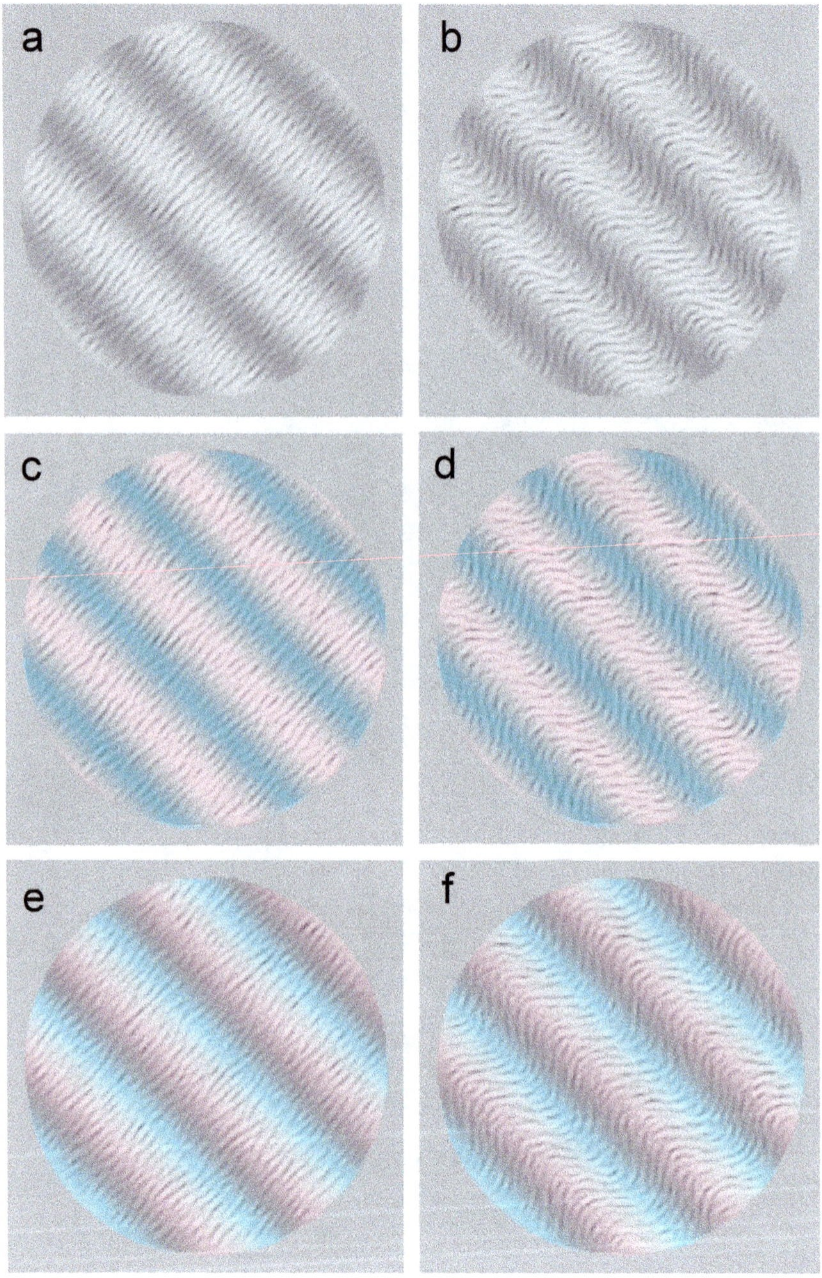

Kingdom *et al.*, Figure 1.

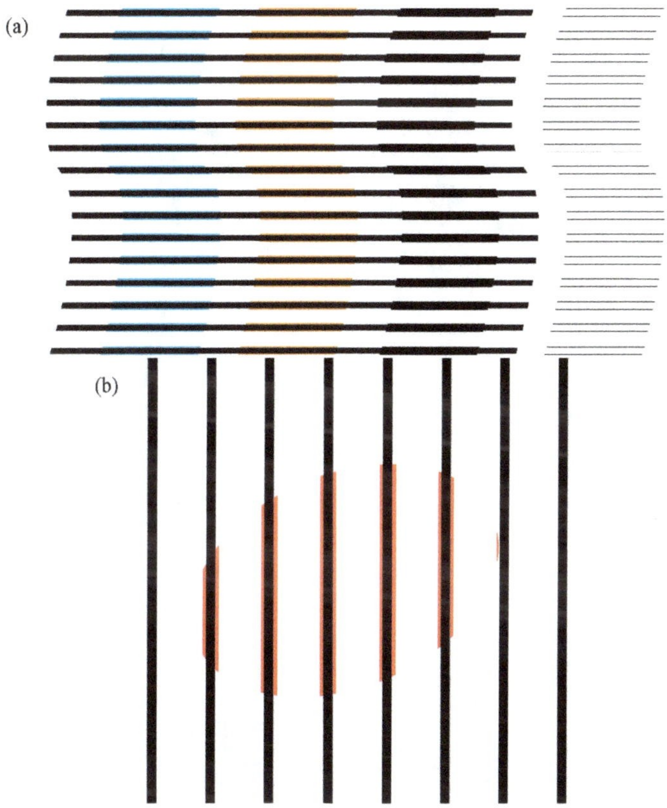

Wollschläger and Faul, Figure 1.

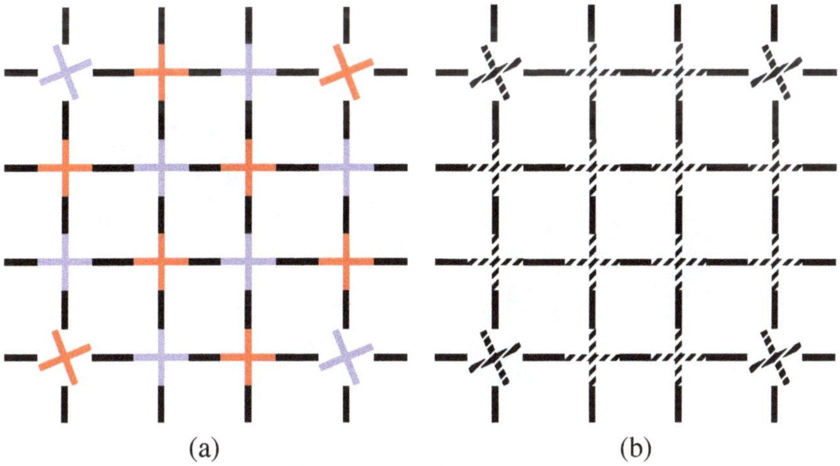

Wollschläger and Faul, Figure 2.

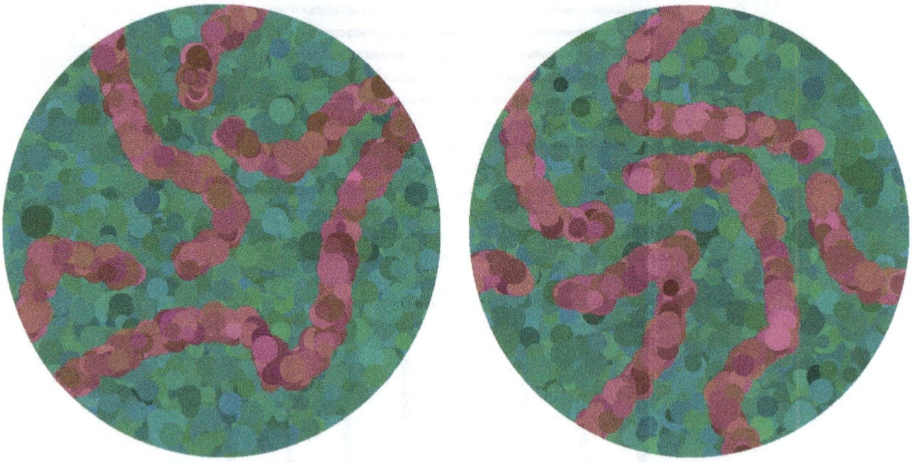

(a) (b)

Wollschläger and Faul, Figure 4.

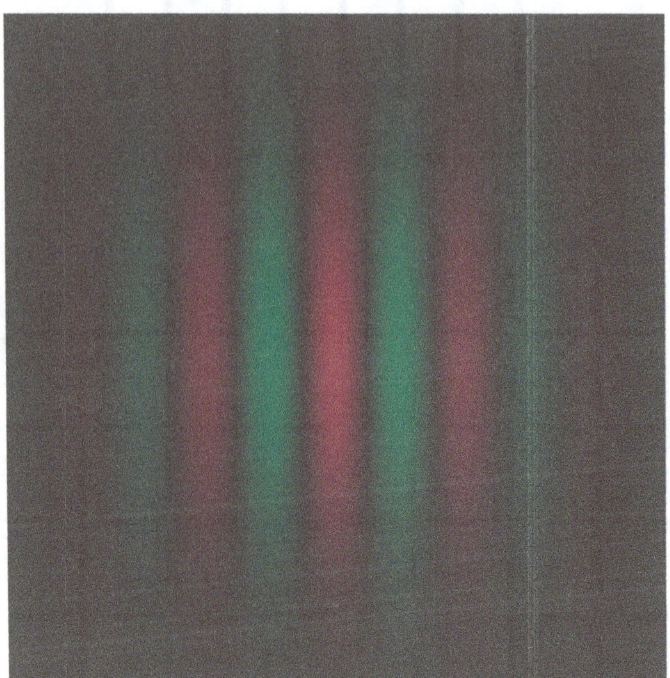

Wollschläger and Faul, Figure 6.

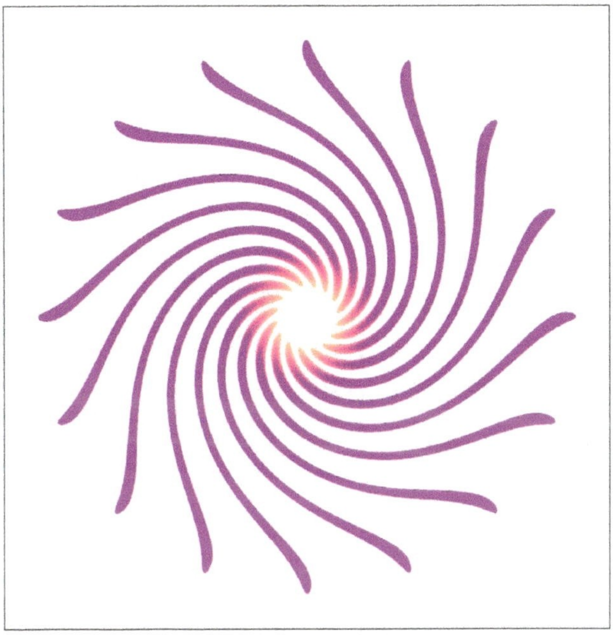

(a)

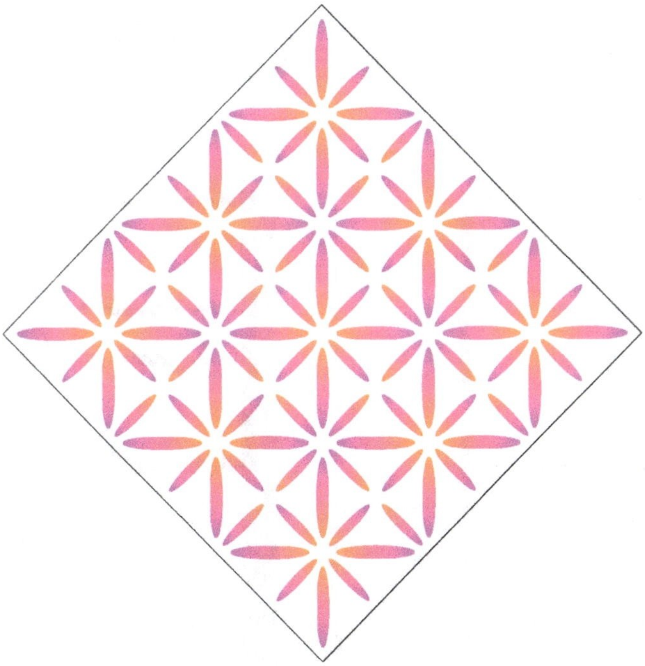

(b)

Todorović, Figure 11.

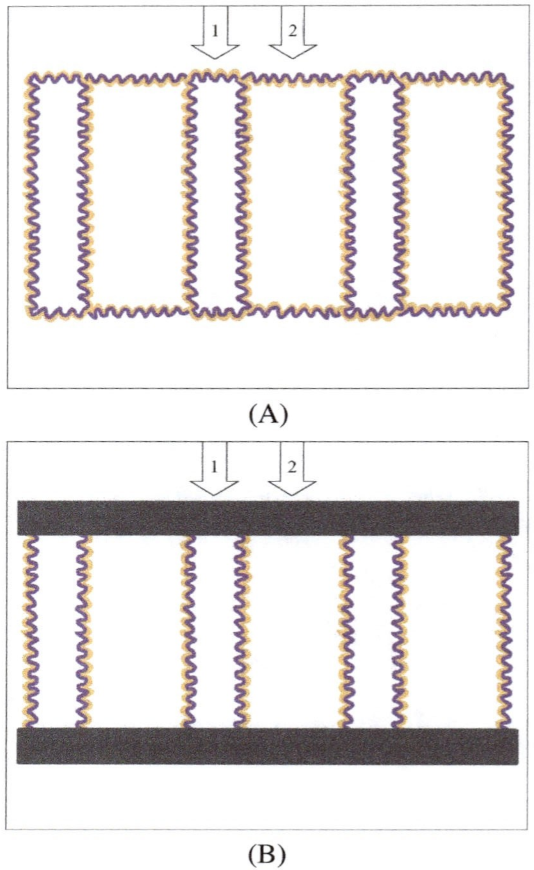

Heydt and Pierson, Figure 1.

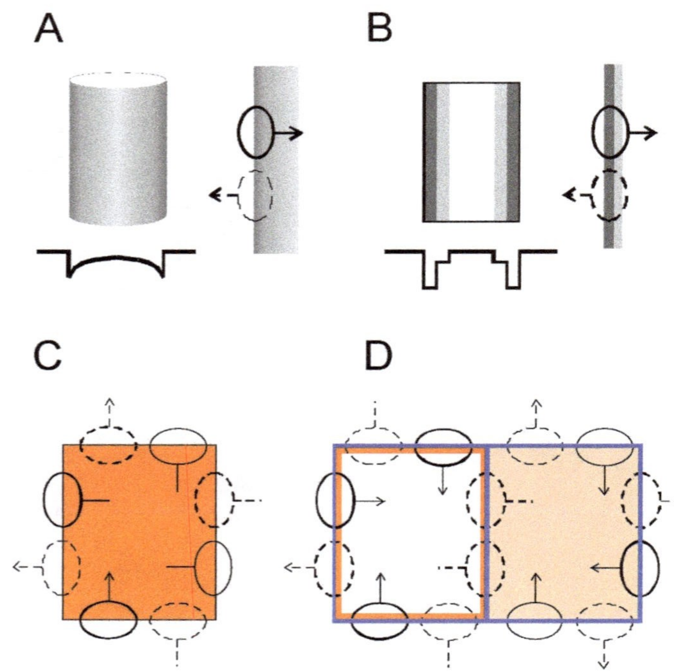

Heydt and Pierson, Figure 8.

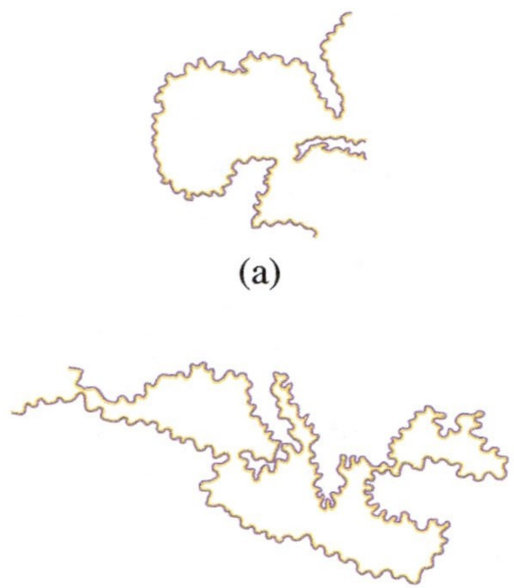

(a) (b)

Pinna and Reeves, Figure 1.

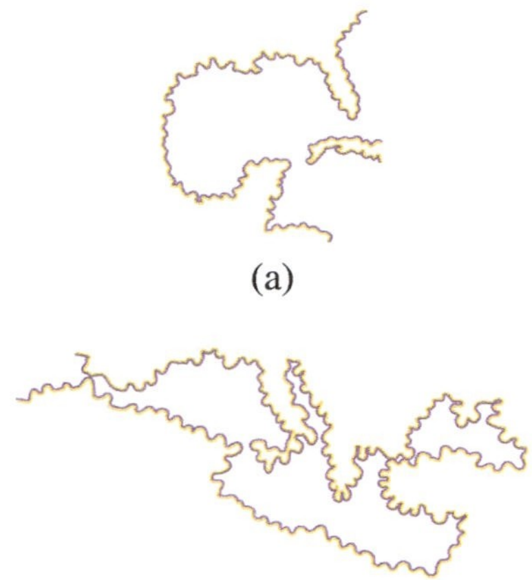

(a) (b)

Pinna and Reeves, Figure 2.

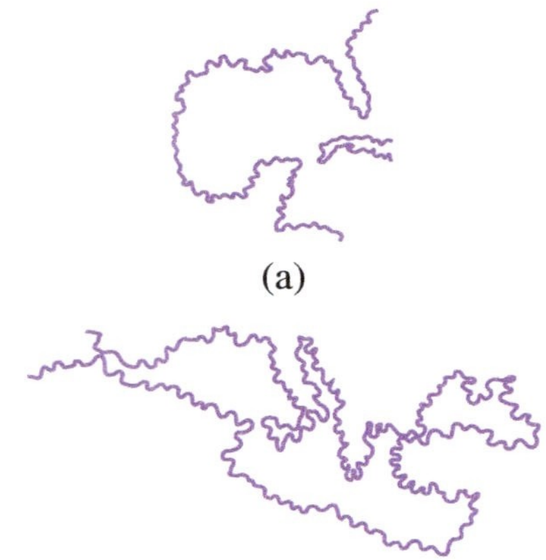

(a) (b)

Pinna and Reeves, Figure 3.

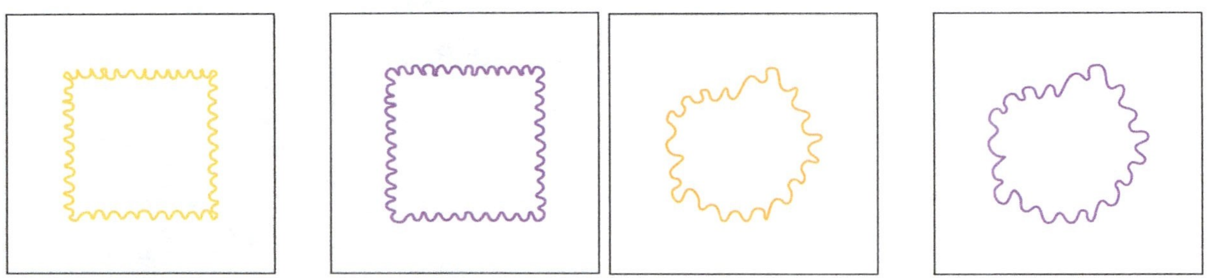

(a) (b)

Pinna and Reeves, Figure 4.

Plate XII

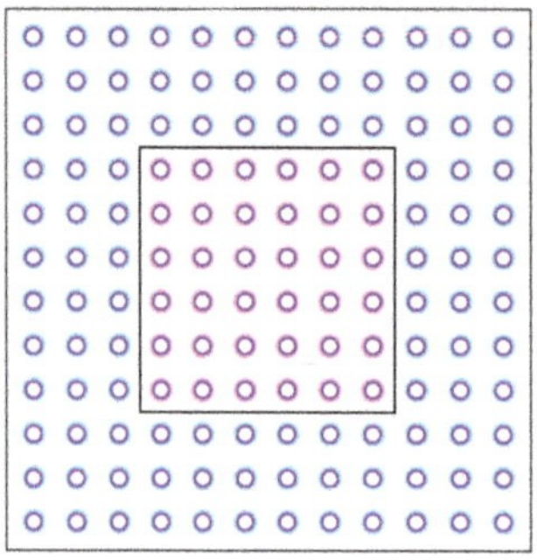

(a)

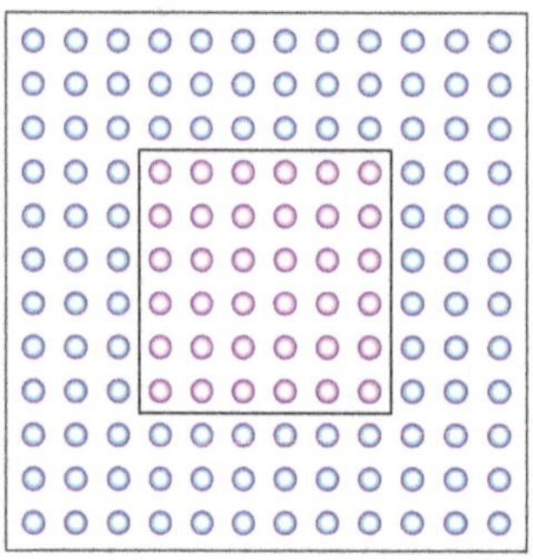

(b)

Pinna and Reeves, Figure 5.

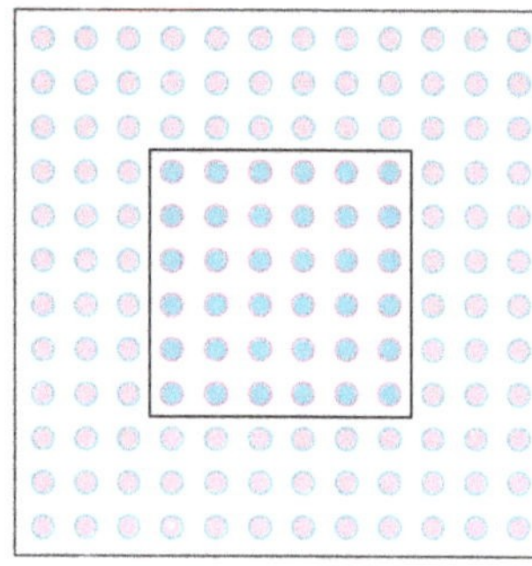

(a)

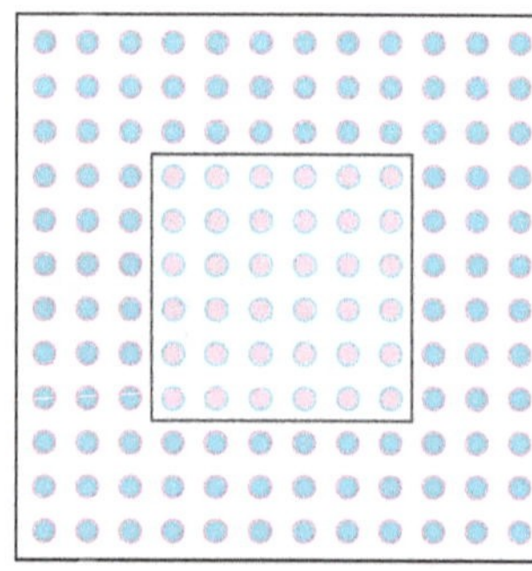

(b)

Pinna and Reeves, Figure 6.

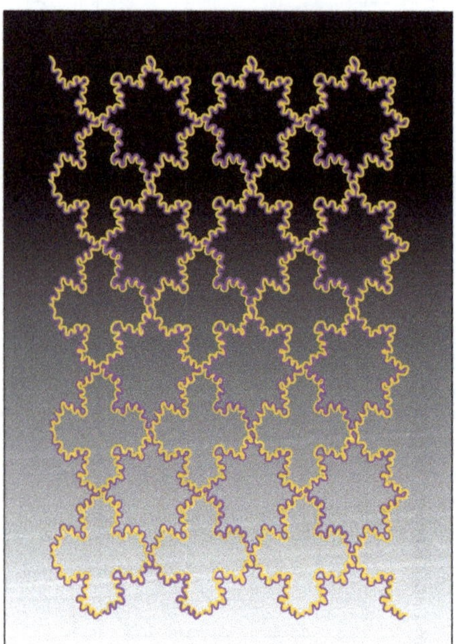

Pinna and Reeves, Figure 7.

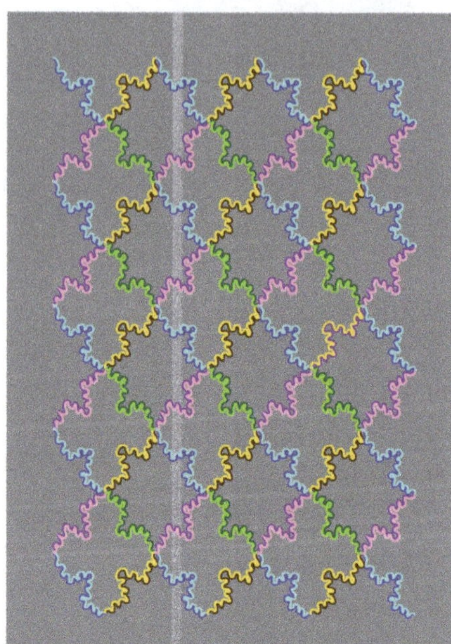

Pinna and Reeves, Figure 9.

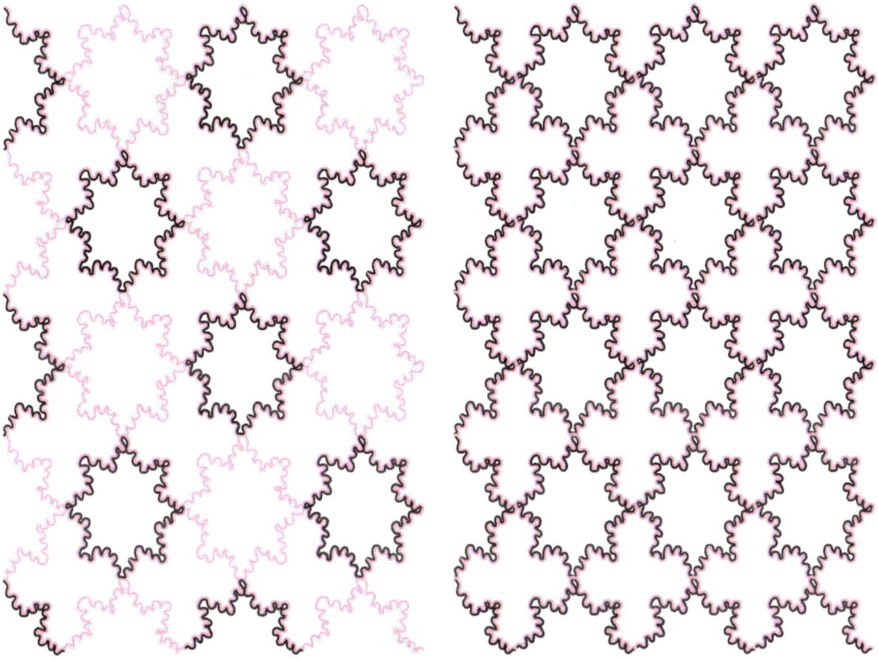

(a) (b)

Pinna and Reeves, Figure 8.

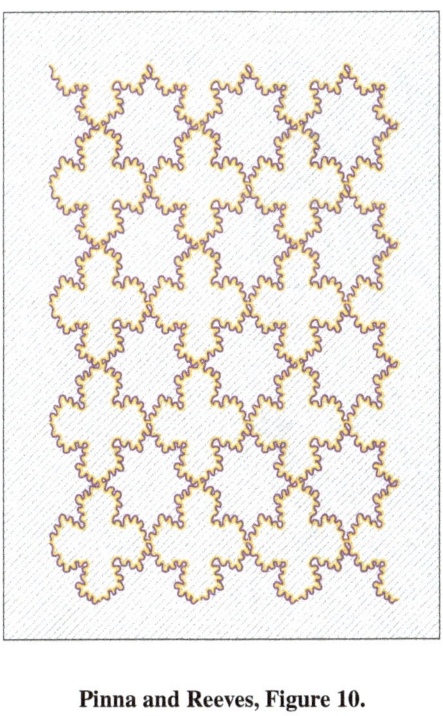

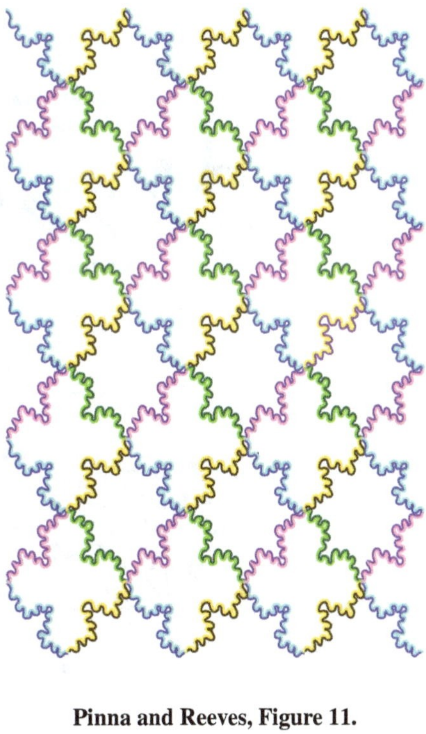

Pinna and Reeves, Figure 10.

Pinna and Reeves, Figure 11.

Plate XIV

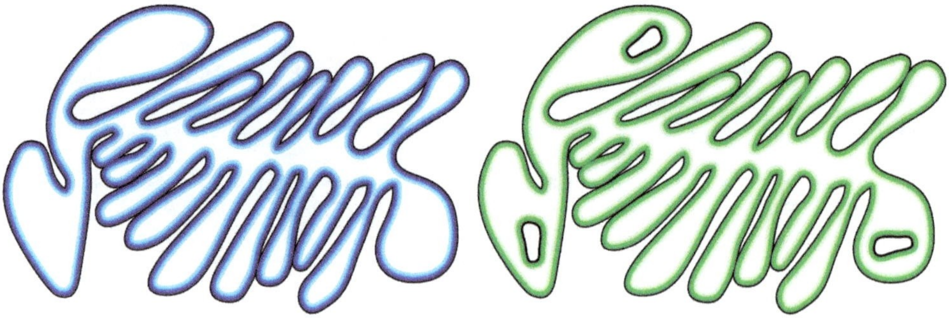

(a) (b)

Pinna and Reeves, Figure 12.

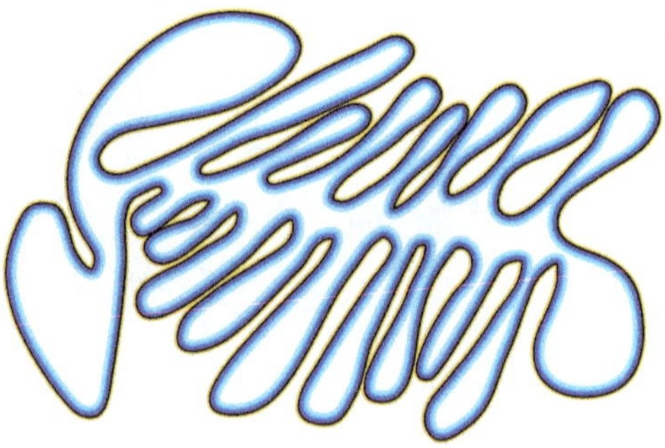

Pinna and Reeves, Figure 13.

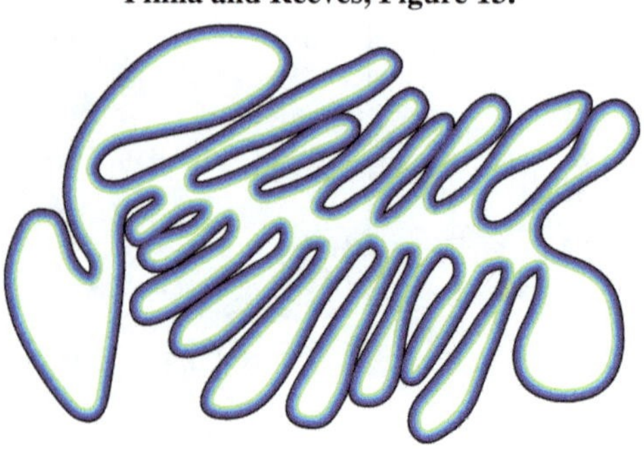

(a)

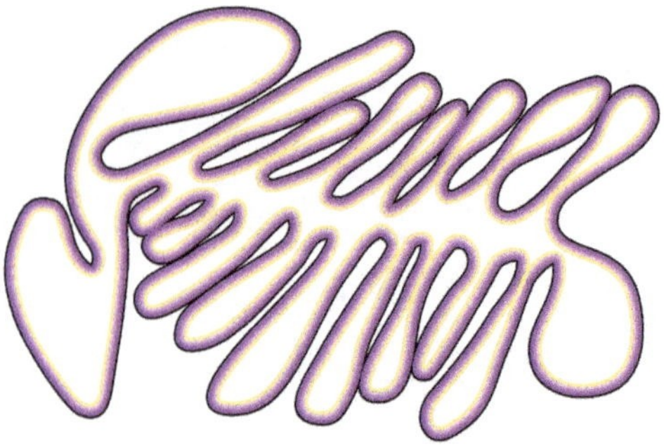

(b)

Pinna and Reeves, Figure 14.

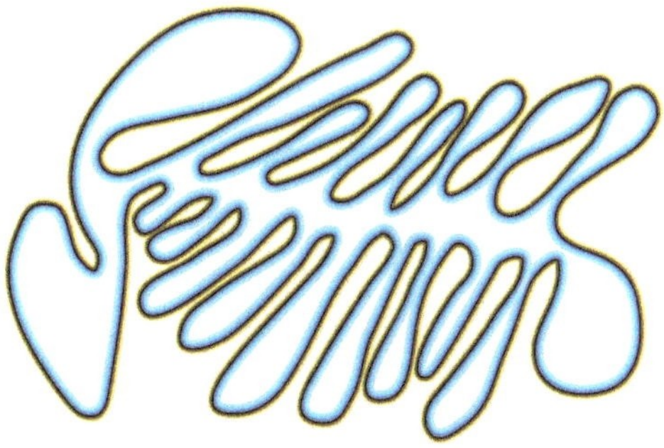

Pinna and Reeves, Figure 15.

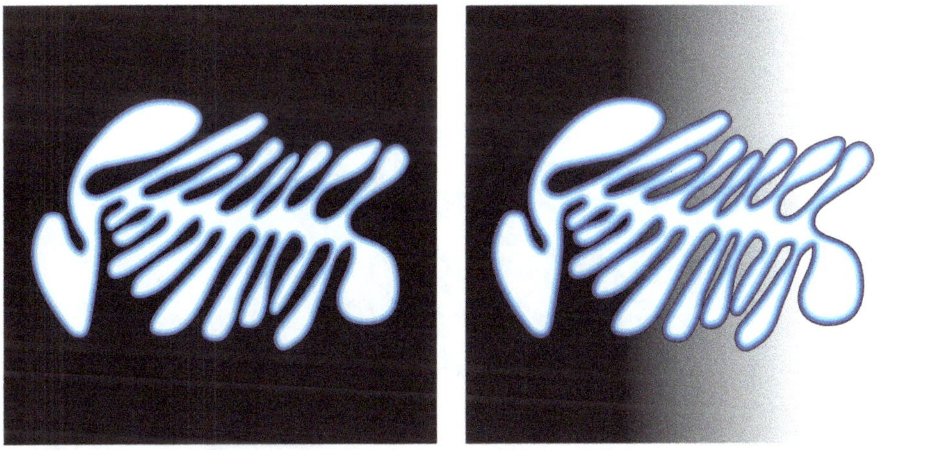

(a) (b)

Pinna and Reeves, Figure 16.

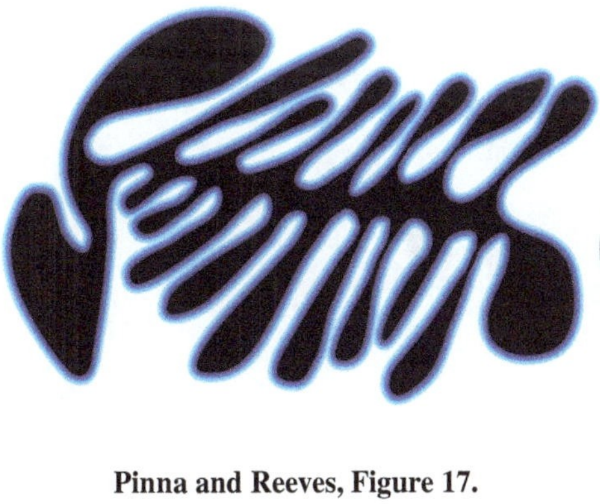

Pinna and Reeves, Figure 17.

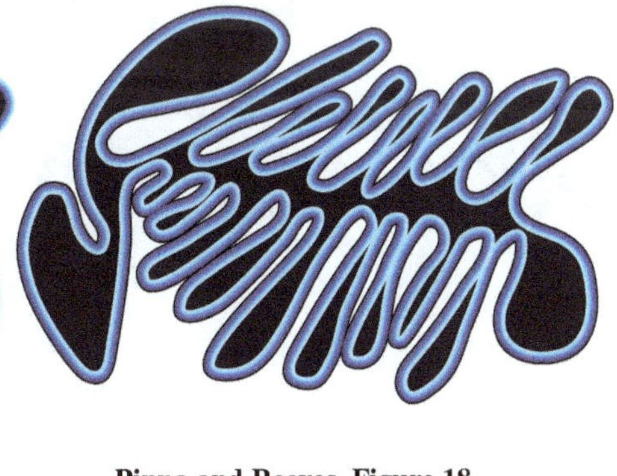

Pinna and Reeves, Figure 18.

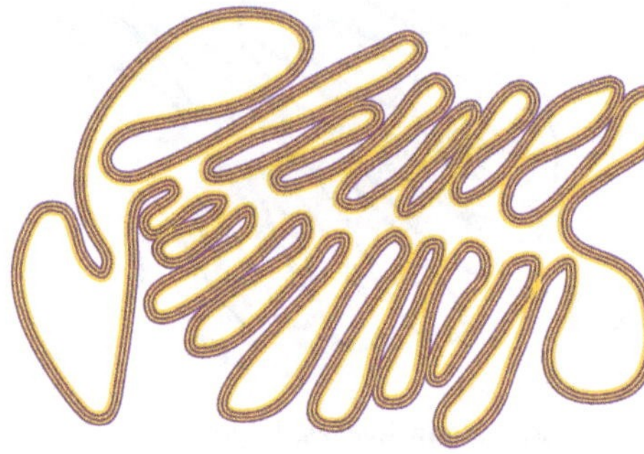

Pinna and Reeves, Figure 19.

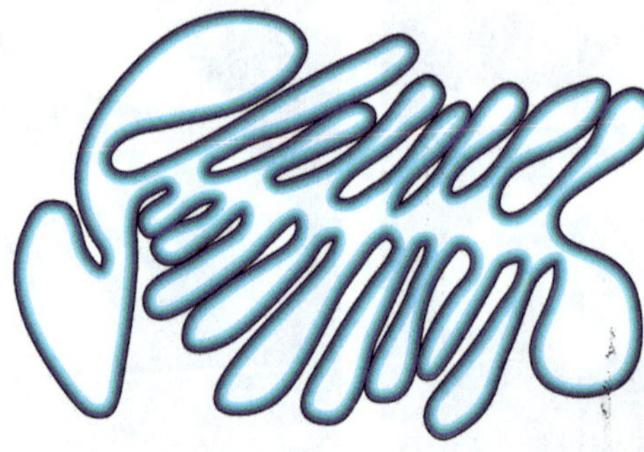

Pinna and Reeves, Figure 20.

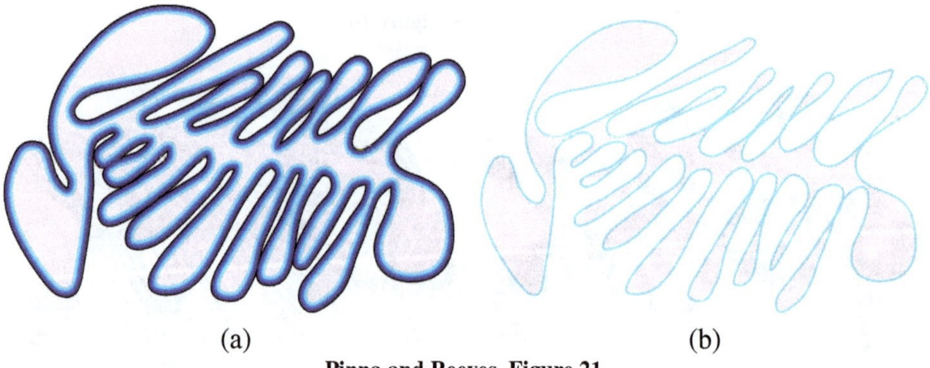

(a) (b)

Pinna and Reeves, Figure 21.

www.ingramcontent.com/pod-product-compliance
Lightning Source LLC
Chambersburg PA
CBHW061202240926
51846CB00006B/1094
* 9 7 8 9 0 0 4 1 5 3 0 6 6 *